DNA and RNA Cleavers and Chemotherapy of Cancer and Viral Diseases

NATO ASI Series

Advanced Science Institutes Series

A Series presenting the results of activities sponsored by the NATO Science Committee, which aims at the dissemination of advanced scientific and technological knowledge, with a view to strengthening links between scientific communities.

The Series is published by an international board of publishers in conjunction with the NATO Scientific Affairs Division

A	**Life Sciences**	Plenum Publishing Corporation
B	**Physics**	London and New York
C	**Mathematical and Physical Sciences**	Kluwer Academic Publishers
D	**Behavioural and Social Sciences**	Dordrecht, Boston and London
E	**Applied Sciences**	
F	**Computer and Systems Sciences**	Springer-Verlag
G	**Ecological Sciences**	Berlin, Heidelberg, New York, London,
H	**Cell Biology**	Paris and Tokyo
I	**Global Environmental Change**	

PARTNERSHIP SUB-SERIES

1.	**Disarmament Technologies**	Kluwer Academic Publishers
2.	**Environment**	Springer-Verlag / Kluwer Academic Publishers
3.	**High Technology**	Kluwer Academic Publishers
4.	**Science and Technology Policy**	Kluwer Academic Publishers
5.	**Computer Networking**	Kluwer Academic Publishers

The Partnership Sub-Series incorporates activities undertaken in collaboration with NATO's Cooperation Partners, the countries of the CIS and Central and Eastern Europe, in Priority Areas of concern to those countries.

NATO-PCO-DATA BASE

The electronic index to the NATO ASI Series provides full bibliographical references (with keywords and/or abstracts) to more than 50000 contributions from international scientists published in all sections of the NATO ASI Series.
Access to the NATO-PCO-DATA BASE is possible in two ways:

– via online FILE 128 (NATO-PCO-DATA BASE) hosted by ESRIN,
Via Galileo Galilei, I-00044 Frascati, Italy.

– via CD-ROM "NATO-PCO-DATA BASE" with user-friendly retrieval software in English, French and German (© WTV GmbH and DATAWARE Technologies Inc. 1989).

The CD-ROM can be ordered through any member of the Board of Publishers or through NATO-PCO, Overijse, Belgium.

Series C: Mathematical and Physical Sciences – Vol. 479

DNA and RNA Cleavers and Chemotherapy of Cancer and Viral Diseases

edited by

Bernard Meunier

Laboratoire de Chimie de Coordination du CNRS,
Toulouse, France

Kluwer Academic Publishers

Dordrecht / Boston / London

Published in cooperation with NATO Scientific Affairs Division

Proceedings of the NATO Advanced Research Workshop on
DNA and RNA Cleavers and Chemotherapy of Cancer and Viral Diseases
Toulouse, France
September 11–14, 1995

A C.I.P. Catalogue record for this book is available from the Library of Congress

ISBN-13: 978-94-010-6596-2   e-ISBN-13: 978-94-009-0251-0
DOI: 10.1007/978-94-009-0251-0

Published by Kluwer Academic Publishers,
P.O. Box 17, 3300 AA Dordrecht, The Netherlands.

Kluwer Academic Publishers incorporates the publishing programmes of
D. Reidel, Martinus Nijhoff, Dr W. Junk and MTP Press.

Sold and distributed in the U.S.A. and Canada
by Kluwer Academic Publishers,
101 Philip Drive, Norwell, MA 02061, U.S.A.

In all other countries, sold and distributed
by Kluwer Academic Publishers Group,
P.O. Box 322, 3300 AH Dordrecht, The Netherlands.

Printed on acid-free paper
Softcover reprint of the hardcover 1st edition 1996

This book contains the proceedings of a NATO Advanced Research Workshop held within the programme of activities of the NATO Special Programme on Supramolecular Chemistry as part of the activities of the NATO Science Committee.

Other books previously published as a result of the activities of the Special Programme are:

WIPFF, G. (Ed.), *Computational Approaches in Supramolecular Chemistry*. (ASIC 426) 1994.
ISBN 0-7923-2767-5

FLEISCHAKER, G.R., COLONNA, S. and LUISI, P.L. (Eds.), *Self-Production of Supramolecular Structures*. From Synthetic Structures to Models of Minimal Living Systems. (ASIC 446) 1994.
ISBN 0-7923-3163-X

FABBRIZZI, L., POGGI, A. (Eds.), *Transition Metals in Supramolecular Chemistry*. (ASIC 448) 1994.
ISBN 0-7923-3196-6

BECHER, J. and SCHAUMBURG, K. (Eds.), *Molecular Engineering for Advanced Materials*. (ASIC 456) 1995. ISBN 0-7923-3347-0

LA MAR, G.N. (Ed.), *Nuclear Magnetic Resonance of Paramagnetic Macromolecules*. (ASIC 457) 1995.
ISBN 0-7923-3348-9

SIEGEL, JAY S. (Ed.), *Supramolecular Stereochemistry*. (ASIC 473) 1995. ISBN 0-7923-3702-6

WILCOX, C.S. and HAMILTON A.D. (Eds.), *Molecular Design and Bioorganic Catalysis*. (ASIC 478) 1996. ISBN 0-7923-4024-8

TABLE OF CONTENTS

Site-specific DNA cleavage

Mechanism of oxidative DNA cleavage

RNA cleavage by RNase H

Hydrolysis of RNA by ribozymes and metal complexes

PREFACE

The past decade has witnessed a burst of activity and interest in the discovery and design of drugs that cleave DNA and RNA with sequence-specificity. This interest stems from the potential of this class of compounds to be useful as therapeutics agents, in particular in the field of the treatment of cancer and viral diseases. Further, a side benefit of such studies is the discovery of novel mechanisms and uses of such agents as tools in the study of structure and function of nucleic acids.

Up to now, no international meeting has been organized to recognize the immense progress that has been made in this field. The field of DNA and RNA cleavage by natural and chemical drugs now includes researchers working with rather dissimilar agents but with common underlying mechanisms of DNA damage. Until recently, these scientists were working in separate, apparently unrelated areas, such as the enediyne antibiotics and their synthetic analogues, bleomycin-metal complexes, metal-drug complexes, ribozymes and ribozyme mimics, and antisense and antigene oligonucleotides, etc. It is now clear that these research areas have in common strategies and targets. Researchers representing these areas worked together at this workshop where these common interests were discussed and scientific ideas modified and criticized. Such a workshop should lead to new research approaches and collaborative interactions, and is expected to significantly enhance the progress in the field of DNA and RNA cleavage.

The workshop was focused on molecular mechanisms and supramolecular aspects (molecular recognition of specific DNA and RNA sequences by cleavers) of nucleic acids cleavage, drug discovery, chemical synthesis of drug analogues, three dimensional structures of drug-DNA complexes and mutagenic mechanisms. It was a special opportunity to bring together chemists, biochemists, pharmacologists and molecular biologists to share their expertise with one another.

Seventy-six scientists attended this NATO Workshop "DNA and RNA cleavers and Chemotherapy of Cancer and Viral Diseases" at the Hotel Mercure St Georges in Toulouse for a four-day meeting with 26 lectures and 25 contributions for the poster session. The present volume contains most of the invited lectures presented during the NATO Workshop and few additional contributions of volunteers. The following fields are covered: (i) DNA cleavage by enediyne molecules, (ii) bleomycin: the paradigm of DNA cleavers based on metal complexes, (iii) site-specific DNA cleavage, (iv) mechanism of oxidative DNA cleavage, (v) RNA cleavage by RNase H and (vi) hydrolysis of RNA by ribozymes and metal complexes.

As Director of this Workshop, I am deeply indebted to my two colleagues, Prof. Irving GOLDBERG of Harvard Medical School and Prof. David S. SIGMAN of the University of California at Los Angeles, for accepting with great enthusiasm to co-organize the meeting. The support of our colleague K. C. NICOLAOU of the Scripps Institute at San Diego in the early stages of the organization of the workshop was also a determining factor.

The local organization was handled by Jean BERNADOU, Marguerite PITIE, Anne ROBERT, Geneviève PRATVIEL, Maryse BEZIAT and all the members of my research group. Their efficient and enthusiastic help is gratefully acknowledged. We also thank the City Council of Toulouse for helping us to welcome all participants.

This fruitful Workshop was made possible with the strong financial support mainly provided by the Scientific Affairs Division of NATO (Supramolecular Chemistry Program) and also by the Région Midi-Pyrénées, Rhône-Poulenc Rorer and the Chemistry Department of CNRS.

Bernard MEUNIER, Toulouse, January 1996

LIST OF CONTRIBUTORS

(listed as appearing in the present volume)

IRVING H. GOLDBERG
Harvard Medical School, Department of Biological Chemistry and Molecular Pharmacology, 250 Longwood Avenue, Boston, Massachusetts 02115, USA

PETER C. DEDON
Whitaker College of Health Sciences and Technology Division of Toxicology, Massachusetts Institute of Technology, Cambridge, Massachusetts 02139-4307, USA

GEORGE A. ELLESTAD
American Cyanamid Company, 401 N. Middletown Road, Bldg. 205/421, Pearl River, New York 10965, USA

NADA ZEIN
Pharmaceutical Research Institute, Bristol -Myers Squibb, P.O. Box 4000, Princeton, New Jersey 08543-4000, USA

YUKIO SUGIURA
Institute for Chemical Research, Kyoto University, Uji, Kyoto 611, Japan

SIDNEY M. HECHT
Department of Chemistry, University of Virginia, Charlottesville, Virginia 22901, USA

RICHARD M. BURGER
The Public Health Research Institute, 455 First Avenue, New York, New York 10016, USA

BERND GIESE
Institut für Organische Chemie der Universität Basel, St Johanns-ring 19, CH-4056 Basel, Switzerland

DAVID SIGMAN
University of California at Los Angeles, Molecular Biology Institute, 405 Hilgard Avenue, Los Angeles, California 90024-1570, USA

PETER E. NIELSEN
Center for Biomolecular Recognition, Department of Medical Chemistry and Genetics, Laboratory B, The Panum Institute, Blegdamsvej 3c, DK-2200 Copenhagen N, Denmark

THOMAS C. BRUICE
Department of Chemistry, University of California at Santa Barbara, Santa Barbara, California 93106, USA

ISAO SAITO
Department of Synthetic Chemistry and Biological Chemistry, Faculty of Engineering, Kyoto University, Kyoto 606-01, Japan

KENNETH T. DOUGLAS
Department of Pharmacy, The University of Manchester, Oxford Road, Manchester M13 9PL, United Kingdom

VALENTIN VLASSOV
Institute of Bioorganic Chemistry, Siberian Division of Russian Academy of Sciences, 8 Lavrentiev Avenue, Novosibirsk, 630090 Russia

BERNARD MEUNIER
Laboratoire de Chimie de Coordination du CNRS, 205 route de Narbonne, 31077 Toulouse cedex, France

STEEN STEENKEN
Max-Planck-Institut für Strahlenchemie, Stifstr. 34-36, D-45470 Mülheim, Germany

JACQUES PIETTE
Laboratoire de Virologie, Institut de Pathologie B23, Université de Liège, B-4000 Liège, Belgium

JEAN-JACQUES TOULME
Laboratoire de Biophysique Moléculaire, Université de Bordeaux II, 146 rue Léo Saignat, 33076 Bordeaux cedex, France

FRITZ ECKSTEIN
Max Planck Institut für experimentelle Medizin, Hermann-Rein Strasse 3, D-37075 Göttingen, Germany

JANE GRASBY
Sheffield University, Chemistry Department, Sheffield S3 7HF, United Kingdom

ROBERT HÄNER
Central Research Laboratories, Ciba-Geigy Limited, CH-4002 Basel, Switzerland

MAKOTO KOMIYAMA
Department of Chemistry and Biotechnology, Faculty of Engineering, The University of Tokyo, Hongo, Tokyo 113, Japan

DARREN MAGDA
Pharmacyclics, 995 E. Arques Avenue, Sunnyvale, California 94086-4593, USA

JAMES K. BASHKIN
Department of Chemistry, Washington University, Campus Box 1134, One Brookings Drive, Saint Louis, Missouri 63130-4899, USA

LIST of PARTICIPANTS
(listed by alphabetic order of the country)

Belgium

Jacques PIETTE
Laboratoire de Virologie, Institut de Pathologie B23, Université de Liège, B-4000 Liège, Belgium

Canada

Jik CHIN
Department of Chemistry, McGill University, Montreal, Canada H3A 2K6

Denmark

Peter E. NIELSEN
Research Center for Medical Biotechnology, Department of Biochemistry, The Panum Institute, Blegdamsvej 3c, DK-2200 Copenhagen N, Denmark

France

François AMALRIC
CNRS, IBCG, 118 route de Narbonne, 31062 Toulouse cedex, France

Christian BAILLY
Institut de Recherches sur le Cancer de Lille, Unité d'Oncohemathologie Moléculaire, Cité Hospitalière, Place de Verdun, 59045 Lille cedex, France

Bernard BARBIER
Centre de Biophysique Moléculaire du CNRS, rue Charles Sadron, 45071 Orléans cedex 2, France

Jean-Paul BEHR
Laboratoire de Chimie Génétique, Faculté de Pharmacie, BP 24, 67401 Illkirch cedex, France

Jean BERNADOU
Laboratoire de Chimie de Coordination du CNRS, 205 route de Narbonne, 31077 Toulouse cedex, France

Pascal BIGEY
Laboratoire de Chimie de Coordination du CNRS, 205 route de Narbonne, 31077 Toulouse cedex, France

Jean CADET
Service d'Etudes des Systèmes et des Architectures Moléculaires, Département de Recherche Fondamentale sur la Matière Condensée, CENG, 17 rue des Martyrs, 38054 Grenoble cedex 9, France

Nadia CHOUINI-LALANNE
IMRCP, Université Paul Sabatier, 118 route de Narbonne, 31062 Toulouse cedex, France

Jean-François CONSTANT
LEDSS 6, Bâtiment 301, Chimie Recherche, Université Joseph Fourier, BP 53X, 38041 Grenoble, France

Jean-Luc DECOUT
LEDSS 5, Bâtiment Chimie Recherche, Université Joseph Fourier, BP 53X, 38041 Grenoble cedex 9, France

Victor DUARTE
Laboratoire de Chimie de Coordination du CNRS, 205 route de Narbonne, 31077 Toulouse cedex, France

Gilles FAVRE
Centre Claudius Regaud, Laboratoire de Ciblage Thérapeutique, 20-24 rue du Pont Saint Pierre, 31052 Toulouse cedex, France

Silvana FRAU
Laboratoire de Chimie de Coordination du CNRS, 205 route de Narbonne, 31077 Toulouse cedex, France

Sylviane GIORGI-RENAULT
Laboratoire de Chimie Thérapeutique du CNRS, Faculté de Pharmacie, 4 avenue de l'Observatoire, 75270 Paris cedex 06, France

Claude HELENE
Muséum National d'Histoire Naturelle, Laboratoire de Biophysique, 43 rue Cuvier, 75231 Paris cedex 05, France

Thierry IMBERT
Laboratoire Pierre Fabre Médicaments, 17 avenue Jean Moulin, 81106 Castres cedex, France

Andreas JAKOBS
Laboratoire de Chimie de Coordination du CNRS, 205 route de Narbonne, 31077 Toulouse cedex, France

Mitsuharu KOTERA
LEDSS, Bâtiment 52, Chimie Recherche, Université Joseph Fourier, BP 53X, 38041 Grenoble cedex 9, France

Roger KRAVTZOFF
Biovector Therapeutics S.A., Parc Technologique du Canal, 10 avenue de l'Europe, 31520 Ramonville St Agne, France

Marc LENG
Centre de Biophysique Moléculaire du CNRS, rue Charles-Sadron, 45071 Orléans cedex 2, France

Jean LHOMME
Université Joseph FOURIER, BP 53X, 38041 Grenoble cedex 9, France

Olivier LORTHIOIR
Centre de Biophysique Moléculaire du CNRS, Rue Charles-Sadron, 45071 Orléans cedex 2, France

Patrick MAILLIET
Rhône-Poulenc Rorer, Centre de Recherche de Vitry-Alfortville, 13 quai Jules Guesde, BP 14, 94403 Vitry sur Seine cedex, France

Béatrice MESTRE
Laboratoire de Chimie de Coordination du CNRS, 205 route de Narbonne, 31077 Toulouse cedex, France

Bernard MEUNIER
Laboratoire de Chimie de Coordination du CNRS, 205 route de Narbonne, 31077 Toulouse cedex
France

Nicole PAILLOUS
IMRCP, Université Paul Sabatier, 118 route de Narbonne, 31062 Toulouse cedex, France

Marguerite PITIE
Laboratoire de Chimie de Coordination du CNRS, 205 route de Narbonne, 31077 Toulouse cedex, France

Geneviève PRATVIEL
Laboratoire de Chimie de Coordination du CNRS, 205 route de Narbonne, 31077 Toulouse cedex, France

Anne ROBERT
Laboratoire de Chimie de Coordination du CNRS, 205 route de Narbonne, 31077 Toulouse cedex, France

Isabelle SASAKI
Laboratoire de Chimie de Coordination du CNRS, 205 route de Narbonne, 31077 Toulouse cedex, France

Sophie SIXOU
Centre Claudius Regaud, Laboratoire de Ciblage Thérapeutique, 20-24 rue du Pont Saint Pierre, 31052 Toulouse cedex, France

Anny SLAMA-SCHWOK
Laboratoire de Spectroscopie Biomoléculaire, UFR de Médecine, 74 rue Marcel Cachin, 93012 Bobigny, France

Annick SPASSKY
Institut de Physique et Chimie, Institut Curie, 11 rue Pierre et Marie Curie,
75231 Paris cedex 05, France

Jean-Jacques TOULME
Laboratoire de Biophysique Moléculaire, Université de Bordeaux II, 146 rue
Léo Saignat, 33076 Bordeaux cedex, France

Jean-Christophe TRUFFERT
Centre de Biophysique Moléculaire du CNRS, rue Charles-Sadron, 45071
Orléans cedex 2, France

Bernard VACHER
Pierre Fabre Médicaments, 17 avenue Jean-Moulin, 81106 Castres cedex,
France

Catherine VERCHERE-BEAUR
LCBB - Bâtiment 420, UPS Orsay, 91405 Orsay cedex, France

Krystyna ZAKRZEWSKA
Laboratoire de Biochimie Théorique, IBPC, 13 rue Pierre et Marie Curie,
75005 Paris, France

Germany

Fritz ECKSTEIN
Max Planck Institut für experimentelle Medizin, Hermann-Rein Strasse 3, D-
37075 Göttingen, Germany

Joachim W. ENGELS
Johann Wolfgang Goethe-Universität, Institut für Organische Chemie, Marie-
Curie-Str. 11, 60439 Frankfurt am Main, Germany

Steen STEENKEN
Max-Planck-Institut für Strahlenchemie, Stifstr. 34-36, D-45470 Mülheim,
Germany

Italy

Chrys CHATGILIALOGLU
I.Co.C.E.A. - C.N.R., Via P. Gobetti 101, 40129 Bologna, Italy

Antonio CLEMENTE
Università Degli Studi di Firenze, Dipartimento di Chimica, Via Gino Capponi 7, 50121 Firenze, Italy

Guido DE GUIDI
Dipartimento di Scienze Chimiche, Università di Catania, Via A. Doria 6, 95125 Catania, Italy

Japan

Makoto CHIKIRA
Department of Applied Chemistry, Chuo University, Kasuga 1-13-27, Bunkyo-ku, Tokyo 112, Japan

Makoto KOMIYAMA
Department of Chemistry and Biotechnology, Faculty of Engineering, The University of Tokyo, Hongo, Tokyo 113, Japan

Isao SAITO
Department of Synthetic Chemistry and Biological Chemistry, Faculty of Engineering, Kyoto University, Kyoto 606-01, Japan

Yukio SUGIURA
Institute for Chemical Research, Kyoto University, Uji, Kyoto 611, Japan

Kazunobu TOSHIMA
Department of Applied Chemistry, Keio University, 3-14-1 Hiyoshi, Kohoku-ku, Yokohama 223, Japan

Portugal

Abel VIEIRA
Instituto Superior Tecnico, Departamento de Engenharia Quimica, Secçao de Quimica Organica, Av. Rovisco Pais, P-1096 Lisboa codex, Portugal

Russia

Valentin VLASSOV
Institute of Bioorganic Chemistry, Siberian Division of Russian Academy of Sciences, 8 Lavrentiev Avenue, Novosibirsk, 630090 Russia

Switzerland

Bernd GIESE
Institut für Organische Chemie der Universität Basel, St Johanns-ring 19, CH-4056 Basel, Switzerland

Robert HÄNER
Central Research Laboratories, Ciba-Geigy Limited, CH-4002 Basel, Switzerland

The Netherlands

M. Vincent M. LAFLEUR
Department of Medical Oncology, Section Molecular Toxicology, Faculty of Medicine, Vrije University, Van der Boechorststr. 7, 1081 BT Amsterdam, The Netherlands

United Kingdom

Kenneth T. DOUGLAS
Department of Pharmacy, The University of Manchester, Oxford Road, Manchester M13 9PL, United Kingdom

Jane GRASBY
Sheffield University, Chemistry Department, Sheffield S3 7HF, United Kingdom

Michael J. WARING
Department of Pharmacology, University of Cambridge, Tennis Court Road, Cambridge CB2 1QJ, United Kingdom

USA

James K. BASHKIN
Department of Chemistry, Washington University, Campus Box 1134, One Brookings Drive, Saint Louis, Missouri 63130-4899, USA

Thomas C. BRUICE
Department of Chemistry, University of California at Santa Barbara, Santa Barbara, California 93106, USA

Richard M. BURGER
The Public Health Research Institute, 455 First Avenue, New York, New York 10016, USA

Christiane CASAS
Department of Chemistry, Yale University, P.O. Box 208107, 225 Prospect Street, New Haven, Connecticut 06511-8118, USA

Peter C. DEDON
Whitaker College of Health Sciences and Technology Division of Toxicology, Massachusetts Institute of Technology, Cambridge, Massachusetts 02139-4307, USA

George A. ELLESTAD
American Cyanamid Company, 401 N. Middletown Road, Bldg. 205/421, Pearl River, New York 10 965, USA

Irving H. GOLDBERG
Harvard Medical School, Department of Biological Chemistry and Molecular Pharmacology, 250 Longwood Avenue, Boston, Massachusetts 02115, USA

John T. GROVES
Department of Chemistry, Princeton University, Princeton, New Jersey 08544, USA

Sidney M. HECHT
Department of Chemistry, University of Virginia, Charlottesville, Virginia 22901, USA

Pascal HOFFMANN
Department of Chemistry, Yale University, P.O. Box 208107, 225 Prospect Street, New Haven, Connecticut 06511-8118, USA

John W. KOZARICH
Merck Research Laboratories, Biochemistry Department, P.O. Box 2000, Rahway, New Jersey 07065-0900, USA

Neocles B. LEONTIS
Department of Chemistry, Overman Hall, Bowling Green State University, Bowling Green, Ohio 43403, USA

Darren MAGDA
Pharmacyclics, 995 E. Arques Avenue, Sunnyvale, California 94086-4593, USA

Jonathan L. SESSLER
Department of Chemistry and Biochemistry, University of Texas, Austin, Texas 78731, USA

David SIGMAN
University of California at Los Angeles, Molecular Biology Institute, 405 Hilgard Avenue, Los Angeles, California 90024-1570, USA

Nada ZEIN
Pharmaceutical Research Institute, Bristol-Myers Squibb, P.O. Box 4000, Princeton, New Jersey 08543-4000, USA

DNA Cleavage by Enediyne Molecules

ENEDIYNES AS PROBES OF NUCLEIC ACID STRUCTURE

I.H. Goldberg, L.S. Kappen, Y.J. Xu, A. Stassinopoulos, X. Zeng, Z. Xi, and C.F. Yang.
Department of Biological Chemistry and Molecular Pharmacology
Harvard Medical School
Boston, Massachusetts 02115, USA

The enediyne antitumor antibiotics are believed to kill cancer cells by damaging cellular DNA. These agents undergo activation by thiols, reducing compounds or spontaneously to form a diradical species that interacts with DNA in a specific way so as to generate oxidative deoxyribose lesions in the form of strand breaks and abasic sites. Distinctive sequence specificities in bistranded lesion formation have been demonstrated for several members of this expanding family of antibiotics. For the past few years my laboratory has been particularly interested in uncovering the structural basis for specific drug interactions with DNAs of differing microstructure and in developing these agents as probes for unusual nucleic acid structures. This interest arose initially from our work with neocarzinostatin chromophore (NCS-chrom) [1,2] and has recently been extended to include the potent enediyne C1027 chromophore (C1027-chrom) [3,4]. The chemical and biochemical basis for the structural specificity of these agents is the subject of this report.

1. Neocarzinostatin

1.1 MECHANISM OF DUPLEX DNA DAMAGE

NCS-chrom binds to DNA by intercalation via its minor groove and upon activation by thiol adduction rearranges via a cumulene intermediate to a diradical species (Scheme 1) that abstracts hydrogen atoms from minor groove-accessible carbons (C-5', C-1' and C-4') of the sugar backbone of each strand of duplex DNA, as reviewed in [1,2]. Molecular oxygen adds onto the carbon-centered radicals generated on the DNA deoxyribose to form peroxy radicals that give rise to the final DNA damage products. Under anaerobic

1

B. Meunier (ed.), DNA and RNA Cleavers and Chemotherapy of Cancer and Viral Diseases, 1–21.
© *1996 Kluwer Academic Publishers. Printed in the Netherlands.*

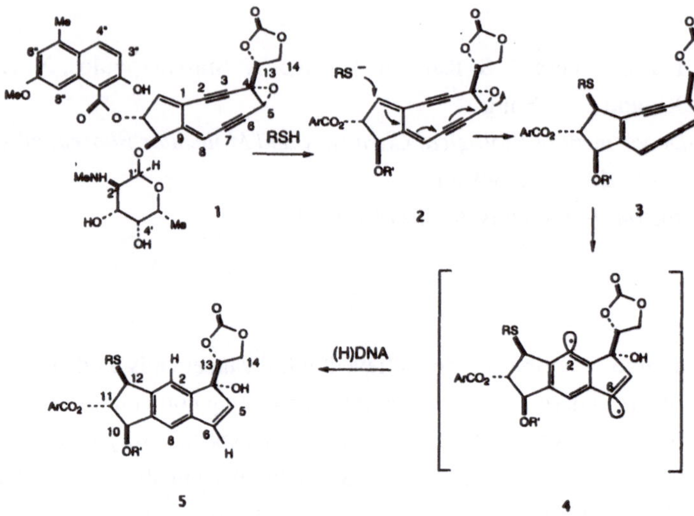

Scheme 1. Proposed Thiol-Dependent Mechanism of NCS-Chrom-Induced Cleavage of Duplex DNA.

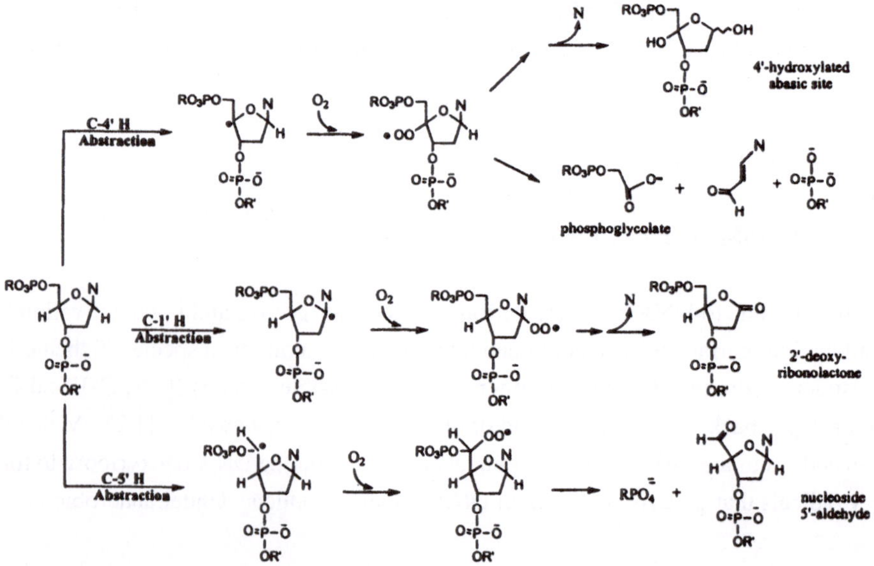

Scheme 2. Proposed Mechanism of NCS-Induced DNA Damage at C-1', C-4', and C-5'.

conditions, nitroaromatic radiation sensitizers can substitute for dioxygen. Three main attack sites and the consequent damage products have been characterized (Scheme 2): (a) Abstraction of a hydrogen atom from the 5'-carbon results mainly in single-strand breaks predominantly at T and A residues with a PO_4 at the 3'- and a nucleoside aldehyde at the 5'-ends of the break. (b) 4'-Attack, which occurs predominantly at the T residues of GT steps, especially at AGT sequences, leads either to an alkali-labile, 4'-hydroxylated abasic site or to a strand break with a phosphoglycolate at the 3'- and a PO_4 at the 5'-termini. (c) Abstraction of a hydrogen atom from the C-1' position of mainly C residues of AGC sequences also results in an alkali-sensitive abasic site having a 2-deoxyribonolactone residue. Deuterium abstraction experiments [5,6], molecular modeling studies [7], and determination of the three-dimensional structure of the complex formed between the thiol-induced post-activated form of the drug, 5 (Scheme 1), and an AGC containing oligonucleotide [8,9] indicate that in the formation of bistranded lesions the radical center at C-2 of NCS-chrom is involved in attack at the C of AGC (and the T of AGT), while that at C-6 abstracts hydrogen from the T two bases to the 3'-side on the complementary strand. The radical center at C-2 is involved only in bistranded lesion formation, since without attack at the C of AGC (or at the T of AGT) by C-2 of NCS-chrom only single-stranded lesions due to C-6 attack on the complementary strand result [10, 11].

Recently we have been able to obtain a high-resolution, three dimensional structure of the glutathione-induced post-activated form of NCS-Chrom (NCSi-glu),which most closely resembles the active diradical species of the drug, and 5'-d(GCAGCGC) · d(GCGCTCC)-3' by ^{1}H NMR [8,9]. This structure (Figure 1) is in basic agreement with the earlier proposed model [7] but offers details of important interactions, not otherwise apparent, that clarify the sequence and attack site specificity of the drug. Drug intercalation is stabilized by stacking interactions between the naphthoate and A3·T12 and G4·C11 base pairs with the long axis of the naphthoate being nearly parallel with the long axis of the G4·C11 base pair. The binding of the postactivated drug to the duplex induces large structural changes at the binding site with a larger base rise than those found in most intercalation complexes with an extended, unwound V-shaped intercalation site and a twist angle accounting for an unwinding of 22˚. Both minor and major grooves are wider and shallower than in canonical B-DNA. The most distinctive feature on the backbone helical contacting interface is the formation of a right-handed hydrophobic surface by the drug, which is complementary to the curvature of the right-handed DNA helix. Of particular interest, the 2'-N-methyl of the drug amino sugar makes contact with A3(H2) at a distance of less than 3Å.

This plus other interactions place the drug sugar in the minor groove. There are many van der Waals interactions between the helical sugar-phosphate backbone of the

4

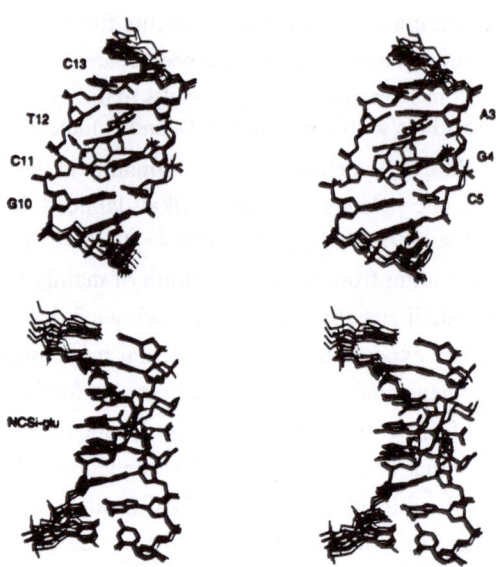

Figure 1. Overlay stereoview (cross-eyed) of five calculated structures and the energy-minimized average structure of the NCSi glu-d(GGAGCGC)·(GCGCTCC) complex. (Top) View into the minor groove, displaying the extended, V-shaped binding site. The C5(1') (top left arrow) and T12(5') (top right arrow) cleavage sites by the C2 and C6 radicals, respectively, are indicated. (Bottom) View displaying the intercalation binding mode and the opening of both the major and minor grooves (left and right side of the drawing, respectively).

C11-T12-C13 segment and the 2'-N-methylfucosamine of the chromophore, as well as between the hydrophobic edge of the tetrahydroindacene and the methoxy group of the inactivated chromophore. The T·A site in the duplex provides the right groove depth and molecular affinity to accommodate the amino sugar of the drug. This result predicts that the binding specificity is determined by the 2'-N-methyl sugar moiety, which recognizes the T·A base pair and assists in the intercalation process. Further, the double-stranded sequence specificity of the drug can be understood in terms of van der Waals and electrostatic interactions involving the orientation of the two amino groups of G4 and G10 on the floor of the minor groove. The structure of the complex shows that the H6 and H2 of the chromophore are proximal to the pro-S H5' of T12 (2.87Å) and H1' of C5 (3.78Å), respectively. This is consistent with the lower frequency of H1' abstraction by the C2 radical and the ease of C2-radical quenching by solvent or the adducted thiol. Further, electrostatic surface measurements demonstrate that the binding of the drug in the minor groove results in a negatively charged surface covering the otherwise relatively neutral minor groove [9]. Thus, the formation of the complex is largely driven by hydrophobic forces and solvation of the polar surfaces of the complex.

Thiol, which has a dual role as an adducting activator of the drug and as a reductant of the DNA damage intermediates, influences the partitioning of the reaction intermediates and hence the final damage products [11,12]. In addition, the structure of the thiol affects the relative proportions of single- and double-strand lesions. While base-specific 5'-attack results mainly in single-strand breaks, sequence-specific 1'- and, to a major degree, 4'-attacks occur as part of a bistranded lesion, which also includes a strand break due to 5'-chemistry on the complementary strand two nucleotides to the 3'-side. Several lines of evidence support the notion that the bistranded lesions that occur at AGC-ACT and AGT-ACT sequences result from the concerted action of a single molecule of NCS-chrom on both DNA strands [10-12].

1.2 EFFECT OF BASE MISMATCHES ON DAMAGE CHEMISTRY

Mechanistic studies, such as those described above, revealed a special property of NCS-chrom: its ability to alter the chemistry of attack (C-1' at the C residue of AGC or C-4' at the T residue of AGT), depending upon the microstructure and the local geometry of the drug-DNA complex. This is further shown by the finding that replacement of a G·C pair with an I·C pair in an oligonucleotide duplex makes a significant difference in the extent of attack at C-1' and at C-4' by NCS-chrom [13]. In addition, it was found that replacement of the G·C base pair 5' to the C of AGC·GCT by a G·T wobble mismatch results in the remarkable switching of the chemistry of damage at the C from C-1' to C-4' [14]. The 1' chemistry is almost eliminated and replaced by 4' chemistry, so that the latter accounts for 64% of the damage (Table 1), mainly in the form of the 4'-hydroxylation product (abasic site) and a smaller amount of the DNA fragment with a phosphoglycolate at the 3' end (strand break). On the complementary strand, the G·T mismatch results in an increase in 4' chemistry at the T residue, but 5' chemistry remains the main mechanism. When a G·A mismatch is inserted 5' to the C, there is a marked decrease in all damage at this site without detectable switching of chemistry.

Table 1. Distribution of NCS-chrom-induced chemistry
at C in AGC with and without a G·T mismatch

Substrate	% of total damage		
	5'	4'	1'
-AGC- . . . -TCG-	15	13	72
-AGC- . . . -TTG-	28	64	8

A G·C to G·T or G·A pair substitution destabilizes the DNA helix much less than other base-pair mismatches [15]. These mismatches result in little alteration in the conformation of the DNA backbone or in the global conformation of the double helix [16, 17]. Displacement of the guanine of the G·T and G·A base pairs into the minor groove results in a loss of the pseudosymmetry about the glycosyl bond that Watson-Crick base pairs have relative to a vector joining the C-1' carbons of the two sugar residues. Of the two mismatches the G·T wobble pair is much more asymmetric, resulting in substantial changes in the DNA minor groove so that the quasiequivalence of certain functional elements, such as the hydrogen bond acceptors present on the purine N-3 and pyrimidine O2 of the Watson-Crick base pair is altered [16, 17]. The marked decrease in all damage 3' to the G·A mismatch suggests that the drug fails to bind at this site, possibly due to alteration at the intercalation site. The projection of the guanine of G·T into the minor groove of DNA [15, 17], where the indacene diradical form of the drug abstracts hydrogen atoms from nearby C-1', C-4', and C-5' deoxyribose sites, might be expected to modify the orientation of the radical centers vis-à-vis these sites. Since C-1' lies deeper in the minor groove than C-4', it is possible that protrusion of the guanine keeps the drug in a more superficial position in the minor groove so that it attacks C-1' poorly but is better situated for reacting with C-4'. This may also account for the relative increase in chemistry at C-5', whose prochiral H_S resides at the outer edge of the minor groove.

In a similar fashion, a series of mismatches has been explored for their effect on the chemistry of damage at the T of AGT·ACT in olgodeoxynucleotides, a site at which 4'-chemistry ordinarily occurs [18]. Placement of a G·T mispair 5' to the T results in a marked increase in 4'-chemistry (from 25% to 62% of total damage), as measured by the formation of breaks with 3'-phosphoglycolate ends and abasic sites due to 4'-hydroxylation. Strikingly, 4'-chemistry is induced at the T on the complementary strand, a site ordinarily restricted to 5'-chemistry. Both stable T·G and unstable T·C mismatches at the attack site itself are associated with marked inhibition of damage at this site. Whereas placement of the relatively stable G·A mismatch on the 5'-side of the T residue (AGT) results in substantial inhibition of damage at the T without shifting of chemistry, the same mismatch at the 3'-side of the attack site decreases damage only slightly but is associated with the appearance of significant 1'-chemistry.

These findings provide further support for the role of minor groove microstructure in determining the chemical mechanisms of DNA damage and underscore the usefulness of NCS-Chrom as a probe of DNA microheterogeneity. Several options exist for the abstraction of hydrogen atoms from minor groove accessible sites on the deoxyribose by the radical centers of activated drug. Relatively minor changes in DNA local geometry result in substantial changes in the pattern of hydrogen atom abstraction. That sequence is not the sole determinant of local geometry is shown by the finding of selective strand scission opposite sites of single-base bulges, independent of the specific sequence [19].

1.3 BISTRANDED LESIONS IN DNA·RNA HYBRIDS

Until recently, the reported damage induced by thiol-activated NCS-chrom has been limited to B-type duplex DNA; lesions in single-stranded DNA have been attributed to the formation of double-stranded regions. Duplex RNA has not been found to be a substrate for this reaction. We have recently found, however, that glutathione-activated NCS-chrom generates staggered double-stranded lesions in DNA·RNA hybrids, involving C-1' hydrogen abstraction from the targeted ribonucleotide on the RNA strand and C-5' chemistry at the targeted deoxyribonucleotide staggered two nucleotides in the 3' direction on the complementary DNA strand [20]. By contrast, the lesions generated on a DNA analogue (DNA$_a$) of the RNA strand (in a DNA·DNA$_a$ duplex) involves C-5', not C-1' chemistry.

In initial experiments it was found that glutathione-activated NCS-chrom generates bistranded lesions in the hybrid formed by yeast tRNA^{phe} and DNA complementary to its 31-mer 3' terminus. To elucidate the chemistry of the RNA cleavage reaction and to show that the lesions are double-stranded, a series of shorter oligoribonucleotides containing the target sequence r(AGAAUUC)·d(GAATTCT) (underlining indicates major attack sites) were studied as substrates. In addition to cleavage at both U residues, major damage was produced in the form of an abasic site at the U residues. Evidence for abasic site formation on the RNA strand was obtained from sequencing-gel analysis and measurement of uracil base release. Substitution of deuterium for hydrogen at the C-1' position of the U residues led to a substantial isotope effect ($k^1H/k^2H = 3$) on the formation of the RNA abasic lesion and the RNA cleavage products, providing conclusive evidence for selective 1'chemistry. On the other hand, the cleavage at the T residues on the complementary DNA strand involved C-5' hydrogen abstraction, as was also true for the T residue in an oligodeoxynucleotide analogue of the RNA strand. Initial evidence for the double-stranded nature of the damage came from experiments in which 2'-O methyluridine was substituted for uridine in the RNA at one or both of the target sites. The site containing the substitution was not a target for cleavage or abasic site formation and the particular T residue, staggered two nucleotides in the 3' direction on the complementary DNA strand, was cleaved significantly less. Presumably, the presence of the substituted uridine moiety altered the structure of the drug binding site sufficiently, so that either the drug failed to bind at all or bound in such a way that attack on neither strand was possible. Thus, these studies were valuable in identifying the DNA double-stranded partner of the RNA attack site.

Direct evidence for double-stranded lesions, however, came from analysis of the products generated from a hairpin oligonucleotide construct in which the RNA and DNA strands were linked by four T residues and contained an internal ^{32}P-label at the 3' end of the RNA strand (Figure 2). The various oligonucleotide products of the drug reaction were identified on polyacrylamide gel electrophoresis by comparison with known markers and with the use of substrates specifically deuterated at C-1' at one or both U residues. The results are summarized in the Figure. Double-stranded cleavages generated fragments R3D9 and R2D8, whereas cleavage at t_{21} or t_{20}, in association with abasic site formation at U8 or U9, respectively, produced fragments a + b, respectively.

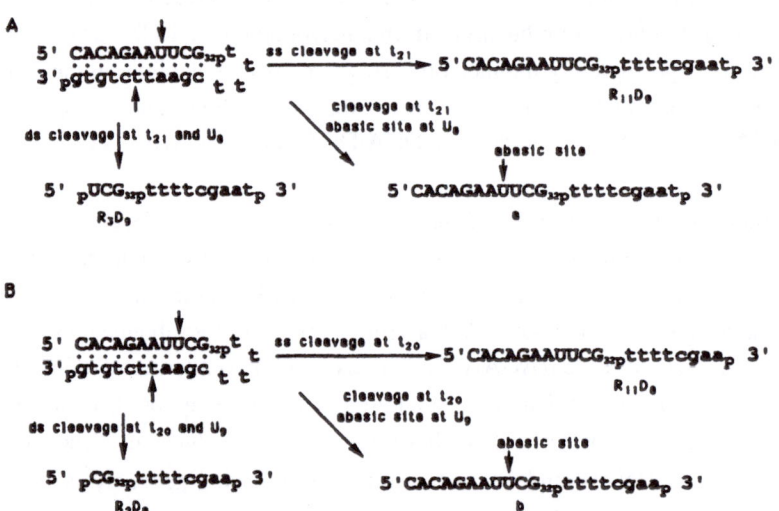

Figure 2. Cleavage products generated from NCS-Chrom treated hairpin RNA-DNA oligonucleotide containing internal ^{32}P-phosphodiester label. RNA oligonucleotide, upper case. DNA oligonucleotide, lower case. (A) Single-stranded (ss) and double-stranded (ds) damage at t_{21} and U8. (B) ss and ds damage at t_{20} and U9.

Single-stranded cleavage at t_{21} or t_{20} formed $R_{11}D_9$ or $R_{11}D_8$, respectively. Proposed chemical mechanisms to account for the RNA cleavage and abasic site formation via C-1' hydrogen abstraction are shown in Scheme 3. The abstraction of a hydrogen from C-1' of ribose to form the ribonic acid is analogous to the 1' chemistry in DNA damage (Scheme 2). The formation of the cleavage product is proposed to involve the generation of **3** from the putative hydroperoxynucleotide (**2**) via base-catalyzed cleavage of the ß-hydroxyperoxide, as cited in [21]. Duff et al. [22] proposed the formation of intermediate **3** from **2** via a Criegee-type rearrangement in the case of C-1' hydrogen abstraction by Fe-bleomycin from a single ribonucleotide in an oligodeoxynucleotide. However, the

Criegee-type mechanism is unlikely to occur for NCS-chrom-mediated damage due to the absence of metal in the reaction or of acidic conditions. Following ß-elimination of **3** the enol phosphate **4** is formed, and the hydrolysis of **4** causes strand cleavage.

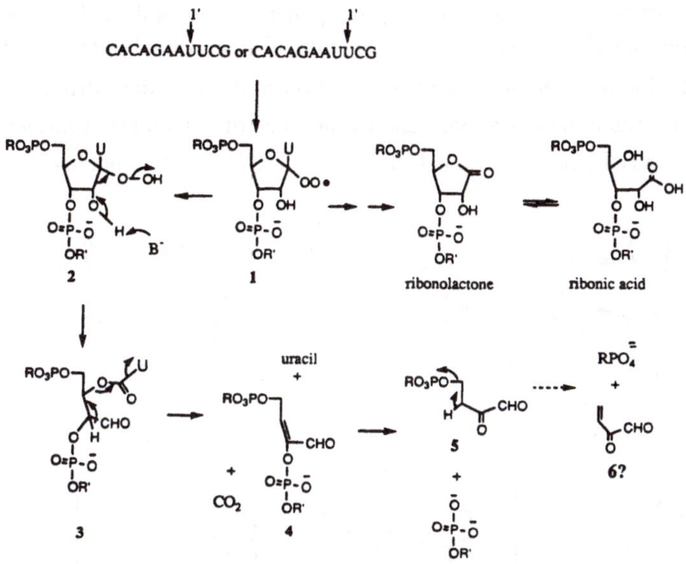

Scheme 3. Proposed 1' Chemistry Mechanism of NCS-Chrom-Mediated Damage on RNA Strand.

It is clear that a DNA·RNA hybrid is a substrate for NCS-chrom-induced damage. This reaction resembles that with duplex DNA, in contrast to that with bulged DNA (see later), in that a thiol is required to generate the appropriate diradical reactive species. Further, the damage produced is primarily bistranded. The chemistry involved in the damage on the RNA strand of the hybrid appears to be limited to C-1' hydrogen abstraction and contrasts with the C-5' chemistry that occurs at the analogous site of a similar DNA·DNA duplex. This difference in chemistry may be due to the wider minor groove in the A-type hybrid compared with the B-type DNA duplex. On the other hand, the lesion on the complementary DNA strand appears to involve primarily C-5' hydrogen abstraction for both the hybrid and DNA·DNA duplex.

1.4 DNA AND RNA BULGE-SPECIFIC DAMAGE

In the absence of thiol, duplex DNA is not a target for damage by NCS-chrom [1], and the drug decomposes in a base-catalyzed reaction under physiological conditions (pH ≥ 6) to a mixture of inactive forms [23]. Recently, however, it has been found that NCS-Chrom can efficiently and selectively cleave DNA in a thiol-independent reaction at a specific bulged structure in a reaction involving general base catalysis [24, 25]. Cleavage was restricted to a target nucleotide at the 3' side of the bulge (T$_{22}$, Figure 3), and was entirely due to 5' chemistry. Determination of the chemical structures of the drug products generated in the presence and the absence of bulged DNA has led to the proposal shown in Scheme 4 for their formation [26, 27]. In this mechanism, the

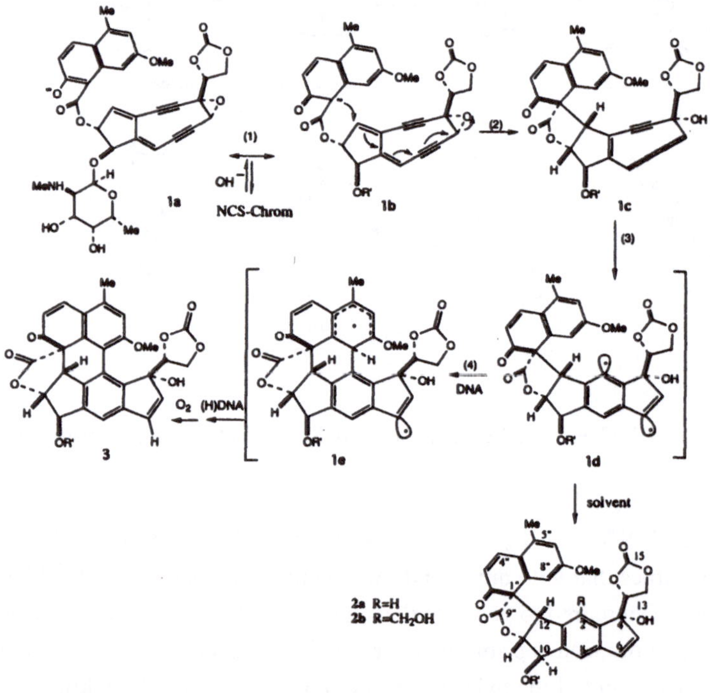

Scheme 4. Proposed Thiol-Independent Mechanism of NCS-Chrom-Induced Cleavage of Bulged DNA.

spirolactone cumulene **1c** is stereoselectively generated via an intramolecular Michael addition at C12 by the enolate anion **1b**, which is a resonance form of the naphthoate anion **1a** of NCS-chrom, resulting in the formation of the biradical **1d**. This series of reactions occurs spontaneously, and in the absence of bulged DNA, the biradical is quenched by other proton sources (including methanol in the solvent) to produce **2a** and **2b**. Drug product **3** is generated only in the presence of a substrate-bulged DNA. Thus, it has been proposed that in the presence of bulged DNA the cumulene **1c**, which has been implicated as the species that searches for the favored DNA binding site is in equilibrium between bound and free forms, which lead to **3** and to **2a** and **2b**, respectively, via **1d**. Intermediate **1e** is formed by quenching of the C2-based radical of **1d** by intramolecular bond formation with C8". In this scheme, **1e** (and its final product **3**) is formed only in the presence of substrate-bulged DNA, which possibly induces a conformational change in the drug (step 4), so as to bring C8" of the naphthoate moiety close to the radical center at C2. As noted later, however, it is not possible to distinguish whether **1d** or **1e** is the actual species involved in C-5' hydrogen abstraction from the bulge site.

Point mutations, deletions, and insertions in the DNA analogue and its complement of the 3'-terminus of yeast tRNA^{phe} show that for a single-stranded DNA to be cleaved by NCS-chrom the DNA must generate a hairpin structure with an apical loop and at least a two-base-pair stem hinged to a region of duplex structure via a bulge containing a target nucleotide at its 3' side (Figure 3) [25]. The size of the loop is not critical so long

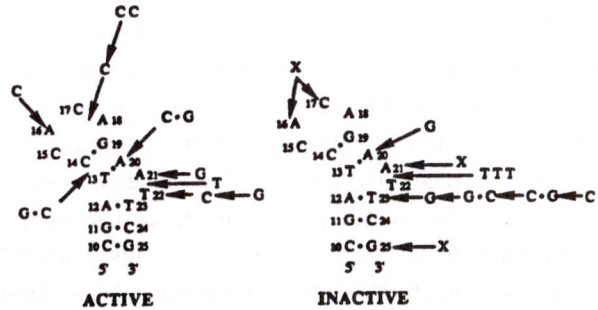

Figure 3. Summary of the point mutations, insertions, and deletions determinming whether oligonucleotide $_{10-25}$ is an active or inactive substrate. Placement position of the arrows indicates whether base(s) is a substitution or insertion. Deletion is indicated by **x**.

as it contains at least three nucleotides; the bulge requires a minimum of two nucleotides but must have fewer than five. With a notable exception involving base-pair changes immediately 3' to the bulge, base changes in the bulge and base-pair changes immediately 5' to the bulge retain substrate activity for NCS-chrom. Conversion of the $T_{23} \cdot A_{12}$ base pair 3' to the bulge (Figure 3) to G·C (or to $A_{23} \cdot T_{12}$) eliminates the substrate property; further, the G·C change results in the absence of **2a** binding (see later). Maintenance of the bulged structure requires stable duplex regions on each side of the bulge. A similar bulged structure, but lacking a loop, formed by the annealing of a linear 8-mer and a 6-mer is an excellent target for cleavage in the thiol-independent reaction. Drugs such as netropsin, which sequester the DNA into nonbulge containing structures inhibit the reaction. In the absence of O_2 strand cleavage is blocked and quantitatively replaced by a presumed drug-DNA covalent adduct.

The most important determinant of competency of a DNA substrate for NCS-chrom-induced site-specific cleavage is the presence of a bulge site containing at least two (and less than five) nucleotides and a sufficient number of base-pairs on either side of the bulge to maintain the bulged structure. The optimal bulge size is probably translated into an optimal extent of bending at the bulge. Since a singly-bulged pyrimidine may loop out from the double helix, in contrast to an ATA-bulge where all bulged bases are stacked within the helix [28], the substrate with a single-bulged nucleotide (Figure 3) may be incompetent because of its position outside the DNA helix and not necessarily because of diminished bending.

The cleavage reaction in the non-thiol reaction is very slow compared with the thiol-dependent reaction (50-fold). This probably is reflective of the activation step, which for glutathione in the presence of DNA is very fast [12]. Nevertheless, the thiol-independent, site-specific cleavage in the bulged DNA is highly efficient and selective. A ratio of about one drug molecule per strand break was calculated under the most optimal reaction conditions. Unlike in the thiol-dependent reaction with duplex DNA where high levels of drug cause breaks at less favored sites, no other cleavage sites are found in the bulged DNA in the non-thiol reaction. This suggests that in the latter reaction DNA damage is restricted to the single site where drug activation is induced. This lesion also differs substantially from the cleavage pattern described earlier at the site of a single-base bulge in duplex DNA in the presence of thiol [19]. In the latter reaction a break is generated on the strand opposite the bulged base at the residue just 3' to the bulge; the bulged residue is not a cleavage target. This interaction between NCS-chrom and single-base bulged DNA is consistent with studies showing that intercalating drugs, including the enediyne dynemicin A, bind selectively to such structures [19, 29-32].

In an effort to identify the active species responsible for hydrogen abstraction from the bulge and to obtain a three-dimensional structure of its complex with bulged DNA we examined the binding of stable drug products, generated in the course of the cleavage reaction, with oligodeoxynucleotides containing the bulged structure [33]. By use of fluorescence quenching, we have found that one drug product, 2a, which is also formed in the absence of bulged DNA and most closely resembles the biradical intermediate (1d) in the cleavage reaction, specifically binds bulged DNA with a K_d in the low micromolar range and competitively inhibits the cleavage reaction. A linear Scatchard plot suggests one mode of binding and since the base-catalyzed cleavage reaction gave a single lesion in high yield, one binding site per bulged DNA can be assumed. ^{1}H NMR determination of the structure formed between 2a and bulged DNA is in progress.

By contrast, the DNA bulge-specific drug end product **3** showed no fluorescence quenching by the bulge-containing oligonucleotide and failed to inhibit **2a** binding. Models of **3** show it to have a highly planar and relatively rigid structure, very different overall from the one expected for **2a**. Also, **2b**, which differs from **2a** only by having a CH_2OH moiety at C2 instead of a hydrogen, does not bind to the 22-mer. A different conformation of the ring system (an alternative, flattened half-chair conformation for ring A of the naphthoate) has been proposed for the **2b** product [27], which will change its overall shape compared to **2a**, especially that of the naphthoate moiety. This might explain how this seemingly small difference results in such a drastic change in binding properties between the two compounds. It seems possible that the DNA bulge structure induces a conformational change in **1d** (or initially in **1c**) from a conformation similar to that of **2a** to one resembling that of **2b** in which C2 and C8" are closer together (see Figure 3 in ref. 27). In such an "averaged" conformation, the exact pucker of the two cis-fused five-membered rings largely dictates the proximity of the two centers C2 and C8", which must be held sufficiently close under the influence of the bulged DNA for the quenching reaction involving the naphthoate and the radical at C2 to occur with C2-C8" bond formation. Further, it is also possible that the shape of the binding site on the bulged DNA is modified by the binding of **2a** --induced fit-- in such a way as to facilitate the quenching reaction. The ability of the C8" of the naphthoate to approach C2 of the core under the influence of the bulged DNA will also depend on its flexibility and whether access to C2 is hindered. Thus, the mere existence of the CH_2OH group may interfere with the proposed conformational change that is proposed to result in the formation of the C2-C8" bond and product **3** during the base-catalyzed reaction on the bulged oligonucleotide.

The described studies indicate that the drug must bind to bulged DNA for it to cleave selectively within the bulge and for it to be converted into product **3**, following quenching of the C2 radical by C8" of the naphthoate moiety. Although it appears that product **3** is always formed in association with the cleavage reaction, it is possible to prepare closely-related bulge-containing oligonucleotides that bind the drug selectively but are not cleavage substrates (Z. Xi, L.S. Kappen and I.H. Goldberg, unpublished data). Such studies will permit further exploration of the role of the DNA bulge conformation in effecting the C2-C8" quenching reaction, since the DNA binding substrate will not be expected to undergo transformation (damage) in the reaction and should be able to function catalytically in product **3** formation. Preliminary experiments support such a role for the DNA and bring to light a new role for DNA as an effector molecule (a deoxyribozyme), that enables carbon-carbon bond formation. Whether this reaction results from a conformational change in **1d**, induced by the bulged DNA (see above), or is a more passive consequence of protection of the radical at C2 from solvent quenching by the DNA, such as described for C-1027-chrom (see later), remains to be clarified.

The experiments on the selective binding of **2a** to the bulged substrate suggest that **1d**, its diradical counterpart, may be, in fact, the species responsible for hydrogen abstraction from the DNA. This wedge-shaped molecule, in which the naphthoate and indacene ring systems are held in a relatively rigid partially-stacked conformation by the spirolactone moiety, might either convert to a precursor of **3** in a relatively concerted reaction by C-5' hydrogen abstraction from the DNA and intramolecular quenching of the C2 radical or undergo the intramolecular reaction first, to form **1e**, which, as shown in Scheme 4, then abstracts the DNA hydrogen. In the latter case, the conformational change involved in **1e** formation, in which a highly planar and rigid molecule is formed, might conceivably be responsible for bringing the radical at C6 closer to the C-5' of the DNA bulge for hydrogen atom abstraction. This drug species and its products are no longer able to bind to the DNA.

NMR studies to elucidate the three-dimensional structure of the **2a**-bulged DNA complex are in progress (A. Stassinopoulos, J. Jie, X. Gao and I.H. Goldberg, unpublished data) and should provide information on the distance of the radical on C6 of **1d** from the C-5' attack site. Preliminary experiments suggest that the wedge-shaped **2a** binds in the bulge space generated by the looping out of the two bulge nucleotides from the DNA helix and stacks with Watson-Crick bases on either side. In fact, **2a** resembles a bisintercalator (or a dinucleotide section of a DNA strand) in which the two overlapping planar ring systems are fixed in relation to each other by a short, rigid spirolactone tether. Thus, it seems possible that synthetic congeners based on **2a** might be developed to have favorable and selective binding affinity for bulged nucleic acid structures. It should also be noted that a related proposal to the one above was made to account for the selective cleavage adjacent to a one-base bulge by thiol-activated NCS-chrom [19] and the cuprous complex of ortho-phenathralene [30]. In these cases, however, a single planar ring system (the napththoate of thiol-activated NCS-chrom) is the moiety believed to be intercalating in the bulge region. A similar model has been proposed for the binding of 9-aminoacridine at a single-base bulge [31].

Recently, in order to use the thiol-independent, DNA bulge-specific reaction as a selective probe of the tertiary structure of naturally occurring bulged structures, we have extended these studies to include high molecular weight single-stranded phage DNAs and long fragments derived therefrom (A. Stassinopoulos and I.H. Goldberg, unpublished data). Bulge-specific cleavage, similar to that described in the model substrates noted above, could be identified in fragments derived from the phage DNA. Similarly, an effort has been made to determine if RNA bulges, including that present in the biologically-relevant single-stranded hairpin HIV-1 transactivation response region (TAR) RNA, are targets for the thiol-independent reaction [34]. In most instances there was little evidence of NCS-chrom-specific cleavage, although in TAR RNA a distinct, but weak, cleavage site could be identified in the central U residue of the three nucleotide bulge. This reaction had all the characteristics of the classical thiol-independent DNA bulge reaction and could be reproduced in an RNA substrate made up of two linear RNA oligomers which can presumably form the same bulged structure. Stronger cleavage was found in the DNA analogue of the TAR RNA. Again, cleavage of the DNA was mainly in the bulge region, although evidence for 4' chemistry was found in addition to 5' chemistry. Given the likely three-dimensional structural difference between the RNA and its DNA analogue, it is not surprising that TAR RNA is a much poorer substrate than its DNA analogue. Perhaps only some small fraction of the TAR RNA bears a structure in space similar to that of the bulged DNA. For that matter, even the DNA analogue of TAR RNA is a relatively poor substrate in the thiol-independent reaction, being only about 10% as effective as the bulged DNA studied earlier [24, 25]. It seems possible that through manipulation of the **2a** structure a compound might be obtained that is speclfic for binding to the TAR RNA bulge. Such studies are in progress.

1.5 NUCLEAR DNA DAMAGE

While most studies have been concerned with enediyne-induced damage of naked DNA, it must be remembered that in the cell nucleus DNA exists in association with nucleosomal structures and chromatin proteins. When DNA damage due to enediynes was assessed in cell nuclei and isolated nucleosome core particles, it was found that NCS-induced damage was restricted to the linker DNA, whereas that due to agents such as calicheamicin, lacking an intercalator moiety, involved the nucleosomal core DNA, with a 10-11 nucleotide periodicity, as well as the linker DNA [35].

2. C1027

C1027, a new highly potent antitumor antibiotic, like neocarzinostatin, consists of a labile chromophore, which is responsible for most of its biological activities, and a noncovalently linked apoprotein [36]. C1027-chrom consists of two functional domains [37], an enediyne core moiety responsible for DNA damage, and benzoxazonilate and sugar moieties, which are presumably involved in the placement of the enediyne moiety in the DNA minor groove. Unlike most other enediynes, C1027 induces damage in DNA in the absence of thiols or reducing compounds. As noted above, NCS-chrom exhibits both types of characteristics. C1027-chrom is presumed to undergo a Bergman-type rearrangement of the enediyne core to form the highly reactive benzenoid diradical

16

species (Scheme 5), which abstracts hydrogen atoms from the deoxyribose of the DNA backbone, as shown in Scheme 2. C1027-chrom generates single-stranded lesions, mainly at A and T residues, and double-stranded lesions, which show a five-nucleotide 3'-stagger of the cleaved residues [3]. Cleavage at the target site on the positive strand involves C-4' chemistry [3,38], whereas that on the negative strand is due to C-5' chemistry [3].

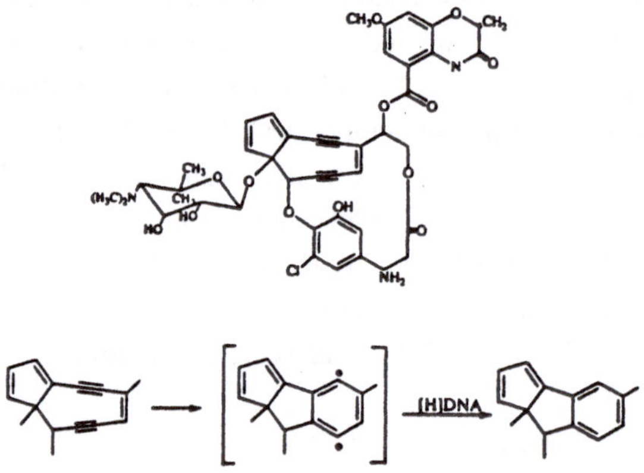

Scheme 5. (A) Chemical structure of C1027 chromophore. (B) Proposed mechanism of activation of C1027 chromophore.

Recently, we have extended the study of the damage reaction using the model $GTTA_1T/ATA_2A_3C$ sequence in oligonucleotides and have obtained conclusive evidence for two types of double-stranded lesions, one involving A_1 and A_2 and the other involving A_1 and A_3 [4]. As shown in Figure 4, hydrogen atom abstraction is mainly from C-4' at A_1 to generate the 4'-hydroxylated abasic product or cleavage with a 3'-phosphoglycolate-ended fragment in both sets of lesions. At A_3, however, mainly 5' chemistry is involved, with the formation of nucleoside 5'-aldehyde-ended fragment. At A_2 the lesion was shown to be an abasic site, formed by selective C-1' chemistry. Lesions at A_2 or A_3 are always part of double-stranded lesions, whereas those at A_1 can also exist as single-stranded ones. Interestingly, the drug radical center that attacks at A_2 or A_3, but not the one involved in attack at A_1, is readily quenched by solvent methanol, so as to produce a single-stranded lesion at A_1. In fact, by using methanol containing carbon-bound deuterium there is a substantial isotope effect on the quenching reaction. In the absence of methanol almost all of the damage at A_1 belongs to double-stranded lesions. These results suggest that in the activated drug-DNA complex the radical center involved in A_1 site attack is better protected by the DNA binding pocket

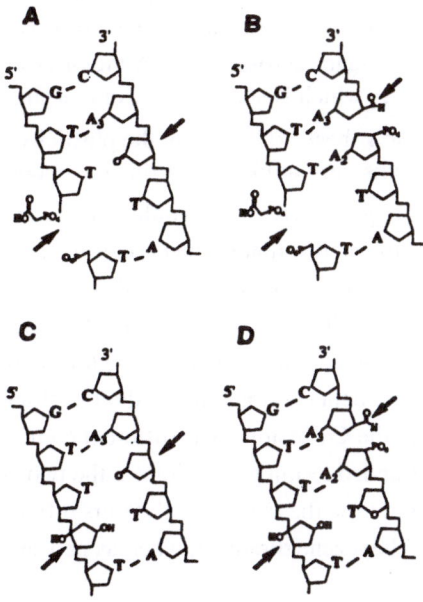

Figure 4. Stylized models of C1027-induced double-stranded lesions in the $GTTA_1T/ATA_2A_3C$ interaction sequence. A and C are the one-nucleotide 3'-staggered double-stranded lesions involving A_1 and A_2; B and D are the two-nucleotide 3'-staggered double-stranded lesions involving A_1 and A_3.

from solvent attack than is the other radical center. A similar finding has been reported in the case of the C6 radical of NCS-chrom complexed with DNA, as cited in [2]. The radical center of C1027-chrom involved in A_2 or A_3 attack also exhibits considerable flexibility in terms of which of the two target nucleotides is attacked. This is shown in experiments where deuterium is substituted at C-1' of A_2, resulting in almost total inhibition of attack at A_2 and substantial shift of the reaction to hydrogen abstraction from C-5' of A_3. This unique shuttling of attack between adjacent nucleotides strongly implies that it is the same drug molecule, with a single binding mode, that has the ability to attack at A_2 or A_3 Consistent with this proposal are the results from quantitative affinity cleavage binding experiments that show virtually identical binding constants for the two types of double-stranded lesions.

The finding that the double-stranded lesion involving A_1 and A_3 is staggered by two nucleotides in the 3' direction is consistent with the intercalation by the benzoxazine moiety of C1027-chrom within the target sites on the complementary DNA strands, as has been demonstrated for NCS-chrom [8,9]. This contrasts with the staggered three nucleotide double-stranded lesion generated by calicheamicin [39], which lacks an intercalating moiety. The intercalated moiety adds the equivalent of a base pair to the intercalation site so that separation of the lesions on each strand by two, rather than three base pairs, represents the shortest distance across the DNA minor groove. A double-stranded lesion involving A_1 and A_2 which is staggered by only a single nucleotide in the 3' direction is most unusual and is best explained by the plasticity of the drug-DNA structure at the A_2, A_3 site. It seems very unlikely that this novel double-stranded lesion, and the drug-DNA complex giving rise to it, would exist as an entirely independent lesion. Further, it is of interest that in the case of neocarzinostatin the double-stranded lesion involves C5'-chemistry on one strand and either C1'- or C4'-chemistry on the other; single-stranded lesions are predominantly associated with the C5' lesion [1,2]. C1027 differs in that C4'-chemistry on one strand is associated with either C5'- or C1'-chemistry on the complementary strand in the double-stranded lesion; single-stranded lesions are mainly due to the C4' lesion. The different combination of chemistries involved in double-stranded lesion generation for the two enediynes may be due to the fact that the distances between the two radical centers differ for the two antibiotics. C1027 involves a 1,4 benezene diradical, whereas neocarzinostatin utilizes a 2,6 indacene diradical.

3. Concluding Remarks

Enediynes are versatile nucleic acid cleaving agents. This property stems from their ready rearrangement under physiological conditions to form diradicals that act as bifunctional reagents able to attack each strand of a nucleic acid duplex: DNA·DNA or DNA·RNA. The attack site specificity stems, to a considerable degree, from the other moieties --sugars, intercalators, etc. -- that are attached to the activatible enedyne. Diradical formation can be triggered by thiols or reducing agents, and, in the case of NCS-chrom, by a base-catalyzed intramolecular mechanism. C 1027-chrom activation is presumed to be thermal. The thiol-independent NCS-chrom reaction generates a diradical species differing substantially in structure from the thiol-induced species and thus in its requirements in the nucleic acid for binding and lesion formation. It is expected that identification of the activated drug structure responsible for bulge-specific cleavage will lead to the design and synthesis of congeners with even more selective nucleic acid bulge damage properties.

As clearly shown for NCS-chrom and C1027-chrom, the flexibility in target site (whether different minor groove carbon-bound hydrogens on the same nucleotide or on adjacent nucleotides) selection is reflective of the nucleic acid microstructure at the attack site. Local geometry, of course, is dependent on sequence and secondary structural features, as influenced by the presence of bulges, mismatches and hybrid structures. Relatively minor structural changes, such as the loss of a 2-amino group on guanine, substitution of a deuterium for a protium, stable wobble mismatches, all lead to marked changes in attack site chemistry. Further, one of the drug radical centers in the drug-nucleic acid complex is particularly susceptible to solvent interactions, resulting in substantial solvent-based alterations in the ratio of double-stranded to single-stranded lesions. The sum total of these diverse features enable the enediynes to act as sensitive probes of nucleic acid structure.

Acknowledgement

The work described in this report was supported by U.S. Public Health Service Grant CA44257 from the National Institutes of Health.

References

1. Goldberg, I.H. (1991) Mechanism of neocarzinostatin action: role of DNA microstructure in determination of chemistry of bistranded oxidative damage. *Accounts of Chemical Research* 24, 191-198.
2. Goldberg, I.H. and Kappen, L.S. (1994) Neocarzinostatin: chemical and biological basis of oxidative DNA damage, in D.B. Borders and T. Doyle (eds.), *Enediyne Antibiotics as Antitumor Agents*, Marcel Dekker, Inc., New York and Basel, pp. 327-362.
3. Xu, Y.-j., Zhen, Y.-S., and Goldberg, I.H. (1994) C1027 chromophore, a potent new enediyne antitumor antibiotic, induces sequence-specific double-strand DNA cleavage. *Biochemistry* 33, 5947-5954.
4. Xu, Y.-J., Xi, Z., Zhen, Y.-S. and Goldberg, I.H. (in press) A single binding mode of activated enediyne C1027 generates two types of DNA double-strand lesions: deuterium isotope induced shuttling between adjacent nucleotide target sites. *Biochemistry*.
5. Meschwitz, S. M. and Goldberg, I. H. (1991) Selective abstraction of ^{2}H from C-5' of thymidylate in an oligodeoxynucleotide by the radical center at C-6 of the diradical species of neocarzinostatin: chemical evidence for the structure of the activated drug-DNA complex. *Proc. Natl. Acad. Sci. USA* 88, 3047-3051.
6. Meschwitz, S. M., Schultz, R. G., Ashley, G. W., and Goldberg, I. H. (1992) Selective abstraction of ^{2}H from C-1' of the C̲ residue in AGC̲•ICT by the radical center at C-2 of neocarzinostatin chromophore: structure of the drug-DNA complex responsible for bistranded lesion formation. *Biochemistry* 31, 9117-9121.
7. Galat, A. and Goldberg, I.H. (1990) Molecular models for neocarzinostatin damage of DNA: Analysis of sequence dependence in 5'GAGCG·5'CGCTC. *Nucleic Acids Res.* 18, 2093-2099.
8. Gao, X., Stassinopoulos, A., Rice, J.S., and Goldberg, I.H. (1995) Structural basis for the sequence specific DNA strand cleavage by the enediyne neocarzinostatin chromophore. Structure of the post-activated chromophore-DNA complex. *Biochemistry* 34, 40-49.

9. Gao, X., Stassinopoulos, A., Gu, J. and Goldberg, I.H. (1995) NMR studies of the post-activated neocarzinostatin chromophore-DNA complex. Conformational changes induced in drug and DNA. Symposium-in-Print on "Recent Advances in DNA Binding Agents." *Bioorg. Med. Chem.* **3**, 795-809.

10. Povirk, L.F., Houlgrave, C.W., and Han, Y.-H. (1988) Neocarzinostatin-induced DNA base release accompanied by staggered oxidative cleavage of the complementary strand. *J. Biol. Chem.* **263**, 19263-19266.

11. Dedon, P.C., Jiang, Z.-W., and Goldberg, I. H. (1992) Neocarzinostatin-mediated DNA damage in a model AGT•ACT site: Mechanistic studies of thiol-sensitive partitioning of C-4' DNA damage products. *Biochemistry* **31**, 1917-1927.

12. Dedon, P.C. and Goldberg, I.H. (1992) Influence of thiol structure on neocarzinostatin activation and expression of DNA damage. *Biochemistry* **31**, 1901-1917.

13. Kappen, L.S., Chen, C.-Q. and Goldberg, I.H. (1988) Atypical abasic sites generated by neocarzinostatin at sequence-specific cytidylate residues in oligodeoxynucleotides. *Biochemistry* **27**, 4331-4340.

14. Kappen, L. S. and Goldberg, I. H. (1992) Neocarzinostatin acts as a sensitive probe of DNA microheterogeneity: Switching of chemistry from C1' to C4' by a G•T mismatch 5' to the site of DNA damage. *Proc. Natl. Acad. Sci. USA* **89**, 6706-6710.

15. Patel, D.J., Kozglowski, S.A., Ikuta, S., and Itakura, K. (1984) Dynamics of DNA duplexes containing internal G·T, G·A, A·C, and T·C pairs: hydrogen exchange at and adjacent to mismatch sites. *Fed. Proc. Fed. Am. Soc. Exp. Biol.* **43**, 2663-2670.

16. Kneale, G., Brown, T., Kennard, O., and Rabinovich, D. (1985) GT base-pairs in a DNA helix: the crystal structure of d(G-G-G-G-T-C-C-C). *J. Mol. Biol.* **186**, 805-814.

17. Hunter, W.N., Brown, T., Kneale, G., Anand, N., Rabinovich, D., and Kennard, O. (1987) The structure of guanosine-thymidine mismatches in B-DNA at 2.5-Å resolution. *J. Biol. Chem.* **262**, 9962-9970.

18. Kappen, L. S. and Goldberg, I. H. (1992) Mismatch-induced switch of neocarzinostatin attack sites in the DNA minor groove. *Biochemistry* **31**: 9081-9089.

19. Williams, L.D. and Goldberg, I.H. (1988) Selective strand scission by intercalating drugs at DNA bulges. *Biochemistry* **27**, 3004-3011.

20. Zeng, X., Xi, Z., Kappen, L.S., Tan, W. and Goldberg, I.H. (in press) Double-stranded damage of DNA-RNA hybrids by neocarzinostatin chromophore: Selective C-1' chemistry on the RNA strand. *Biochemistry* .

21. Giese, B., Beyrich-Graf, X., Erdmann, P., Giraud, L., Imwinkelried, P., Muller, S.N., and Schwitter, U. (1995) Cleavage of single-stranded 4'-oligonucleotide radicals in the presence of O_2. *J. Am. Chem. Soc.* **117**, 6146-6147.

22. Duff, R.J., de Vroom, E., Geluk, A., and Hecht, S. M. (1993) Evidence for C-1' hydrogen abstraction from modified oligonucleotides by Fe-bleomycin. *J. Am. Chem. Soc.* **115**, 3350-3351.

23. Napier, M.A., Holmquist, B., Strydom, D.J. and Goldberg, I.H. (1981) Neocarzinostatin chromophore: Purification of the major active form and characterization of its spectral and biological properties. *Biochemistry* **20**, 5602-5608.

24. Kappen, L.S. and Goldberg, I.H. (1993) DNA conformation induced activation of an enediyne anticancer drug for site-specific cleavage. *Science* **261**, 1319-1321.

25. Kappen. L.S. and Goldberg, I.H. (1993) Site-specific cleavage at a DNA bulge by neocarzinostatin chromophore via a novel mechanism. *Biochemistry* **32**, 13138-13145.

26. Hensens, O.D., Helms, G.L., Zink, D.L., Chin, D.-H., Kappen, L.S., and Goldberg, I.H. (1993) Bifunctional involvement of the hydroxy naphthoate moiety in the activation of neocarzinostatin chromophore in DNA-mediated site-specific cleavage. *J. Am. Chem. Soc.* **115**, 11030-11031.

27. Hensens, O.D., Zink, D.L., Chin, D.-H., Stassinopoulos, A., Kappen, L.S., and Goldberg, I.H. (1994) Spontaneous generation of a biradical species of neocarzinostatin chromophore: role in DNA bulge-specific cleavage. *Proc. Natl. Acad. Sci. USA* **91**, 4534-4538.

28. Rosen, M.A., Shapiro, L., and Patel, D.J. (1992) Solution structure of a trinucleotide A-T-A bulge loop within a DNA duplex. *Biochemistry* **31**, 4015-4026.

29. Nelson, J.W. and Tinoco, I. Jr. (1985) Ethidium binds more strongly to a DNA double helix with a bulged cytosine than to a regular double helix. *Biochemistry* **24**, 6416-6421.

30. Williams, L.D., Thivierge, J., and Goldberg, I.H. (1988) Specific binding of o-phenanthroline at a DNA structural lesion. *Nucleic Acid Res.* **16**, 11607-11615.

31. Woodson, S.A. and Crothers, D.M. (1988) Binding of 9-aminoacridine to bulged-base DNA oligomers from a frame-shift hot spot. *Biochemistry* **27**, 8904-8914.

32. Kusakabe, T., Maekawa, K., Ichikawa, A., Uesugi, M., and Sugiura, Y. (1993) Conformation-selective DNA strand breaks by dynemicin: a molecular wedge into flexible regions of DNA. *Biochemistry* **32**, 1169-11675.

33. Yang, C.F., Stassinopoulos, A. and Goldberg, I.H. (1995) Specific binding of the biradical analog of neocarzinostatin chromophore to bulged DNA: Implications for thiol-independent cleavage. *Biochemistry* **34**, 2267-2275.

34. Kappen, L.S. and Goldberg, I.H. (1995) Bulge-specific cleavage in transactivation response region RNA and its DNA analogue by neocarzinostatin chromophore. *Biochemistry* **34**, 5997-6002.

35. Yu, L., Goldberg, I.H. and Dedon, P.C. (1994) Enediyne-mediated DNA damage in nuclei is modulated at the level of the nucleosome. *J. Biol. Chem.* **269**, 4144-4151.

36. Otani, T., Minami, Y., Marunaka, T., Zhang, R. and Xie, M.Y. (1988) A new macromolecular antitumor antibiotic, C-1027 II. Isolation and physico-chemical properties. *J. Antibiot.* **41**, 1580-1585.

37. Yoshida, K., Yoshinori, M., Azuma, R., Saeki, M., and Otani, T. (1993) Structure and cycloaromatization of a novel enediyne, C-1027 chromophore. *Tetrahedron Lett.* **34**, 2637-2640.

38. Sugiura, Y. and Matsumoto, T. (1993) Some characteristics of DNA strand scission by macromolecular antitumor antibiotic C-1027 containing a novel enediyne chromophore. *Biochemistry* **32**, 5548-5553.

39. Zein, N., Sinha, A.M., McGahren, W.J., and Ellestad, G.A. (1988) Calicheamicin γ_1^{I}: an antitumor antibiotic that cleaves double-stranded DNA site specifically. *Science* **240**, 1198-1201.

THE INTRINSIC FLEXIBILITY AND DRUG-INDUCED BENDING OF CALICHEAMICIN DNA TARGETS

AARON SALZBERG, PUNAM MATHUR and PETER DEDON
Division of Toxicology
Massachusetts Institute of Technology, 16-336
Cambridge, MA, USA 02139

Abstract. Calicheamicin γ_1^I is an enediyne antitumor antibiotic that causes radical-mediated bistranded DNA cleavage. We recently proposed that calicheamicin recognizes a structural discontinuity at the 3'-ends of oligopurine sequences and that drug binding further alters DNA structure [Yu *et al.* (1995) *Biorg. Med. Chem.* **6**, 729-741]. We now present evidence to support the hypothesis that calicheamicin bends DNA at its target sequences. Gel migration studies of DNA polymers containing phased calicheamicin recognition sequences demonstrate that these species do not possess intrinsic curvature. However, the same sequences readily form small circles (<189 base pairs) which suggests that calicheamicin binding sites may be flexible. Addition of calicheamicin ε, the aromatized form of calicheamicin γ_1^I, shifts the population of circles to smaller sizes (<149 bp). When the binding sites are placed out of phase with the helical repeat, the shift to smaller circle sizes is significantly reduced. These results suggest that drug binding increases the probability of cyclization by bending the DNA rather than improving the helical alignment of the ends. In circle closure assays with a 273 bp construct containing phased calicheamicin recognition sequences, a drug concentration of 10 μM increased the probability of cyclization more than 10-fold. Together, these results suggest that calicheamicin causes significant bending of its DNA targets. This targeting of specific structural features of DNA may have significant biological implications for calicheamicin activity *in vivo*.

1. Introduction

The enediyne antitumor agent calicheamicin has received significant attention both for its potency as a cytotoxic agent [1,2] and for the chemistry of its DNA damage [3,4].

Like other members of the enediyne family, calicheamicin produces DNA damage *via* a diradical intermediate that, when positioned in the minor groove, abstracts deoxyribose hydrogen atoms (Fig. 1) [3,4]. The sugar radicals undergo oxygen-dependent reactions to form abasic sites or strand breaks with resulting DNA damage products dependent on the position of the abstracted hydrogen. Single molecules of calicheamicin produce mainly double-stranded lesions, with damage sites on each strand staggered by 2 base pairs (bp) in a 3'-direction [2,5,6].

B. Meunier (ed.), DNA and RNA Cleavers and Chemotherapy of Cancer and Viral Diseases, 23–36.

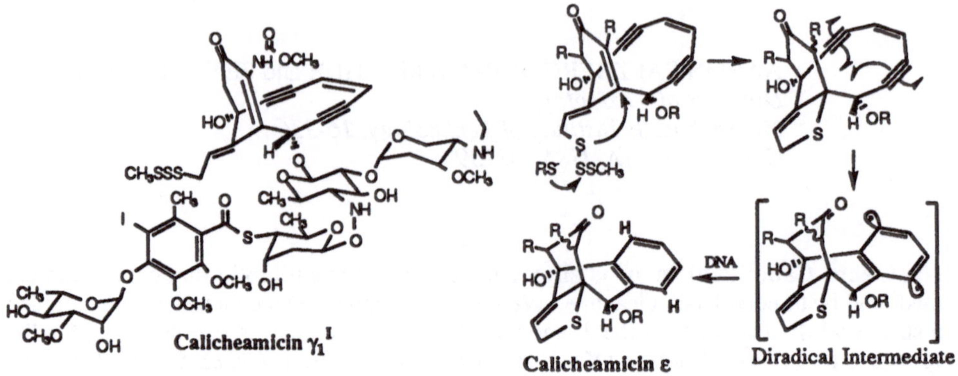

Figure 1. Structure and activation of calicheamicin

There has been significant progress in defining the elements of calicheamicin target recognition. It is now generally accepted that the apparent tetrapurine sequence selectivity displayed by calicheamicin reflects recognition of sequence-dependent local DNA conformation and dynamics rather than a direct reading of specific base sequences [2,7-17]. Isotope transfer studies, hydroxyl radical footprinting, and NMR solution structures of the calicheamicin/DNA complex all suggest that calicheamicin binds to target sequences with its carbohydrate tail extending toward the 5'-end of the tetrapurine strand [10,13,15,16,18,19]. The tail, which appears to adopt a right-handed screw that complements the minor groove [13,15,16,20], is responsible for the sequence-selectivity and is a major contributor to the binding energetics [8,9,12,21].

Several investigations suggest that both calicheamicin and its DNA binding site act together to affect target selection. According to NMR studies, the relatively rigid carbohydrate side chain [22] is proposed to induce a widening of the minor groove along the purine•pyrimidine tract binding site [13,15,16]. On the basis of circular dichroism studies, Ellestad and coworkers have proposed that calicheamicin causes overwinding of the DNA helix [23]. We have also observed that calicheamicin increases the apparent negative superhelical density of plasmid DNA (L. Yu and P. Dedon, unpublished). As yet, however, there is no unified model for the calicheamicin target recognition process.

Recent studies of calicheamicin-induced damage in nuclei and nucleosomes have shed light on the structure of drug binding sites. We observed that calicheamicin is capable of binding to the bent and dynamically-constrained DNA of the nucleosome core, with cleavage sites positioned where the minor groove faces away from the histone core [24,25]. In studies of calicheamicin-induced damage in reconstituted nucleosomes, we observed three novel features of calicheamicin target selection: (1) that bending of nucleosome DNA increased drug-induced cleavage at one site; (2) that there is little correlation between minor groove width *per se* and calicheamicin target selection; and (3) that calicheamicin appears to recognize the structural discontinuity at the 3'-ends of purine tracts [25]. The last point is supported by a review of the literature, as shown in Figure 2.

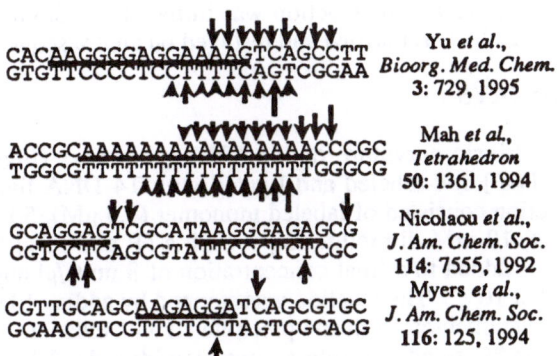

Figure 2. Examples of calicheamicin-induced DNA damage at the 3'-ends of purine tracts.

These observations led us to propose a model for calicheamicin interaction with its DNA targets [25]. In this model, the drug recognizes a structural perturbation (*i.e.*, bend or hinge) lying 3' to the purine tract at or beyond the purine-pyrimidine junction. We now present experimental results consistent with the hypothesis that calicheamicin binds to flexible DNA sequences and bends the DNA.

2. Materials and Methods

2.1. MATERIALS

Calicheamicins $\gamma_1 I$ and ε were generously provided by Dr. George Ellestad, Wyeth-Ayerst Research, and were stored in methanol at -80°C. Enzymes were obtained from New England Biolabs and all other chemicals were reagent grade.

2.2. SYNTHESIS AND CLONING OF DUPLEX OLIGOMERS THAT CONTAIN CALICHEAMICIN BINDING SITES

Three 21 bp duplex DNA constructs (Figure 3) were prepared by annealing complementary oligonucleotide strands. These monomers were ligated together with T4 DNA ligase to produce a population of polymers of integral monomer lengths. The polymers were then ligated into a modified pUC19 cloning vector containing a single *Sap*I site in the polylinker region (A. Salzberg, P. Dedon, unpublished). Individual clones were characterized by DNA sequencing and polyacrylamide gel electrophoresis.

2.3. LABELING OF BINDING SITE DNA FRAGMENTS AND CALICHEAMICIN $\gamma_1 I$ DAMAGE REACTIONS

To assess calicheamicin-induced damage, cloned DNA constructs were digested with *Eco*RI, 5'-[32P] end-labeled with T4 polynucleotide kinase and γ-[32P]-ATP, purified on G25 Quick-spin Columns (Boehringer), and digested with *Hind*III [26]. The labeled fragment containing the insert was purified by elution from a 10% polyacrylamide gel [26]. Damage reactions consisted of labeled DNA (~50,000 cpm), 30 µg/ml of calf thymus DNA, calicheamicin $\gamma_1 I$ (0-30 nM), 1% methanol, 10 mM glutathione, 50 mM

HEPES, 1 mM EDTA, pH 7. The reaction was initiated by adding calicheamicin γ_1^I. After 1 hr at 0°C, the DNA was purified and resolved on an 8% sequencing gel.

2.4. LIGATION ASSAYS

To determine the intrinsic curvature of the DNA monomers (Figure 3), the 21 bp constructs were 5'-[32P] end-labeled and ligated with T4 DNA ligase to form linear polymers. The reaction consisted of labeled monomer (7.5 µM), 50 mM Tris-HCl, pH 7.8, 10 mM MgCl$_2$, 10 mM dithiothreitol, 1 mM ATP and 50 µg/ml bovine serum albumin. Ligase was added to a final concentration of 8 units/µl and the mixture was incubated overnight at 16°C. The reaction was stopped by adding EDTA, and the DNA was extracted with phenol/chloroform and precipitated with ethanol. The DNA products were then resolved on 8% nondenaturing polyacrylamide gels. Ligated *Bam*HI linkers (New England Biolabs) were used as "straight" DNA controls [27].

To assess the formation of DNA circles, labeled binding site constructs were reacted as before, except at lower DNA concentrations (4.5 pM), in the presence of varying amounts of calicheamicin ε; methanol was added to the control reactions. Identification of circular DNA was accomplished either by two-dimensional gel electrophoresis [28] or by digestion of linear fragments with *Bal*31 exonuclease prior to one-dimensional electrophoresis [29]. The specific circle sizes where determined by denaturing gel analysis of DNA eluted from the nondenaturing gel [28].

In both sets of experiments, the quantity of radioactivity in individual bands was determined by phosphorimager analysis (Molecular Dynamics). In the circularization studies, the radioactivity in each band was normalized according to circle size and presented as a fraction of the total radioactivity in all the bands considered.

2.5. CIRCLE CLOSURE ASSAYS

Circle closure assays were performed to assess the effect of calicheamicin ε on the rate of DNA cyclization. A 273 bp polymer of monomer #9 was 5'-[32P] end-labeled, purified twice on G-50 Quick-spin columns, and combined with cold polymer at a ratio of less than 1:20 [30]. The DNA was ligated at 20°C in a reaction containing 12 µg/ml of DNA; 50 mM Tris-HCl, pH 7.8; 10 mM MgCl$_2$; 10 mM dithiothreitol; 1 mM ATP; 50 µg/ml bovine serum albumin; 1.3% v/v MeOH; 0 or 10 µM calicheamicin; and 0.01 units/µl T4 DNA Ligase. At various times, 10 µl aliquots were removed and quenched by adding 5 µl of 100 mM EDTA, 0.04% w/v bromophenol blue, and 5% glycerol, and heating at 65°C for 10 minutes. Samples were cooled to room temperature and loaded directly onto a 4% agarose gel (2% NuSeive GTG and 2% high-melting agarose). The quantity of radioactivity in individual bands was determined by phosphorimager analysis (Molecular Dynamics).

3. Results

3.1. PREPARATION OF DUPLEX DNA CONSTRUCTS

We have undertaken studies to define the curvature and flexibility of calicheamicin DNA target sequences and to assess the effect of drug binding on DNA conformation. The hypothesis is that calicheamicin targets flexible DNA sequences and either bends the DNA or traps it in a bent conformation.

To investigate the structural characteristics of calicheamicin binding sites, polymers were prepared with the target sequences aligned in phase with the helical repeat of B-DNA. This phasing ensures that any bending or anisotropic flexibility introduced by the binding sequences will add constructively and lead to macroscopic changes in DNA conformation that can be observed by gel electrophoresis or circularization assays.

As shown in Figure 3, we have prepared 21 bp duplex oligonucleotides that contain two calicheamicin binding sites: AGGA•TCCT [3] and ACAA•TTGT [10,14]. These sequences have been shown to be high affinity binding sites for both calicheamicins γI and ε [10,14]. We also prepared a construct with two A-tracts in phase with the helical repeat [28]. This serves as a control for intrinsically curved DNA.

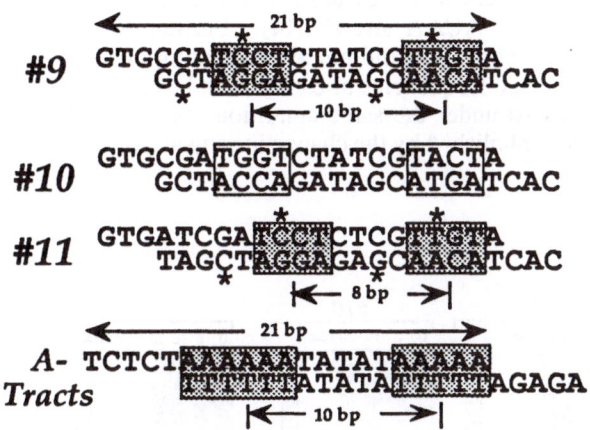

Figure 3. Calicheamicin binding site constructs. Shaded boxes encompass the binding sites (#9, #10, #11) or curved DNA sequences (A-Tracts). Asterisks denote sites of cleavage produced by calicheamicin γ.

In monomer #9, the binding sites are separated from each other by 10 bp. When the monomer is self-ligated, these sites are, on average, aligned with the 10.5 bp helical repeat of B-DNA. The sites are further constructed with the purine-rich sequences on the same strand to ensure that the drug binds with the same orientation at each site [10,18,19]. Should these sites possess intrinsic curvature or become curved upon drug binding, polymers assembled from monomer #9 will develop planar curvature. Construct #10 contains binding sites that have been disrupted by exchanging the central two base pairs on opposite strands. Polymers constructed from this monomer serve as a control for a similar DNA sequence that does not contain drug binding sites. Finally, construct #11 contains the same binding site sequences as #9, except that the sites are now separated by 8 bp. These out-of-phase polymers test whether or not increased cyclization results from alterations in helical alignment rather than DNA bending. If the drug affects DNA cyclization by altering the helical alignment of the DNA ends, then polymers consisting of construct #11 will behave in a similar manner to those of monomer #9. If instead, the drug bends the DNA, then the out-of-phase polymer would adopt a three-dimensional "zig-zag" conformation rather than a planar curve and will not cyclize as efficiently as the in-phase polymer.

3.2. CALICHEAMICIN-INDUCED DAMAGE IN THE BINDING SITE CONSTRUCTS

To ensure that the DNA monomers possessed the predicted binding sites, cloned polymers of each construct were subjected to reactions with calicheamicin γ_1^L. The resulting calicheamicin cleavage products were then separated and analyzed by polyacrylamide gel electrophoresis. In these studies, we make the reasonable assumption that the frequency of drug-induced damage correlates with the affinity of either calicheamicin γ_1^I or ε for the binding sites. The results of this study are shown in Figure 4. In both constructs #9 and #11, calicheamicin γ_1^I produced damage only at the expected sites. Consistent with previous studies, damage at the AGGA site occurred more frequently than the ACAA site [10,14]. Furthermore, the spacing between binding sites in the two constructs did not affect binding of the drug, as suggested by the same relative cleavage frequencies at each site in both constructs. The absence of damage in construct #10 under the same conditions as #9 and #11 indicates that the binding sites have been abolished by the change in sequence.

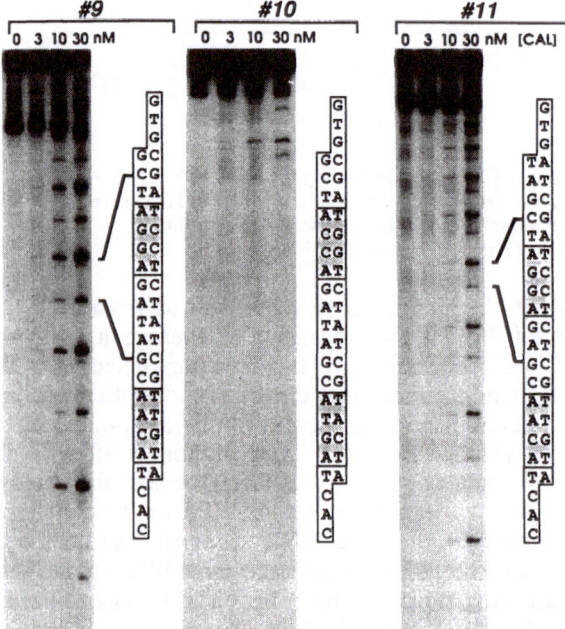

Figure 4. Damage produced by calicheamicin γ in the binding site constructs. Cloned polymers of constructs #9, #10, and #11 were end-labeled, treated with calicheamicin γ (0-30 nM), and resolved on a sequencing gel. Sequence diagrams indicate location of the damage sites in the constructs.

3.3. OLIGOMERS CONTAINING AGGA AND ACAA BINDING SITES DO NOT POSSESS SIGNIFICANT STATIC CURVATURE

To assess whether calicheamicin binding sites are intrinsically curved, we examined the migration of the DNA constructs on nondenaturing polyacrylamide gels. These studies are premised upon the anomalously slow migration of curved DNA sequences on polyacrylamide gels when compared to "straight" sequences of the same size, possibly due to the trapping of curved conformations in the gel matrix [27,31-33]. Since these effects become more pronounced as the DNA length increases, the 21 bp binding site constructs were first polymerized, by self-ligation, to form longer molecules exhibiting observable macroscopic curvature. The family of polymers was then resolved on an 8% nondenaturing polyacrylamide gel. The mobility of these fragments relative to "straight" DNA, in this case polymers of 10 bp *Bam*HI linkers, is expressed as the ratio of apparent length to true length in base pairs (R_L).

An autoradiograph of the results with construct #9 is shown in Figure 5, while the results for all of the constructs are depicted graphically in Figure 6. None of the binding site constructs exhibited significant migration anomalies relative to the straight DNA standard (Figure 6). These results stand in striking contrast to the retarded migration of the A-tract construct. As expected for a DNA sequence possessing intrinsic curvature, the anomalous mobility of the A-tract polymers increases as the chain length grows [27]. We thus conclude that the constructs containing AGGA and ACAA binding sites do not possess significant static curvature.

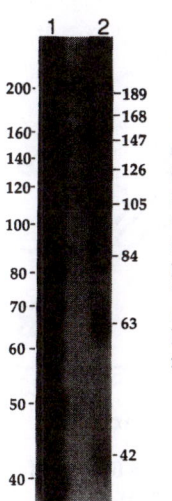

Figure 5. Polyacrylamide gel migration of construct #9 polymers. DNA fragments resulting from ligation of construct #9 (lane 2) were resolved on an 8% native polyacrylamide gel along with polymers of a 10 bp *Bam*HI linker (lane 1). Fragment sizes (bp) are noted in the margins.

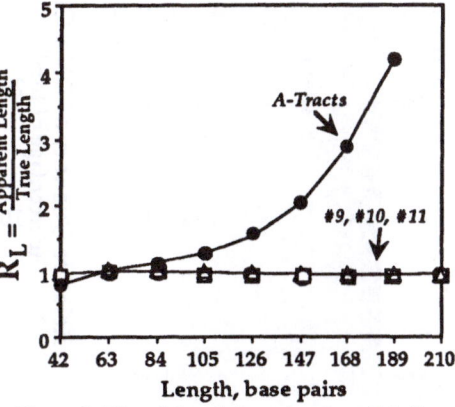

Figure 6. Plot of the relative migration of binding site construct polymers on polyacrylamide gels *versus* the length of the DNA fragments.

3.4. BINDING OF CALICHEAMICIN-ε CAUSES A SHIFT IN THE POPULATION OF CIRCLES FORMED IN THE LIGATION ASSAY

The results of the preceding studies suggested that the calicheamicin binding site constructs did not possess intrinsic curvature. However, this did not rule out the possibility that the binding sites may be flexible sequences. When curved or anisotropically flexible DNA sequences are repeated in phase with the helical repeat of

DNA, the two ends of the DNA polymer will be in proximity to each other more frequently than in "straight" DNA [27,28,30,34]. Experimentally, this can be observed by covalently "trapping" the ends when they are near each other with DNA ligase. The resulting circles can then be resolved from linear polymers by two-dimensional gel electrophoresis [28] or by removing linear fragments with an exonuclease (*e.g.*, *Bal*31) [29].

We performed such a study with calicheamicin binding site construct #9. For these studies, we used calicheamicin ε, the postactivated form of calicheamicin γ1I (Figure 1), since the presence of dithiothreitol in the ligase stock can activate calicheamicin γ1I to produce double-strand breaks in the DNA. Previous studies indicate that both forms of calicheamicin bind to the same DNA sequences [10,14], although the ε form binds with lower affinity [23].

An autoradiograph of the results is shown in Figure 7, and the results are depicted graphically in Figure 8. In the absence of calicheamicin ε, the population of circle sizes occurs with a mode of 189 bp. This relatively small circle size is consistent with a flexible DNA sequence, but further studies are necessary to define the flexibility of construct #9 relative to other DNA sequences.

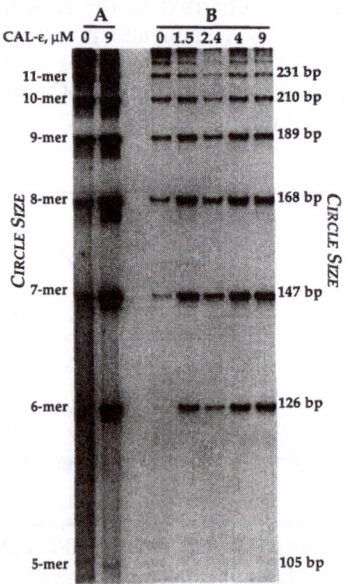

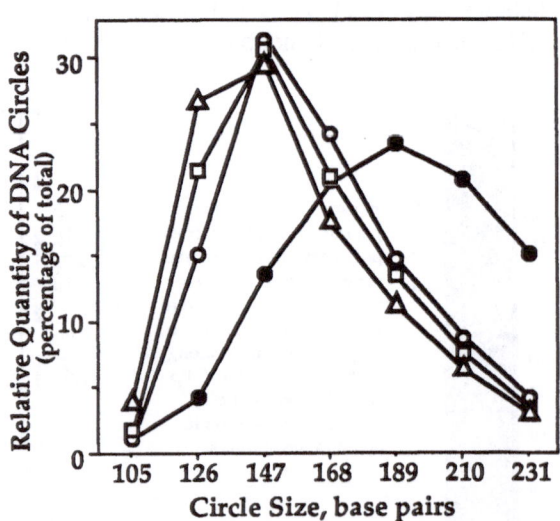

Figure 7. Effect of calicheamicin ε on the formation of DNA circles. Construct #9 (± calicheamicin ε) was treated with DNA ligase and the products were digested with *Bal*31 to remove linear DNA. The DNA was then resolved on a 10% polyacrylamide gel. A: undigested DNA; B: DNA circles remaining after *Bal*31 digestion.

Figure 8. Distribution of DNA circles as a function of calicheamicin ε concentration. The quantity of DNA circles from gels similar to that shown in Figure 7 is plotted *versus* the circle size. Calicheamicin ε concentrations: ●, 0 μM; O, 1 μM; □, 4 μM; Δ, 12 μM.

The addition of 1 μM calicheamicin ε (~5 drug molecules/binding site) shifts the population distribution of circle sizes towards a new mode of 147 bp. Raising the drug concentration shifts the distribution further towards the left, increasing the production of smaller circles. The drug does not appear to affect the activity of the ligase as suggested by the formation of linear polymers in all cases (see the undigested samples, "A," in Figure 7).

Three factors could account for this observed increase in the cyclization probability of the construct #9 polymers: (1) the addition of calicheamicin to the reaction mixture may act to increase the probability of cyclization regardless of the DNA sequence; (2) the binding of calicheamicin to the DNA construct #9 introduces phased bends into the polymer, bringing the ends closer together and increasing the probability of cyclization; and/or (3) the binding of calicheamicin alters the helical rotation of the polymer such that the ends are more favorably aligned for cyclization.

To test for sequence-independent effects and the contribution of helical rotation to cyclization, we repeated the ligation studies with binding site construct #11. In DNA polymers constructed by the self-ligation of monomer #11, the two calicheamicin binding sites are out of phase with the helical repeat. If sequence plays no role in the increased cyclization probability, then this polymer would behave as monomer #9 in the presence of calicheamicin. A similar response would be expected if calicheamicin binding altered DNA structure solely by changing the helical repeat, since the realignment of the ends is independent of the phasing of the damage sites.

The results of the studies are shown in Figure 9. While there is a slight shift to smaller circle sizes in the presence of similar concentrations of calicheamicin ε, drug binding to construct #11 did not produce the large shift seen previously when binding sites were in phase with the helical repeat. This result suggests that calicheamicin either causes significant bending of the DNA or traps it in a bent conformation.

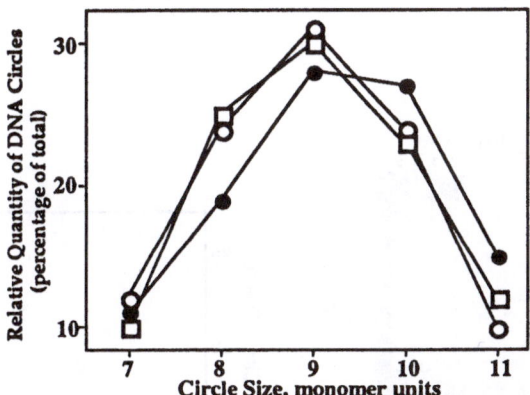

Figure 9. Effect of calicheamicin ε on the distribution of DNA circles formed from construct #11. The quantity of DNA circles from a gel similar to that shown in Figure 7 is plotted *versus* the circle size. Calicheamicin concentrations: ●, 0 µM; ○, 0.5 µM; □, 10 µM.

3.5. CALICHEAMICIN-ε SIGNIFICANTLY INCREASES THE RATE OF CIRCLE FORMATION

The preceding studies indicate that calicheamicin bends DNA upon binding. However, obtaining quantitative information from the self-ligation of 21 bp DNA constructs is hampered by our inability to identify all the products resulting from the continued polymerization of a specific size polymer. We are therefore unable to determine the probability of cyclization, j, which is defined as the ratio of K_c to K_a, where K_c and K_a are the equilibrium constants for cyclization and bimolecular association, respectively. Expressed in molar concentration units, j can be thought of as the concentration of one end of the polymer about the other [29]. Knowing how the probability of cyclization changes with respect to the phasing of calicheamicin binding sites and the degree of drug binding might allow us to determine the relative contributions of flexibility, bending, and/or helical rotation. Furthermore, by testing other constructs, we might be able to identify the locus of flexure/bending within the molecule.

To address these issues, we have begun to perform cyclization assays to determine the probability of cyclization for specific polymers. Using methods developed by several researchers [29,30,35-37], we ligate a single short (<500 bp) DNA fragment under conditions that allow us to measure the disappearance of substrate and the appearance of products so that the relative reaction rates, and therefore the probability of cyclization, can be determined [29,30,35-37].

The results of such a study with calicheamicin ε binding to a 273 bp polymer of construct #9 are shown in Figure 10. The data is presented graphically in Figure 11. In the absence of calicheamicin, the probability of cyclization is ~10^{-7} M. This value is larger than that determined by other groups for similarly sized DNA fragments [34,36,37], perhaps due to the three bp overhangs (see Figure 3), but less than that expected for a polymer consisting of phased A-tracts [29,38]. Upon addition of calicheamicin ε, however, the rate of circle formation increases dramatically. At 10 μM drug (~10 molecules of drug/binding site), the j-factor increases over 10-fold. (This assumes that the rate of dimerization, measured with a 105 bp polymer of construct #9 that failed to circularize, remains constant when calicheamicin is added.)

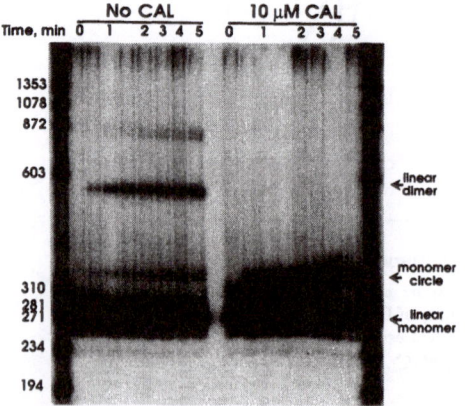

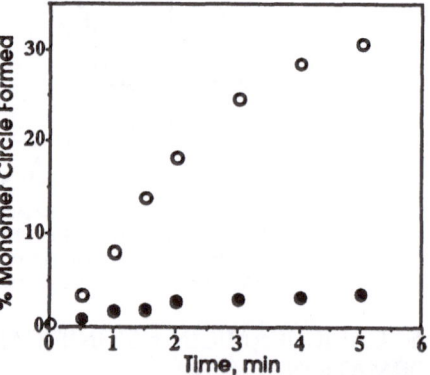

Figure 10. Effect of calicheamicin ε on the rate of circularization of the 273 bp polymer of construct #9. Ligation products were resolved on a 4% agarose gel and the bands quantitated by phosphorimager analysis. The outermost lanes contain labeled size markers and sizes (bp) are indicated in the margin.

Figure 11. Graphical analysis of calicheamicin-induced circle formation. The quantity of 273 bp DNA circles from gels similar to that shown in Fig. 10 are plotted *versus* time (min). Calicheamicin ε concentrations: ●, 0 μM; ○, 10 μM.

We have presented evidence consistent with the bending of DNA by calicheamicin. While further study is needed to define the magnitude of bending, the extent of helical rotation, and the flexibility of the binding site constructs, the present studies reveal several important features of calicheamicin target recognition.

It is clear from these and other studies that calicheamicin recognizes the conformation and dynamics of oligopurine sequences. The NMR structural studies performed by the Kahne [15,16] and Nicolaou groups [13] revealed that calicheamicin caused changes in deoxyribose conformation consistent with widening of the minor groove. Kahne and coworkers further argued that the carbohydrate side chain, extending in a 3'-direction along the purine tract, presses against the pyrimidine strand to produce this widening [15,16]. They suggested that the "flexibility" of the purine•pyrimidine tract is a major determinant of drug binding. This model is not inconsistent with our observations of calicheamicin-induced DNA bending, although we would expect that the major contribution to bending would arise outside the purine tract.

However, the apparent preference of calicheamicin for the 3'-ends of purine tracts adds another facet to the target recognition process. Tullius and coworkers observed that calicheamicin damage is most frequent at the 3'-end of poly A-tracts. They proposed that calicheamicin binding is determined by a narrow minor groove and the structural discontinuity at the 3'-end of the A-tract [11]. While we propose that minor groove width *per se* is not a major determinant of calicheamicin binding [25], we have extended this model to include the structural and dynamic properties at the 3'-end of mixed purine tracts. Our present observations suggest that calicheamicin prefers to bind to DNA sequences that are intrinsically curved (A-tracts) or that can be readily bent (3'-ends of mixed purine sequences). However, it is speculative at this point to interpret the small circles formed by the calicheamicin binding site construct #9 (126-189 bp) as evidence that mixed sequence purine tracts are hyperflexible sequences.

While our preliminary data suggests that calicheamicin-induced circle formation is primarily due to bending of the helix along its axis, it is likely that there is some component of helical rotation that occurs upon drug binding. This hypothesis is supported by two studies. First, Ellestad and coworkers observed circular dichroism changes consistent with helical overwinding upon binding of calicheamicin ε to duplex oligonucleotides containing an AGGA sequence [23]. Second, we have observed that calicheamicin $\gamma_1{}^I$ increases the negative superhelical density of plasmid DNA (L. Yu, P. Dedon, unpublished). However, both of these observations could also be explained by local bending along the helix axis. Further studies are underway to resolve these issues of bending and helical rotation.

5. Acknowledgments

The authors are grateful to Danika LeDuc and Arash Maneshipour for the preparation and analysis of cloned DNA constructs, William LaMarr for his technical advice and support, Dr. George Ellestad (Wyeth-Ayerst Research) for the generous gifts of calicheamicin $\gamma_1{}^I$ and ε, and to other members of the Dedon Lab for critical review of the manuscript. This work was supported by the following sources: an NIH grant to PD (CA54642), a NIEHS training grant to the Division of Toxicology, MIT (ES07020; AS), and the Samuel A. Goldblith Professorship (PCD).

6. References

1. Zhao, B., Konno, S., Wu, J.M., and Oronsky, A.L. (1990) Modulation of nicotinamide adenine dinucleotide and poly(adenosine diphosphoribose) metabolism by calicheamicin γ1 in human HL-60 cells, *Cancer Lett.* **50**, 141-147.
2. Zein, N., Sinha, A.M., McGahren, W.J., and Ellestad, G.A. (1988) Calicheamicin γ1[I]: an antitumor antibiotic that cleaves double-stranded DNA site specifically, *Science* **240**, 1198-1201.
3. Lee, M.D., Ellestad, G.A., and Borders, D.B. (1991) Calicheamicins: Discovery, structure, chemistry, and interaction with DNA, *Acc. Chem. Res.* **24**, 235-243.
4. Dedon, P.C., and Goldberg, I.H. (1992) Free-radical mechanisms involved in the formation of sequence-dependent bistranded DNA lesions by the antitumor antibiotics bleomycin, neocarzinostatin, and calicheamicin, *Chem. Res. Tox.* **5**, 311-332.
5. Kishikawa, H., Jiang, Y.-P., Goodisman, J., and Dabrowiak, J.C. (1991) Coupled kinetic analysis of cleavage of DNA by esperamicin and calicheamicin, *J. Am. Chem. Soc.* **113**, 5434-5440.
6. Dedon, P.C., Salzberg, A.A., and Xu, J. (1993) Exclusive production of bistranded DNA damage by calicheamicin, *Biochemistry* **32**, 3617-3622.
7. Ding, W.-d., and Ellestad, G.A. (1991) Evidence for a hydrophobic interaction between calicheamicin and DNA, *J. Am. Chem. Soc.* **113**, 6617-6620.
8. Drak, J., Iwasawa, N., Danishefsky, S., and Crothers, D.M. (1991) The carbohydrate domain of calicheamicin γ1[I] determines its sequence specificity for DNA cleavage, *Proc. Natl. Acad. Sci. USA* **88**, 7464-7468.
9. Li, T., Zeng, Z., Estevez, V.A., Baldenius, K.U., Nicolaou, K.C., and Joyce, G.F. (1994) Carbohydrate-minor groove interactions in the binding of calicheamicin γ1[I] to duplex DNA, *J. Am. Chem. Soc.* **116**, 3709-3715.
10. Mah, S.C., Townsend, C.A., and Tullius, T.D. (1994) Hydroxyl radical footprinting of calicheamicin. Relationship of DNA binding to cleavage., *Biochemistry* **33**, 614-621.
11. Mah, S.C., Price, M.A., Townsend, C.A., and Tullius, T.D. (1994) Features of DNA recognition for oriented binding and cleavage by calicheamicin, *Tetrahedron* **50**, 1361-1378.
12. Nicolaou, K.C., Tsay, S.-C., Suzuki, T., and Joyce, G.F. (1992) DNA-carbohydrate interactions. Specific binding of the calicheamicin γ1[I] oligosaccharide with duplex DNA, *J Am. Chem. Soc.* **114**, 7555-7557.
13. Paloma, L.G., Smith, J.A., Chazin, W.J., and Nicolaou, K.C. (1994) Interaction of calicheamicin with duplex DNA: Role of the oligosaccharide domain and identification of multiple binding modes, *J. Am. Chem. Soc.* **116**, 3697-3708.
14. Walker, S., Landovitz, R., Ding, W.D., Ellestad, G.E., and Kahne, D. (1992) Cleavage behavior of calicheamicin γ1 and calicheamicin T, *Proc. Natl. Acad. Sci. USA* **89**, 4608-4612.
15. Walker, S., Murnick, J., and Kahne, D.J. (1993) Structural characterization of a calicheamicin-DNA complex by NMR, *J. Am. Chem. Soc.* **115**, 7954-7961.
16. Walker, S.L., Andreotti, A.H., and Kahne, D.E. (1994) NMR characterization of calicheamicin γ1[I] bound to DNA, *Tetrahedron* **50**, 1351-1360.
17. Zein, N., McGahren, W.J., Morton, G.O., Ashcroft, J., and Ellestad, G.A. (1989) Exclusive abstraction of nonexchangeable hydrogens from DNA by calicheamicin γ1[I], *J. Am. Chem. Soc.* **111**, 6888-6890.

18. De Voss, J.J., Townsend, C.A., Ding, W.-d., Morton, G.O., Ellestad, G.A., Zein, N., Tabor, A.B., and Schreiber, S.L. (1990) Site-specific atom transfer from DNA to a bound ligand defines the geometry of a DNA-calicheamicin γ1I complex, *J. Am. Chem. Soc.* **112**, 9669-9670.

19. Hangeland, J.J., De Voss, J.J., Heath, J.A., Townsend, C.A., Ding, W.-d., Ashcroft, J.S., and Ellestad, G.A. (1992) Specific abstraction of the 5'(S)- and 4'-deoxyribosyl hydrogen atoms from DNA by calicheamicin γ1I, *J. Am. Chem. Soc.* **114**, 9200-9202.

20. Hawley, R.C., Kiessling, L.L., and Schreiber, S.L. (1989) Model of the interactions of calicheamicin γ1 with a DNA fragment from pBR322, *Proc. Natl. Acad. Sci. USA* **86**, 1105-1109.

21. Aiyar, J., Danishefsky, S.J., and Crothers, D.M. (1992) Interaction of the aryl tetrasaccharide domain of calicheamicin γ1I with DNA: Influence of aglycon and methidiumpropyl-EDTA.iron(II)-mediated DNA cleavage, *J. Am. Chem. Soc.* **114**, 7552-7554.

22. Walker, S., Valentine, K.G., and Kahne, D.J. (1990) *J. Am. Chem. Soc.* **112**, 6429.

23. Krishnamurthy, G., Ding, W.-d., O'Brien, L., and Ellestad, G.A. (1994) Circular dichroism studies of calicheamicin-DNA interactions: Evidence for a calicheamicin-induced DNA conformational change, *Tetrahedron* **50**, 1341-1349.

24. Yu, L., Goldberg, I.H., and Dedon, P.C. (1994) Enediyne-mediated DNA damage in nuclei is modulated at the level of the nucleosome, *J. Biol. Chem.* **269**, 4144-4151.

25. Yu, L., Salzberg, A.A., and Dedon, P.C. (1995) New insights into calicheamicin-DNA interactions derived from a model nucleosome system, *Bioorg. Med. Chem.* **3**, 729-741.

26. Ausubel, F.M., Brent, R., Kingston, R.E., Moore, D.D., Seidman, J.G., Smith, J.A., and Struhl, K. (1989) *Current Protocols in Molecular Biology*, John Wiley and Sons, New York.

27. Hagerman, P.J. (1990) Sequence-directed curvature of DNA, *Annu. Rev. Biochem.* **59**, 755-781.

28. Ulanovsky, L., Bodner, M., Trifonov, E.N., and Choder, M. (1986) Curved DNA: Design, synthesis, and circularization, *Proc. Natl. Acad. Sci. USA* **83**, 862-866.

29. Crothers, D., Drak, J., Kahn, J., and Levene, S. (1992) DNA bending, flexibility, and helical repeat by cyclization kinetics, *Meth. Enz.* **212**, 3-30.

30. Taylor, W.H., and Hagerman, P.J. (1990) Application of the method of phage T4 DNA ligase-catalyzed ring-closure to the study of DNA structure. II. NaCl-dependence of DNA flexibility and helical repeat, *J. Mol. Biol.* **212**, 363-376.

31. Crothers, D.M., Haran, T.E., and Nadeau, J.G. (1990) Intrinsically Bent DNA, *J. Biol. Chem.* **265**, 7093-7096.

32. Diekmann, S. (1992) Analyzing DNA curvature on polyacrylamide gels, *Meth. Enz.* **212**, 30-46.

33. Kahn, J.D., Yun, E., and Crothers, D.M. (1994) Detection of localized DNA flexibility, *Nature* **368**, 163-166.

34. Shore, D., Langowski, J., and Baldwin, R.L. (1981) DNA flexibility studied by covalent closure of short fragments into circles, *Proc. Natl. Acad. Sci. USA* **78**, 4833-4837.

35. Hagerman, P.J., and Ramadevi, V.A. (1990) Application of the method of phage T4 ligase-catalyzed ring-closure to the study of DNA structure. I. Computational analysis., *J. Mol. Biol.* **212**, 351-362.
36. Shore, D., and Baldwin, R.L. (1983) Energetics of DNA twisting. II. Topoisomer analysis, *J. Mol. Biol.* **170**, 983-1007.
37. Shore, D., and Baldwin, R.L. (1983) Energetics of DNA twisting. I. Relation between twist and cyclization probability, *J. Mol. Biol.* **170**, 957-981.
38. Koo, H.-S., Drak, J., Rice, J.A., and Crothers, D.M. (1990) Determination of the extent of DNA bending by an adenine-thymine tract, *Biochemistry* **29**, 4227-4234.

RECENT BIOPHYSICAL AND BIOLOGICAL STUDIES ON THE MECHANISM OF ACTION OF CALICHEAMICIN

GIRIJA KRISHNAMURTHY, WEIDONG DING AND GEORGE A. ELLESTAD
Wyeth Ayerst Research
Pearl River, New York, 10965

Abstract. Circular dichroism spectroscopy and gel cleavage methods were used to characterize the site-specific DNA interactions and binding constants of the enediyne -containing calicheamicin γ_1^I and cycloaromatized ε. Evidence from circular dichroism studies suggests that the binding of calicheamicin to DNA induces an optically detectable conformational change of B-form DNA. The optical changes are due to the specific interaction of calicheamicin with cleavable / tight binding-sequences and are not observed for noninteracting sites. The gel cleavage method was used to determine the binding constants and evaluate their salt dependence for the three high affinity sites, TCCT, TTGT and ATCT in a pBR322 restriction fragment. The salt dependence of the binding constants and their evaluation in terms of polyelectrolyte theory suggest that the sequence specific interactions are dominated by nonionic rather than electrostatic forces. Calicheamicin γ_1^I inhibits mitochondrial function in yeast suggesting that similar effects on mammalian cells are responsible for the delayed, lethal toxicity in mice.

Introduction

Calicheamicin γ_1^I (1) is a member of the enediyne class of antitumor antibiotics which exhibit unusually potent antitumor activity. It is believed that this activity correlates with these agents' ability to cause extensive double-stranded lesions in the targeted DNA. The basic features of the DNA cleavage chemistry are outlined in scheme 1. Activation of the drug is initiated by reductive cleavage of the allylic trisulfide followed by an intramolecular Michael addition to the eneone moiety. Electrocyclization then occurs via the formation of a transient p-benzyne diradical. It is this non-diffusible, carbon centered diradical, when properly positioned in the DNA minor groove, that causes lesions by the abstraction of proximal hydrogen atoms from the deoxyribose backbone. The sugar radicals then undergo oxygen-dependent reactions that result in strand breaks or abasic sites.

Calicheamicin is unique among the enediynes in that it exhibits remarkable DNA-cleaving specificity for a molecule of only 1367 Daltons. Its binding and cleavage of homopyrimidine/homopurine-containing sequences such as TCCT/AGGA, CTCT/GAGA[1] and TTTT/AAAA[2] has been studied in some detail. It is now generally agreed that the sequence specific interactions in the minor groove are due primarily to

37

B. Meunier (ed.), DNA and RNA Cleavers and Chemotherapy of Cancer and Viral Diseases, 37–51.
© 1996 *Kluwer Academic Publishers. Printed in the Netherlands.*

1 Calicheamicin γ_1^I

recognition and binding therein of the carbohydrate tail portion of the molecule. To date several groups have characterized these minor groove interactions using the unmodified molecule,[3-9] truncated oligosaccharide analogs,[10] modified thiobenzoate analogs[11] and cycloaromatized calicheamicin ε.[9] These studies indicate the following important factors in the sequence specific recognition process (i) minor groove binding is unidirectional, with the aglycone aligned toward the 5' end of the pyrimidine strand while the oligosaccharide binds toward the 3' end of the pyrimidine-containing strand[3,4] (ii) the observed sequence selectivity is due to the isohelical shape of the thiobenzoate oligosaccharide tail[5,12] (iii) a truncated form of the molecule containing only the aglycone and sugars A and E produces a lower cleavage specificity than that of the intact molecule[2] (iv) most but not all binding specificity resides in the carbohydrate tail portion of the molecule[10,13] (v) the iodinated thiobenzoate moiety contributes significantly to the overall binding energy.[7,11,14]

More recently Townsend and coworkers showed that calicheamicin recognizes and cleaves DNA sequences with a narrow minor groove, sequences that contain structural discontinuities deviating from regular B-form DNA and those regions with a propensity for drug-induced helix deformation.[15] This group also demonstrated by hydroxyl footprinting experiments that calicheamicin-DNA binding occurs only at sequences that are cleaved. NMR evidence for drug induced conformational changes suggests expansion of the minor groove to accomodate the drug.[5,6] Dedon's group[16] has extended these studies to nucleosome as well as to other naked DNA targets. They used the *Xenopus borealis* 5S gene and have identified several new features of the DNA/calicheamicin interaction. In general, calicheamicin mediated DNA damage was reduced by incorporation of the DNA into a nucleosome. However, in one instance they observed an enhanced cleavage site and attributed this to a drug-induced bending in the target site. They also observed that the minor groove width *per se* is probably not a major determinant of drug target selection. The generality of the homopurine/homopyrimidine recognition element was confirmed, but with the additional requirement that the purine tract be interrupted at the 3'-end by a pyrimidine(s).

One of the features of the calicheamicin/DNA interaction that is still of some debate is that of the role of the aminosugar in binding and cleavage. The amino group is the only positively charged functionality in the molecule at physiological pH and one would expect this to play a nonspecific role in electrostatic attraction with the negatively charged DNA backbone. But it might also facilitate the reductive cleavage of the

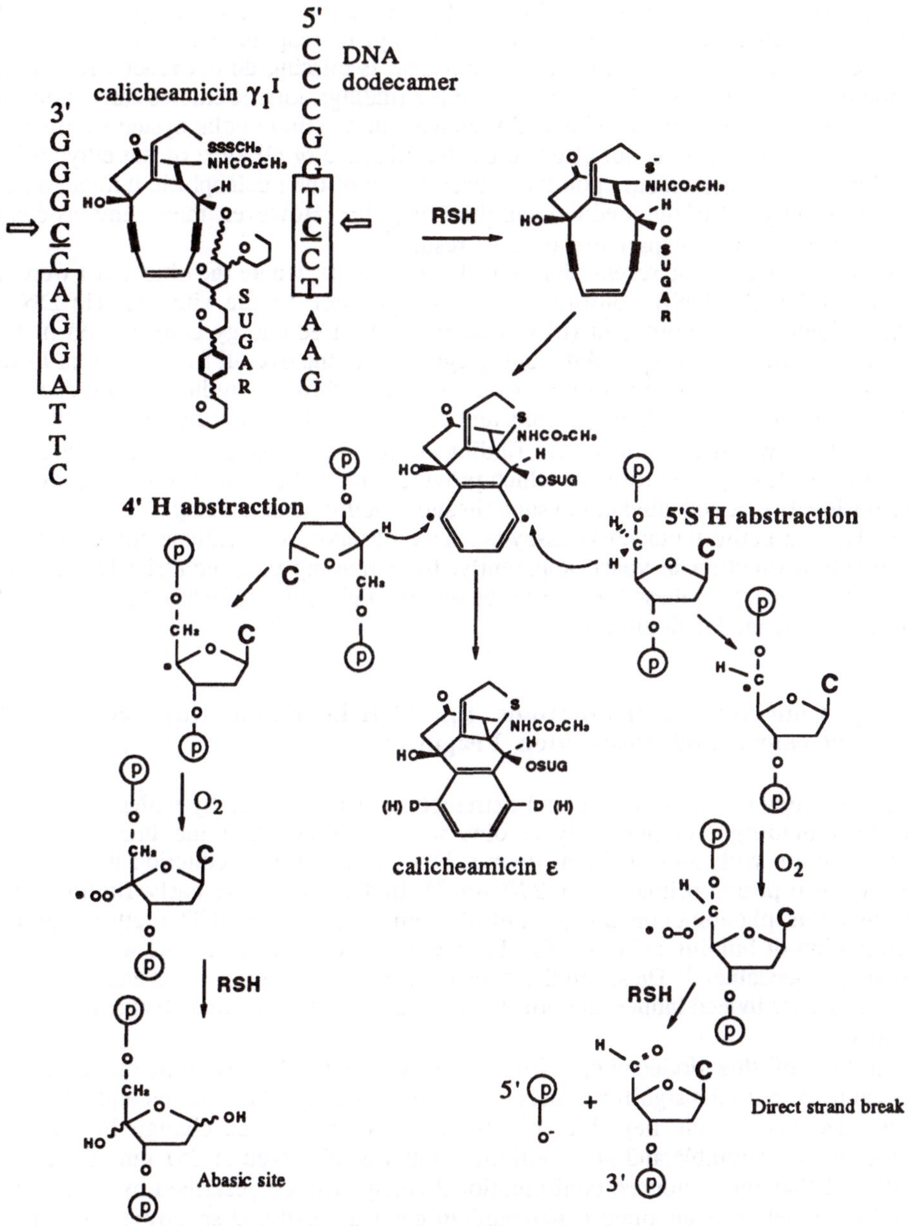

Scheme 1. Summary of Calicheamicin Binding and Cleavage

trisulfide moiety by electrostatically lowering the pK_a of the incoming thiol nucleophile.[17] It has been shown with disulfide exchange reactions in proteins that there is a linear correlation between the pK_a of the attacking thiolate anion and the rate of disulfide reduction.[18,19] Kahne's ROSEY NMR results[5,12] showed that the ethylamino group is proximal to the allylic methyltrisulfide.

In acetonitrile there is strong evidence that the amino group does play a role in facilitating disulfide exchange reactions with alkyl mercaptans since the acetylated derivative and the analog in which the aminosugar is missing do not react without the addition of an organic base.[20] In contrast to these findings, kinetic studies carried out by Townsend's group using aminoethanethiol indicate that in more polar systems reductive cleavage of the trisulfide is not affected by the presence or absence of the ethylamino group but is solely dependent upon the concentration of the nucleophilic thiolate anion as determined by pH of the medium and thiol pK_a.[21,22] However, there remains some disagreement as to the intepretation of these results.[23]

Here, we describe some recent biophysical experiments, where the objectives were to (i) characterize the DNA conformational changes due to the site-specific DNA-calicheamicin(s) interaction and (ii) to determine the free energy contribution of the protonated ethylamino group and the bulky aglycone to the overall free energy of DNA association. First, we report results of the circular dichroism studies on site-specific DNA binding of calicheamicins γ_1^I and ε and compare the binding constants for the two analogs. Then we discuss our recent findings on the salt effects on binding/cleavage using affinity cleavage experiments which provide some indication as to the importance of the positively charged ethylamino sugar in site-specific DNA binding.

Finally, in a petite mutagensis assay we have observed that calicheamicin impairs mitochondrial function in yeast, apparently, by damaging mitochondrial DNA. We discuss the relevance of these findings to the observed biological effects of delayed death in mice unchallenged with tumors.

Binding Constant Determinations for Calicheamicin Interaction with DNA Dodecamer and Restriction Fragments.

CD spectroscopy is a noninvasive and extremely sensitive means by which to monitor drug/DNA binding. In our early studies we determined that the interactions of calicheamicin ε with sonicated calf thymus DNA resulted in a reduction in the DNA dichroic absorption, primarily at 270 nm.[24] In light of these early findings, we examined this phenomenon more carefully and extended the CD method for the determination of binding affinities for the interaction of calicheamicins with a TCCT containing dodecamer.[9] These studies showed that the binding of calicheamicin ε and γ_1^I to DNA does indeed induce an optically detectable conformational change of B-form DNA (figure 1).

Titration of this dodecamer with calicheamicin ε, the aromatized analog of calicheamicin, caused a significant decrease in the positive CD at 266 nm and a slightly smaller decrease in the negative minimum at 240 nm. The change in dichroic absorption was saturable.and an isoelliptic point was observed at 251 nm suggesting that the calicheamicin induced conformational change can be described by a two-state transition model. The unbound calicheamicin ε has a small CD spectrum throughout this region except at 271 nm where its contribution is zero.

Figure 2 shows the binding curves obtained using the CD changes at 271 nm and 219 nm. The small CD signal due to calicheamicin ε at 219 nm was subtracted for the binding curves generated at 219 nm. The binding constants of $6.9 \times 10^4 M^{-1}$ (271 nm) and $5.3 \times 10^4 M^{-1}$ (219 nm) were estimated from a nonlinear least squares fit of the data.

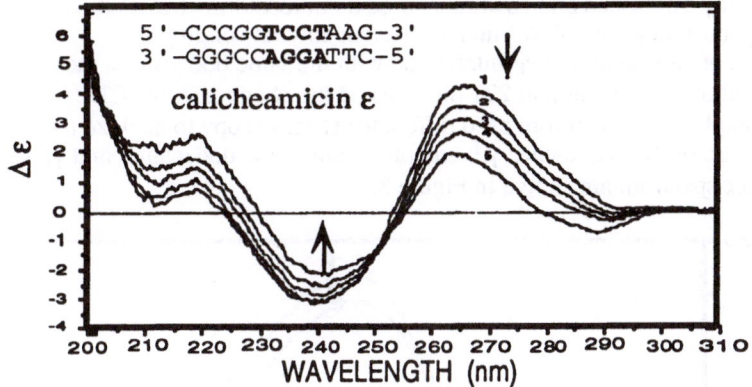

Figure 1. CD spectra of the dodecamer 5'-d(CCCGG**TCCT**AAG) (23 µM of duplex) in the absence (1); and presence of calicheamicin ε at 11 µM (2); 23 µM (3); 46 µM (4); 131 µM (5). The concentration-scaled CD contribution of the drug has been subtracted from the recorded spectra. (taken from reference 9)

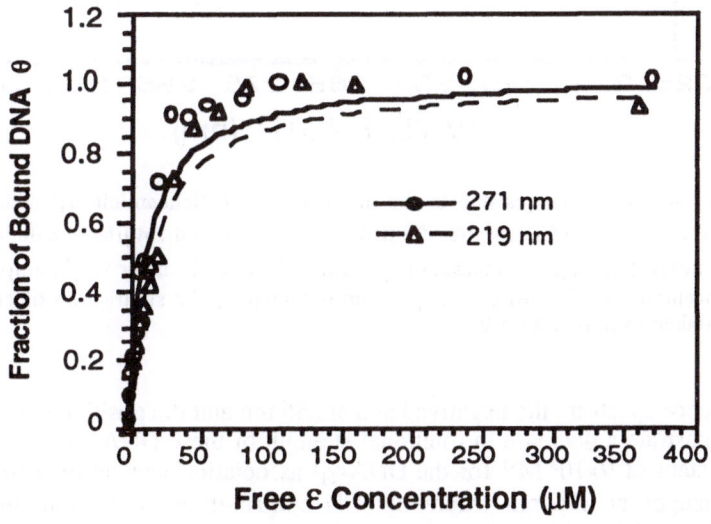

Figure 2. Binding isotherms of calicheamicin ε interaction with the TCCT dodecamer. Increasing concentrations of the drug were titrated into duplex DNA and the resultant CD intensity changes at 219 nm (triangle) and 271 nm (circle) were used to calculate the fractional DNA saturation values. The solid line represents the theoretical binding curve generated by Kaleidagraph. Analysis of the binding curves provide K_a values of 5.3 and 6.9 (x 10^4 M^{-1}) at 219 nm and 271 nm respectively. The χ^2 was 0.14 at 219 nm and 0.08 at 271 nm. The correlation coefficient was 0.97 at both wavelengths. (taken from reference 9)

More recently Townsend and coworkers used a microcalorimetric method and estimated the binding constant as $6 \times 10^4 \, M^{-1}$ under different solvent conditions.[14]

In the case of calicheamicin $\gamma_1{}^I$ interaction with the same dodecamer, the large inherent CD of the unbound $\gamma_1{}^I$ analog at 271 nm masked the changes in the CD spectrum of the resultant complex. We therefore used difference spectroscopy to analyze the CD spectra. The CD spectra of the dodecamer-$\gamma_1{}^I$ complex, sum of the dodecamer and $\gamma_1{}^I$ spectra and the difference spectrum are shown in Figure 3.

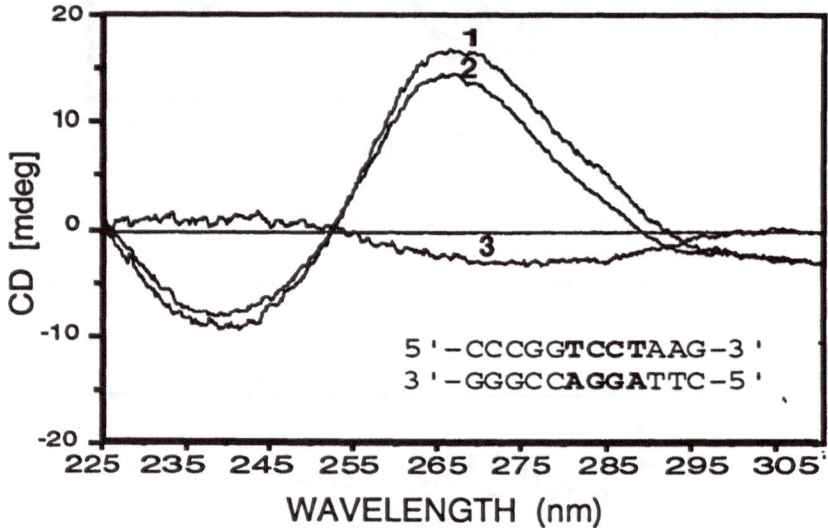

Figure 3. CD spectral changes due to the interaction of calicheamicin $\gamma_1{}^I$ with the TCCT dodecamer. The sum of the CD spectra of 3 μM of dodecamer and 2.4 μM of calicheamicin $\gamma_1{}^I$ assuming no interaction (1); the measured spectrum of the dodecamer - $\gamma_1{}^I$ complex at these indicated concentrations (2); difference spectrum obtained by the subtraction of curve 2 from curve 1 (3). (taken from reference 9)

In the difference spectrum the negative band at 266 nm and the positive band at 240 nm clearly demonstrate a decrease in rotational strength of these DNA CD extremums. A binding constant of $9 \times 10^5 \, M^{-1}$ for the DNA-$\gamma_1{}^I$ association was derived from the CD intensity changes at 236 nm where the CD contribution of the unbound drug is negligible (figure 4).

The 20 fold increased stability of the DNA-$\gamma_1{}^I$ complex relative to the DNA-ε complex suggests that the enediyne segment of the aglycone contributes significantly to the overall binding energy. The role of the aglycone in the sequence specific DNA recognition has been independently addressed by two groups. Danishefsky, Crothers and coworkers observed that the aglycone cleaves DNA nonspecifically. The same group similarly observed in their footprinting studies that preincubation of the tetrasaccharide

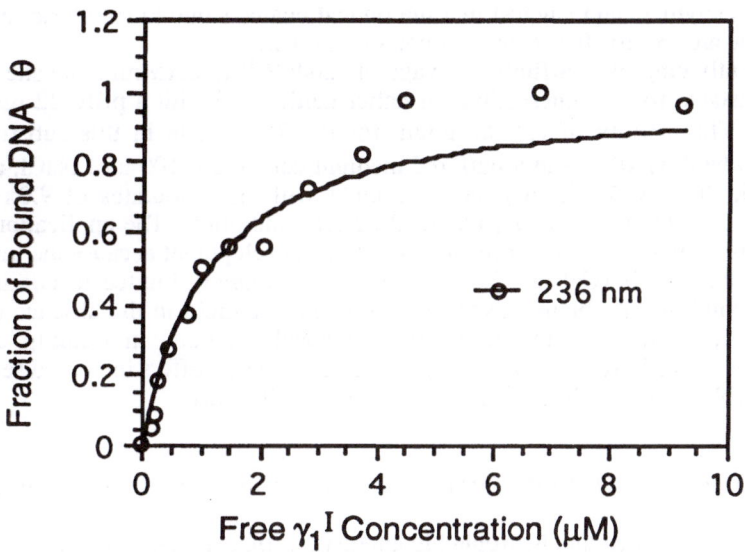

Figure 4. Binding isotherm of calicheamicin γ_1^I interaction with the TCCT dodecamer. Increasing concentrations of the drug were titrated with titrated into dodecamer and the fractional DNA saturation values were calculated from the CD intensity changes at 236 nm. The solid line represents the theoretical binding curve generated by Kaleidagraph. Curve fit analyses of the binding isotherm provided a K_a value of $9x10^5$ M^{-1}. The χ^2 value was 0.08 and correlation coefficient 0.97. (taken from reference 9)

domain followed by incubation with aglycone leads to sequence-selective cleavage not observed in the absence of the carbohydrate.[10] Our CD results suggest that the increased stability of the DNA-γ_1^I complex is perhaps due to a better fit of the enediyne-containing aglycone in the minor groove compared to the aromatized analog.

The observed decrease in the intensity of the positive CD band caused by calicheamicin is strongly reminiscent of a DNA structure brought about by a reduction in the degree of hydration caused by temperature, alcohols and high salt.[25-29] Certain DNA binding proteins have been shown to cause a similar decrease of the positive extremum and this reduction has been attributed to an increase in the winding angle of the DNA helix.[30,31] Calicheamicin γ_1^I was also titrated with poly(dG-dC)$_2$ polymer (used as a carrier in titration experiments with DNA containing TCCT sites to avoid spurious DNA cleavage activity[8]) to compare the CD changes in this substrate with that of the TCCT sequence-containing oligomer. Poly(dG-dC)$_2$-calicheamicin mixtures did not show an ellipticity decrease at 270 nm and 236 nm. The absence of a CD change with a poly(dG-dC)$_2$ substrate is consistent with the fact that the observed ellipticity changes observed with an oligomer containing a principal recognition site are due to a specific drug interaction in the minor groove. We also titrated tRNA^{phe} from *E.coli* with calicheamicin γ_1^I and similarly found that no CD spectral changes were induced consistent with a recent observation that calicheamicin does not cleave tRNA^{phe}.[32] These CD results corroborate our earlier inference that calicheamicin does not tightly bind G-C copolymer[8] and suggest that the optical changes due to calicheamicin-DNA complexation are specific for cleavable / bound complexes.

We recently employed affinity cleavage methods[33-35] to determine the site specific binding constants for the interactions of calicheamicin γ_1^I with a pBR322 restriction fragment.[8] The affinity of calicheamicin for the TCCT site in this substrate was estimated to be 8.0×10^5 M[-1] in a buffered medium containing 100 mM NaCl, at 23 °C and pH 7.5. This value is in good agreement with our estimates of 9.0×10^5 M[-1] obtained by the CD method under identical buffer conditions. The application of this method for binding constant determinations is a key development because the traditional spectroscopic methods (UV and fluorescence) are extremely limited in this instance. Significant solubility problems associated with calicheamicin in the absence of DNA have made it difficult to use traditional physical methods in binding constant determinations. We have found that the gel cleavage method offers increased sensitivity and is adaptable for use with restriction fragments as substrates.

Site Specificities and Salt Effects on DNA-Calicheamicin Interactions

Earlier, we demonstrated that hydrophobic interactions play a major role in the sequence specific interactions of DNA-calicheamicin interactions.[29] In this context we were interested in determining the electrostatic contribution of the protonated amine in the aminosugar to the overall binding free energy. Specifically, we used gel cleavage methods to determine whether counterion release and ion pair formation between the positively charged ethylamino group and the oppositely charged phosphate backbone contribute significantly to the tightness of the interaction. The gel cleavage data were analyzed in terms of the polyelectrolyte theory.[36] The Gibbs free energy change required to bind calicheamicin to the TCCT site was determined as a function of salt at pH 8.1 and 0 °C and at pH 7.5 at 23 and 0 °C. These salt dependence studies provided evidence that the association of calicheamicin with DNA is dominated by nonionic forces rather than electrostatic interactions.[8] The TCCT site binding isotherms at pH 8.1 and 0 °C were investigated as a function of NaCl concentrations of 50, 100, 125, 150, 250, 500 mM and 1 M. Between 150 mM and 125 mM NaCl, the TCCT sequence displays a sharp transition in ΔG from -7.6 to 8.9 Kcal mol[-1] (Figure 5). However, between 150 mM and 1 M NaCl calicheamicin binding to the TCCT site is insensitive to salt. Below 125 mM NaCl, the ΔG values for calicheamicin binding to the TCCT sequence are again essentially invariant as they are from 150 mM to 1 M. The TCCT site binding free energies at 23 °C and pH 7.5 also showed identical trends with tight binding at 50 mM NaCl and weaker binding at salt concentrations of 100 mM and above (Table 1).

Although the origin of the transition between 125 and 150 mM for the TCCT site is not clear, it is possible that the observed free energy differences at this site are due to coupling of the drug/DNA binding equilibrium to salt-dependent DNA conformational changes. The binding free energies of calicheamicin interaction with two other cleavable sites, ATCT and TTGT, were also investigated as a function of salt. Figure 6 shows the binding isotherms for the three sites at 125 mM NaCl (pH 8.1, 0 °C). At 125 mM NaCl, the site binding free energies of the ATCT and TTGT sites are weaker than the TCCT site. In figure 7, the site binding free energies for the ATCT, TTGT and TCCT.

sites are plotted as a function of the NaCl concentration between 50 mM and 1 M NaCl. The site binding free energies are -7.8 and -7.9 Kcal mol^{-1} for the ATCT and TTGT sites respectively and are essentially insensitive to variations in NaCl concentration.

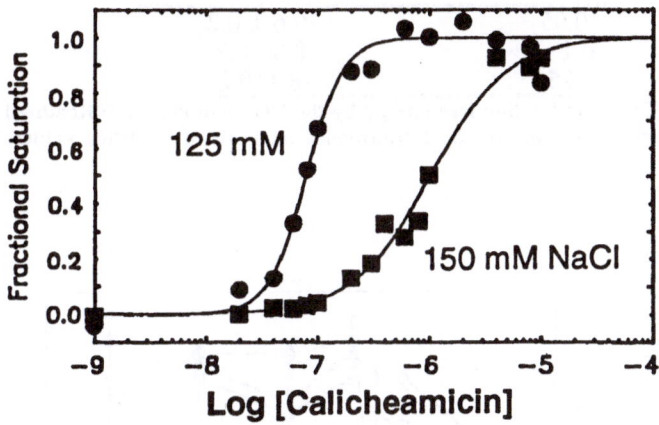

Figure 5. Calicheamicin-TCCT site binding isotherms at 125 (●) and 150 mM (■) NaCl (pH 8.1 at 0 °C). The solid sigmoidal curve in each instance shows the best fit of the data to the Hill equation using nonlinear least-squares analysis. (taken from reference 8)

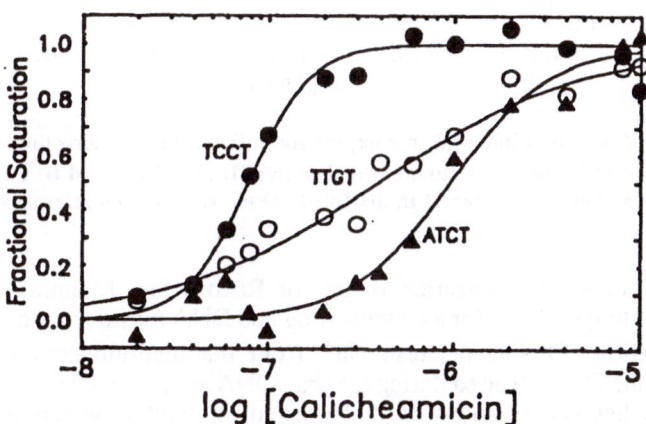

Figure 6. Binding isotherms of calicheamicin interacting with TCCT (●), TTGT (O) and ATCT (Δ) sequences (125 mM NaCl, pH 8.1 and 0 °C). The symbols represent the fractional saturation data and the solid lines connecting the data points are due to nonlinear least-squares fits of the data to the Hill equation. (taken from reference 8)

Table 1. TCCT site binding free energies as a function
of NaCl concentration at 23 °C, pH 7.5[a].

NaCl (M)	ΔG^b (Kcal.mol^{-1})
0.05	-9.6 ± 0.2
0.1	-8.0 ± 0.2
0.25	-8.1 ± 0.2

a The TCCT site binding free energy by the CD method is -8.0 Kcal/mol (0.1 M NaCl).
b The errors represent the 65% confidence intervals of the fitted values.

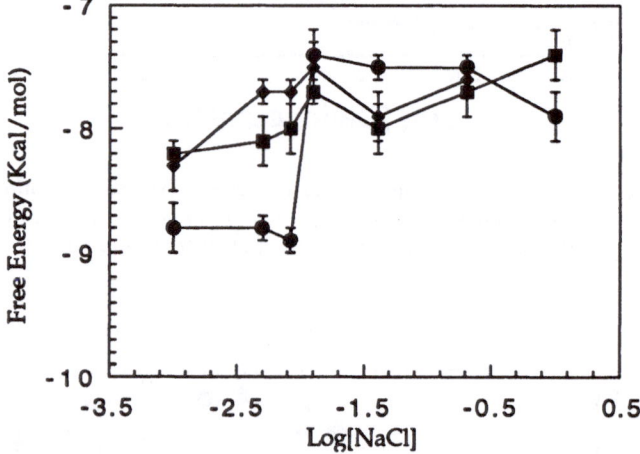

Figure 7. Plot of the site binding free energies for calicheamicin interacting with TCCT(●), TTGT (■) and ATCT (◆) as a function of log [NaCl] at pH 8.1 and 0 °C. The error bars represent the 65% confidence interval in the fitted value. (taken from reference 8)

Based on the ion-condensation theory of Record and Manning, the observed equilibrium binding constant for a cationic drug and DNA may decrease with increasing salt concentration. This is because Na$^+$ from the medium competes with Na$^+$ counterions of the DNA released during the drug/DNA association thereby decreasing the interaction. It has been demonstrated that a plot of $\log K_{obs}$ versus $\ln[MX]$ gives a straight line with a slope of $-Z\psi$ if no other processes are linked to the DNA binding equilibrium. Z is equivalent to the number of counterions released during drug binding and ψ is 0.88 which is the fractional neutralization of a phosphate due to condensed counterions. The observed salt effects on the TCCT, ATCT and TTGT sequences suggest that in the case of the monocationic calicheamicin, counterion release is not a major driving force in the energetics of the complexation.

Previously, the DNA interactions of small cationic molecules such as netropsin, irehdiamine, and daunomycin were shown to be stabilized by ionic interactions between the positively charged amine group and the negatively charged DNA with the concomitant release of sodium ions. Unlike these ligands, calicheamicin consists of a secondary amine with a bulky deoxy sugar and alkyl group substituents. Furthermore, the ethylamino sugar in calicheamicin is part of a relatively rigid molecule[12] and Patel's recent NMR experiments[37] have now shown that this sugar binds externally to the polyphosphate backbone. The diminished importance of electrostatic interactions in this instance may be due to (i) restricted rotation that would have otherwise provided optimal contact with the DNA polyphosphate backbone, as well as (ii) positioning of the aminosugar external to the minor groove. Our observations do not exclude the role of ethylamino group in nonspecific electrostatic association with the DNA and as a hydrogen bond donor along the phosphate backbone.

Another interesting observation to come out of these studies was that the individual site loading energies appear to include contributions from cooperative interactions.[38] Best-fit Hill coefficients (n_h) were determined for each of the binding curves and were found to be greater than one, especially for the TCCT sequence. This is not unique for minor groove binding drugs and has been well documented for distamycin[39] and 4',6'-diamino-2-phenylindole.[40,41]

Calicheamicin Affects Mitochondrial Functions in Yeast.

Because of reports of delayed toxicity associated with another highly potent antitumor antibiotic, CC-1065, calicheamicin γ_1^I was injected at various doses into normal mice which were then held for possible mortality over a period of four months. Calicheamicin displayed delayed, lethal toxicity at a dose of 1.2 mg/kg, which became pronounced only after 60 days. Higher doses exhibited greater acute toxicity. Although the cause of the lethality was unclear, the greatly reduced levels of ATP and NAD^+ appeared to account for the more acute aspects of cell death.[42] Functional mitochondria are required for the production of ATP and mitochondrial DNA damage can cause cellular energy deficiencies with an attendant impairment of cellular activities.[43] Thus the reduced levels of ATP and NAD^+ are likely a result of mitochondrial DNA damage caused by calicheamicin.

In order to examine this aspect, we used a genetic approach, the mitochondrial 'petite' mutagenesis assay, to determine the *in vitro* effects of calicheamicin on mitochondrial DNA in *Saccharomyces cerevisiae*. This assay is based upon the premise that in wild type strains, intact mitochondrial DNA is required for yeast growth on a nonfermentable carbon substrate such as glycerol which can only be utilized by fully functional respiratory enzymes. The genetic assay was set up by treating yeast with calicheamicin in a liquid nutrient medium consisting of glucose. Calicheamicin treated and untreated yeast were then plated out on media containing either glucose or glycerol. Calicheamicin treated *S. Cereviseae* failed to grow on glycerol containing plates ('petite' mutants), while the untreated control yeast produced colonies both on glucose and glycerol containing media. This inhibitory effect was observed at calicheamicin concentrations as low as 25 ng/ml. These results suggest that calicheamicin treatment of yeast culture impaired mitochondrial functions.

Mitochondrial DNA is perhaps more sensitive to damage than nuclear DNA possibly due to at least two reasons (i) mtDNA is not associated with histones and is therefore more susceptibile to DNA damage[43] (ii) mtDNA lacks the excision repair mechanism which characterizes the cellular response to DNA damage.[44] It is also conceivable that calicheamicin binds mitochondrial DNA more tightly than genomic DNA because of the circular nature of mitochondrial DNA which could be an important additional factor in the mutagenesis event. This is based on some early studies in which plasmid DNA was observed to be more susceptible to calicheamcin damage than linear fragments.[1] Potent inducers of petites in yeast have been shown to act at the mitochondrial level in mammalian cells.[45] Similarly, we believe that the delayed toxicity in mice due to calicheamicin is related to impaired mitochondrial function.

References

1. Zein, N., Sinha, A.M., McGahren, W.J., and Ellestad, G.A. (1988) Calicheamicin γ_1^I: An antitumor antibiotic that cleaves double-stranded DNA site specifically, *Science* **240**, 1198-1201.

2. Walker, S., Landovitz, R., Ding, W.-d., Ellestad, G.A. and Kahne, D. (1992) Cleavage behavior of calicheamicin γ_1^I and calicheamicin T, *Proc. Natl. Acad. Sci. U.S.A.* **89**, 4608-4612.

3. Devoss, J.J., Townsend, C.A., Ding, W.-d., Morton, G.O., Ellestad, G.A., Zein, N., Tabor, A.B., and Schrieber, S.L. (1990) Site-specific atom transfer from DNA to a bound ligand defines the geometry of a DNA-calicheamicin complex, *J. Am. Chem. Soc.* **112**, 9669-9670.

4. Hangeland, J.J., Devoss, J.J., Heath, J.A., Townsend, C.A., Ding, W.-d., Ashcroft, J.S., and Ellestad, G.A. (1992) Specific abstraction of the 5'(S)- and 4'-deoxyribosyl hydrogen atoms from DNA by calicheamicin γ_1^I, *J. Am. Chem. Soc.* **114**, 9200-9202.

5. Walker, S., Murnick, J and Kahne, D., (1993) Structural characterization of a calicheamicin-DNA complex by NMR, *J. Am. Chem. Soc.* **115**, 7954-7961.

6. Paloma, L.G., Smith, J.A., Chazin, W.J., and Nicolaou, K.C. (1994) The interaction of calicheamicin with duplex DNA: Role of the oligosaccharide domain and identification of multiple binding modes, *J. Am. Chem. Soc.* **116**, 3697-3708.

7. Hawley, R.C., Kiessling, L.L., and Schreiber, S.L., (1989) Model of the interactions of calicheamicin γ_1^I with a DNA fragment from pBR322, *Proc. Natl. Acad. Sci. USA*, **86**, *1105-1109.*

8. Krishnamurthy, G., Brenowitz, M.D., and Ellestad, G.A. (1995) Salt dependence of calicheamicin-DNA site-specific interactions, *Biochemistry* **34**, 1001-1010.

9. Krishnamurthy, G., Ding, W.-d., O'Brien, L.,& Ellestad, G. A. (1994) Circular dichroism studies of calicheamicin-DNA interaction: Evidence for calicheamicin-induced DNA conformational change, *Tetrahedron* **50**, 1341-1349.

10. Aiyar, J., Danishefsky, S.J., and Crothers, D.M., (1992) Interaction of the aryl tetrasaccharide domain of calicheamicin γ_1^I with DNA: Influence on aglycone and methidium-EDTA-iron(II)-mediated DNA cleavage, *J. Am. Chem. Soc.* **114**, 7552-7554.

11. Li, T., Zeng, Z., Estevez, V.A., Baldenius, K.U., Nicolaou, K.C. and Joyce, G.F. (1994) Carbohydrate - minor groove interactions in the binding of calicheamicin γ_1^I to duplex DNA, *J. Am. Chem. Soc.* **116**, 3709-3715.

12. Walker, S., Valentine, K.G., and Kahne, D. (1990) Sugars as DNA binders: A comment on the calicheamicin oligosaccharide, *J. Am. Chem. Soc.*, **112**, 6428-6429.

13. Nicolaou, K.C., Tsay, S.-C., Suzuki, T., and Joyce, G.F., (1992) DNA-carbohydrate Interactions. Specific binding of the calicheamicin γ_1^I oligosaccharide with duplex DNA, *J. Am. Chem. Soc.* **114**, 7555-7557.

14. Chatterjee, M., Mah, S.C., Tullius, T.D., and Townsend, C.A. (1995) Role of the aryl iodide in the sequence-selective cleavage of DNA by calicheamicin. Importance of thermodynamic binding *vs* kinetic activation in the cleavage process, *J. Am. Chem. Soc.* **117**, 8074-8082.

15. Mah, S.C., Townsend, C.A., and Tullius, T.D. (1994) Hydroxyl radical footprinting of calicheamicin. Relationship of DNA binding to cleavage, *Biochemistry* **33**, 614-621.

16. Yu, L., Salzberg, A.A., and Dedon, P.C. (1995) New insights into calicheamicin-DNA interactions derived from a model nucleosome system, *Bioorganic & Medicinal Chemistry* **3**, 729-741.

17. Keillor, J.W., and Brown, R.S. (1992) Attack of zwitterionic ammonium thiolates on a distorted anilide as a model for the acylation of papain by amides. A simple demonstration of a bell-shaped pH/rate profile, *J. Am. Chem. Soc.* **114**, 7983-7989.

18. Creighton, T.E. (1975) Interactions between cysteine residues as probes of protein conformation: The disulfide bond between cys-14 and cys-38 of the pancreatic trypsin inhibitor, *J. Mol. Biol.* **96**, 767-776.

19. Snyder, G.H., Cennerazzo, M.J., Karalis, A.J., and Field, D. (1981) Electrostatic influence of local cysteine environments on disulfide exchange kinetics, *Biochemistry* **20**, 6509-6519.

20. Ellestad, G.A., Hamann, P.R., Zein, N., Morton, G.O., Siegel, M.M., Pastel, M., Borders, D.B., and McGahren, W.J. (1989) Reactions of the trisulfide moiety in calicheamicin, *Tetrahedron Lett.* **30**, 3033-3036.

21. Cramer, K.D. and Townsend, C.A. (1991) Kinetics of trisulfide cleavage in calicheamicin-assessing the role of the ethylamino group, *Tetrahedron Lett.* **36**, 4635-4638.

22. Chatterjee, M., Cramer, K.D., and Townsend, C.A. (1993) Role of the aryl iodide in the sequence-selective cleavage of DNA by calicheamicin. Importance of thermodynamic binding *vs* kinetic activation in the cleavage process, *J. Am. Chem. Soc.* **115**, 3374-3375.

23. Myers, A.G., Cohen, S.B., and Kwon, B.M (1994) A study of the reaction of calicheamicin γ_1^I with glutathione in the presence of double stranded DNA, *J. Am. Chem. Soc.* **116**, 1255-1271.

24. Zein, N., Poncin, M., Nilakantan, R., and Ellestad, G.A. (1989) Calicheamicin γ_1^I and DNA: Molecular recognition process responsible for site-specificity, *Science* **244**, 697-699.

25. Johnson, B.B., Dahl, K.S., Tinoco, I.,Jr., Ivanov,V.I., and Zhurkin, V.B., (1981) Correlations between deoxyribonucleic acid structural parameters and calculated circular dichroism spectra, *Biochemistry* **20**, 73-78.

26. Bokma, J.T., Johnson, W.C. Jr., and Blok, J. (1987) CD of the Li-salt of DNA in ethanol/water mixtures: Evidence for the B- to C-form transition in solution, *Biopolymers* **26**, 893-909.

27. Fareed, A-e., Varani, G., Walker, G.T., and Tinoco, I. Jr., (1988) The TFIIIA recognition fragment d(GGATGGGAG).d(CTCCCATCC) is B-form in solution, *Nucleic Acids Res.* **16**, 3559- 3572.

50

28. Kilkuskie, R., Wood, N., Ringquist, S., Shinn, R., and Hanlon, S. (1988) Effects of charge modification on the helical period of duplex DNA, *Biochemistry*, **27**, 4377-4386.

29. Ding, W.-d and Ellestad, G.A. (1991) Evidence for hydrophobic interaction between calicheamicin and DNA, *J. Am. Chem. Soc.* **113**, 6617-6620.

30. Fried, M.G., Wu, H-M., and Crothers, D.M. (1983) CAP binding to B and Z forms of DNA, *Nucleic Acid Res.* **11**, 2479-2494.

31. Lyubchenko, Y.L., Shlyakhtenko, L.S., Appella, E., and Harrington, R.E. (1993) CA runs increase DNA flexibility in the complex of 1 Cro protein with the O_R3 site, *Biochemistry* **32**, 4121-4127.

32. Battigello, J-M. A., Cui, M., Roshong, S., and Carter, B.J. (1995) Enediyne-mediated cleavage of RNA, *Bioorganic & Medicinal Chemistry* **3**, 839-849.

33. Brenowitz, M., Senear, D.F., Madeline, S.A., and Ackers, G.K. (1986) "Footprint" titrations yield valid thermodynamic isotherms, *Proc. Natl. Acad. Sci. USA* **83**, 8462-8466.

34. Senear, D.F., Brenowitz, M., Shea, M.A., and Ackers, G.A. (1986) Energetics of cooperative protein-DNA interactions: Comparison between quantitative deoxyribonuclease footprint titration and filter binding, *Biochemistry* **25**, 7344-7354.

35. Singleton, S.F. and Dervan, P.B. (1992) Thermodynamics of oligodeoxyribonucleotide-directed triple helix formation: An analysis using quantitative affinity cleavage titration, *J. Am. Chem. Soc.* **114**, 6956-6965.

36. Record, M.T., Jr., Lohman, T.M. and deHaseth, P.L (1978) Ion effects on ligand-nucleic acid interaction, *J. Mol. Biol.* **107**, 145-158.

37. Ikemoto, N., Kumar, R.A., Ling, T-t., Ellestad, G.A., Danishefsky, S.J., and Patel, D.J. (1995) Calicheamicin-DNA complexes: Warhead alignment and saccharide recognition of the minor groove, in press, *Proc. Natl. Acad. Sci. USA*.

38. Ackers, G.K., Shea, M.A., and Smith, F.R. (1983) Free energy coupling within macromolecules, *J. Mol. Biol.* **170**, 223-242.

39. Hogan, M.H., Dattagupta, N., and Crothers, D.M. (1979) Transmission of allosteric effects in DNA, *Nature* **278**, 521-524.

40. Wilson, W. D., Tanious, F.A., Barton, H.J., Jones, R.L., Fox,K., Wydra, R.L., and Strekowski, L. (1990) DNA sequence dependent binding, *Biochemistry* **29**, 8452-8461.

41. Eriksson, S., Kim, S.K., and Norden, B., (1993) Binding of 4',6'-Diamidino-2-phenylindole (DAPI) to AT regions of DNA: Evidence for an allosteric conformational change, *Biochemistry* **32**, 2987-2998.

42. Lee, M.D., Durr, F.E., Hinman, L.M., Hamann, P.R., and Ellestad, G.A., (1993) Advances in Medicinal Chemistry, in B. E. Maryanoff and C. A. Maryanoff (eds.), *The Calicheamicins* , Jai Press pp 31-66.

43. Shinega, M.K., Hagen, T.M., and Ames, B.N. (1994) Oxidative damage and mitochondrial decay in aging, *Proc. Natl. Acad. Sci. USA* **91**, 10771-10778.

44. Linnane , A. W., Marzuki, S., Ozawa, T., and Tanaka, M., (1989) Mitochondrial DNA mutations as an important contribution to aging and degenerative diseases,. *The Lancet* i, 643-644.

45. Heude, M. (1988) The induction of rho- mutants by UV or γ-rays is independent of the nuclear recombination repair pathways in *Saccharomyces cerevisiae*, *Mutation Research* **194**, 151-163.

46. Ferguson, L.R., and von Borstel, R.C., (1992) Induction of the cytoplasmic 'petite' mutation by chemical and physical agents in *Saccharomyces cerevisiae, Mutation Research* **265**, 103-148.

KEDARCIDIN AND MADUROPEPTIN, TWO NOVEL CHROMOPROTEINS WITH POTENT ANTITUMOR ACTIVITIES.

N. ZEIN
Oncology Drug Discovery
Bristol-Myers Squibb
P.O.Box 4000
Princeton, N.J. 08558-4000
U.S.A.

The enediyne antitumor antibiotics are a growing family of natural products isolated from soils that were collected in diverse parts of the world. As a class, the enediynes are the most potent antitumor agents ever isolated.

Calicheamicin γ_1^I for example is 1,000 times more potent in mice models than adriamycin, one of the most clinically effective antitumor agents. The enediyne moiety, commonly referred to as the warhead, contains two acetylenic groups conjugated to a double bond or incipient double bond within a 9- or a 10-membered ring. To date, the 9-membered enediyne antibiotics have been isolated as a complex with a highly acidic apoprotein; they are referred to as the chromoproteins and include actinoxanthin, auromomycin, neocarzinostatin, kedarcidin, C1027 and maduropeptin. This is in contrast to calicheamicin, esperamicin and dynemicin, agents containing 10-membered enediyne rings. To date, no 10-membered enediyne was isolated with an associated apoprotein [1].

The remarkable biological properties of the enediyne class of antibiotics are believed to result from their unique molecular structure: a structure that appears to be geared towards DNA destruction. The structural skeleton could be viewed as composed of three interrelated components: a DNA binding moiety, a triggering device and an enediyne functionality. Upon binding to DNA, activation of the trigger leads to the cyclization of the enediyne moiety to form a highly reactive diradical species. This species abstracts hydrogen atoms from the DNA backbone and leads to strand scission.

In this chapter, we will focus on the latest studies conducted on the kedarcidin chromophore, the maduropeptin chromophore and one of the maduropeptin chromophore derivatives (structures **1**, **2** and **3** respectively);

B. Meunier (ed.), DNA and RNA Cleavers and Chemotherapy of Cancer and Viral Diseases, 53–63.
© 1996 *Kluwer Academic Publishers. Printed in the Netherlands.*

54

kedarcidin and maduropeptin are two 9-membered enediyne containing chromoproteins recently isolated and characterized at Bristol-Myers Squibb [2-8].

| **1** | **2** | **3** |

1. Kedarcidin

1.1. KEDARCIDIN CHROMOPHORE

The isolation and structure elucidation of the kedarcidin chromophore presented a challenge to the natural product chemists as it is very unstable in most organic solvents. This problem was overcome through a series of chemical degradation and derivatization experiments. Considerable substructural information was obtained by selective methanolysis. This allowed the identification of two glycosidic units, L-mycarose and a new aminosugar, kedarosamine. Chemical reduction of the chromophore afforded a stable derivative and allowed for the structure elucidation of the central core portion [7, 8].

An examination of the structure (1) reveals a number of unique moieties exquisitely deployed within the molecule. For example, kedarcidin chromophore is the first known example of a natural product containing a 2'-chloroazatyrosine unit as a structural fragment. This chloroazatyrosine unit is amide-linked to a naphthoyl moiety, and also bridges the C-14 dienediyne core, forming a 17-membered macrocyclic lactone. Also attached to the core are the two glycosidic units. The core incorporates a highly strained nine-membered enediyne ring system and an epoxide ring allylic to a double-bond in a cyclopentene ring.

1.1.1. *Cleavage Chemistry*
Kedarcidin chromophore exhibits a marked preference for cleaving supercoiled duplex plasmid DNA to give mostly single strand breaks. The chemistry believed to be responsible for cleavage is outlined in Scheme 1. The cycloaromatization is

initiated by attack of a thiol on the less-hindered side of C-12 followed by bond migration and opening of the epoxide ring. Subsequent cyclization of the enediyne gives the 1,4-benzenoid diradical, (Scheme 1, **3**), the species believed to cause DNA damage. The opening of the epoxide reduces the strain energy developed in the transition state leading to cycloaromatization. This mechanism of action contrasts with that of the neocarzinostatin chromophore, in which the epoxide ring opening generates a cumulene intermediate that aromatizes to a C-2/C-6 indacene diradical.

Cleavage experiments with several 5'-end-labeled pBR322 and pUC18 restriction fragments show that the kedarcidin chromophore cleaves DNA site-specifically in a single-stranded manner. A prominent cleavage site is the 3' nucleotide (N=T, C, G, A) adjacent to the TCCT tetramer, as seen in Figure 1A. Another preferred site is TCGTN, where N is a C or a G. Secondary cleavage sites are also observed, e.g., the 5-mers TCATN, ACGCN and TCTAN ($\leq 50\%$ of the TCCTN band), with the extent of cleavage at a given site depending on the flanking sequences [3].

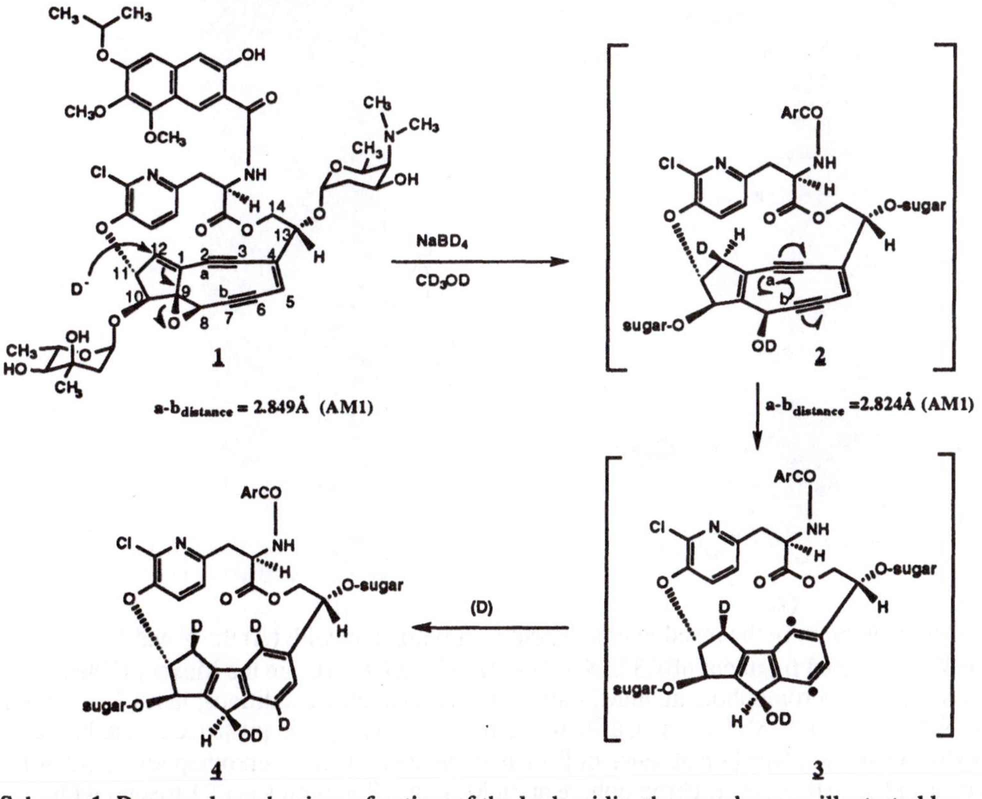

Scheme 1. Proposed mechanism of action of the kedarcidin chromophore as illustrated by NaBD4 reduction.

As observed with calicheamicin γ_1^I, bleomycin and neocarzinostatin, DNA cleavage studies with the 3'-end-labeled pBR322 SalI/BamHI fragment suggest that strand breakage is principally due to 4'-hydrogen atom abstraction along with a small portion of 5'-H abstraction from the targeted DNA deoxyribose sugars Figure 1B [5, 9-14]. Unequivocal chemical evidence for the proposed 4'-H abstraction will have to await detailed NMR studies such as those performed on calicheamicin.

Competition experiments with netropsin, show that kedarcidin cleavage sites are either modified or eliminated upon preincubation of the DNA with the known minor groove binder. These observations suggest that the kedarcidin chromophore acts in the minor groove of DNA.

By excluding oxygen from the reaction mixture, cleavage is dramatically inhibited. Moreover, strand-scission is not affected when the reaction is carried out in presence of excess superoxide dismutase or catalase, Figure 1A. Reduced

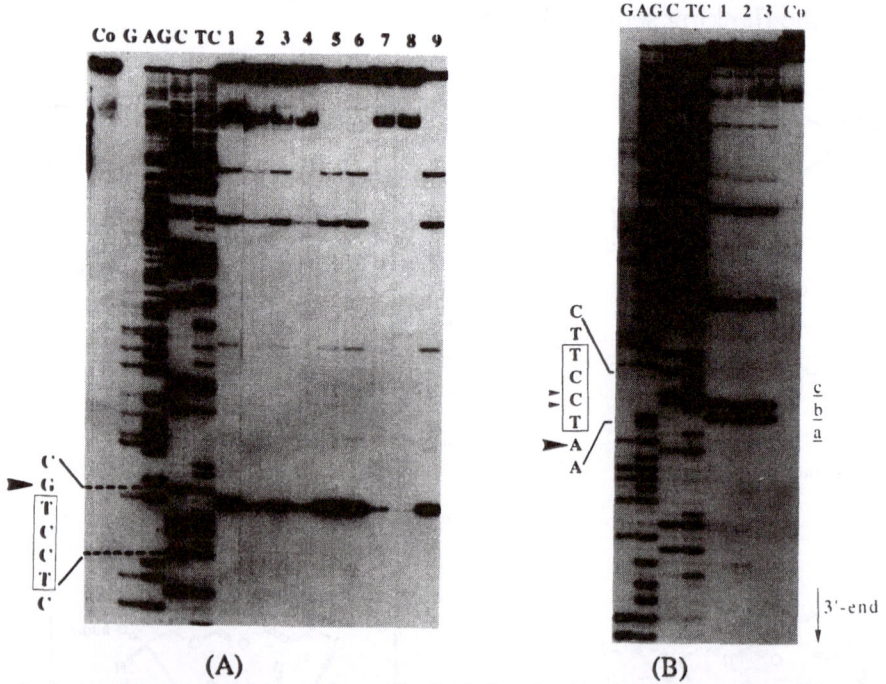

(A)   (B)

Fig. 1. Autoradiogram of the reaction of kedarcidin chromophore with (A) the 5' end-labeled and (B) the 3' end-labeled fragment pBR322 SalI/BamHI. G, AG, C, TC are the Maxam-Gilbert lanes. (A) Lane 1, chromophore at 5mM. Lane 2, under anaerobic conditions; lanes 3, 4, 7 with 50mM NaCl and 10mM MgCl$_2$ and CaCl$_2$ respectively; lanes 5, 6, in presence of catalase and superoxide dismutase; lane 8 in absence of β-mercaptoethanol; lane 9, chromophore at 50mM. Co is control lane. (B) Lane 1, chromophore at 5mM. Lanes 2,3, products of 1 treated with NaOH and NaBH$_4$ respectively as described in text. Co is control lane.

oxygen species are, therefore, not involved in the cleavage reaction. These results along with the specificity of the cuts argue against a diffusible species such as hydroxyl radicals and support instead a nondiffusible, carbon-centered radical produced during the aromatization of the enediyne moiety, as proposed in scheme 1, **3**.

The similarity of the kedarcidin chromophore naphthoyl moiety with that of siderophores led us to evaluate the chelation capacity of the chromophore for certain cations and the effect of these cations on strand cleavage. As seen in Fig. 1(A), addition of 10mM $CaCl_2$ or $MgCl_2$ to the reaction mixtures resulted in at least 90% inhibition of the DNA cleavage. In contrast, addition of 100mM $CaCl_2$ to a DNA/esperamicin A_1 or DNA/calicheamicin γ_1^I reaction mixtures had no effect on their cleaving properties. Worthy of notice is the absence of the siderophore-like moiety in esperamicin A_1 and calicheamicin γ_1^I [3, 15, 16].

NMR studies in presence of increasing amounts of $CaCl_2$ suggest that the Ca^{2+} ion is localized in the region of the naphthoic acid moiety as shown in structure **1'**, Figure 2. In the chelated form, kedarcidin presumably cannot associate with the DNA binding site thus, preventing DNA cleavage. These results suggest that the naphthoyl moiety is involved in DNA binding.

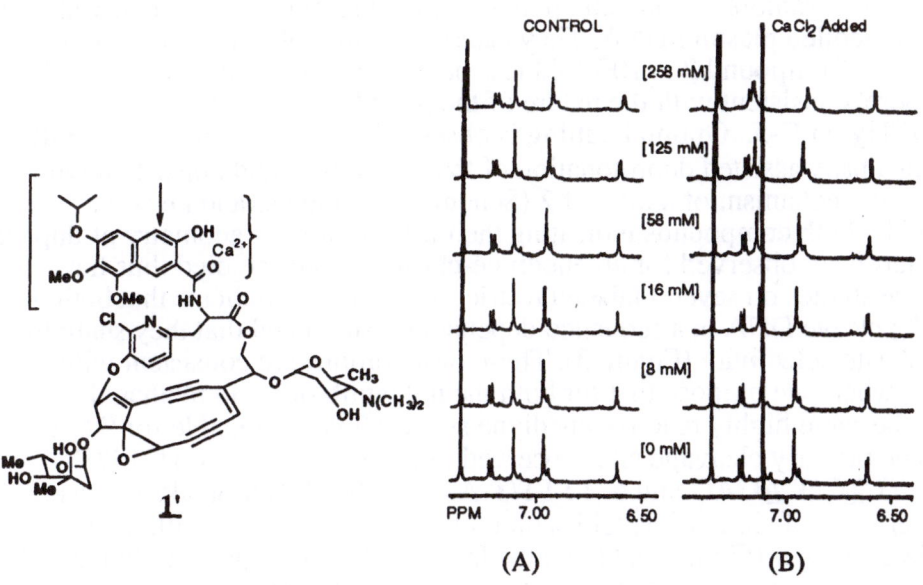

Fig. 2. Expanded NMR spectra of the kedarcidin chromophore with increasing amounts of (A) D_2O (Control) (B) $CaCl_2$ in D_2O.

2. Maduropeptin

2.1. MADUROPEPTIN CHROMOPHORE

Unlike kedarcidin, maduropeptin chromophore does not appear to be easily removed from its associated protein and a new process to extract the chromophore from the protein had to be developed. Conditions necessary to denature the protein and release the maduropeptin chromophore led to the isolation of solvent artifacts where nucleophilic addition to the double bond of the enediyne (at C-5) occured (Scheme 2, **2**). These solvent-artifacts were crucial in the determination of the structure and the mechanism of action of the highly labile parent compound **1**(Scheme 2). Methanol-adduct **2** (Scheme 2) exhibits good antitumor and antibacterial activities. However, these activities are 100-fold less than those of **1** (Scheme 2) [2].

Examination of the proposed structure of the chromophore suggests that it is biogenetically related to the enediyne chromophores that have been characterized thus far, having the same 14-unbranched core carbons.

2.1.1. *Cleavage Chemistry*

Maduropeptin chromophore and compound **2** (Scheme 2) favor covalently closed circular supercoiled plasmid DNA. They cause a mixture of double and single-strand breaks. Compound **2** is 100-fold less potent than maduropeptin chromophore, consistent with the nature of the poor-leaving group in **2** (Scheme 2), i.e. $-OCH_3-$ at C-5. Optimal cutting is observed in basic conditions (pH 8) suggesting a base-assisted deprotonation of the amide NH, and consistent with the proposed mechanism of action of **2** (Scheme 2). Single strand DNA is unaffected by both compounds indicating their affinity for the geometry of duplex DNA, as has been observed for all enediyne-chromophores studied thus far.

Cleavage studies on several labeled restriction fragments indicate that both MDP and **2** cleave DNA in a sequence-dependent fashion and that they share the same DNA site-selectivity (Figure 3). These observations are consistent with Scheme 2, where we propose that for both the holoantibiotic and methanol-adduct **2**, the same highly reactive enediyne species **3** is responsible for DNA cutting. The primary cleavage sites observed in our studies are, 5' TCTT/3' AGAA, 5'TCTC/3' AGAG and 5' TTTT/3' AAAA. The DNA breaks at these sites appear to be of similar intensities on both strands and occur with a two-nucleotide 3'-stagger (Figure 3). Given the fact that the cleavage experiments are run under single-hit kinetics, these results suggest that the DNA damage at those sites could result from bistranded lesions. 3'-end labeled DNA studies (Figure 3B), using restriction fragments and oligomers suggest that in the case of the observed primary cleavage sites, MDP and 2 (Scheme 2) cause DNA breaks by abstracting hydrogen from the C-4' site on the deoxyribose sugars [6].

The nature (DS versus SS), the chemistry (position of the hydrogen atom on the targeted deoxyribose sugar) and the magnitude of cleavage at the secondary sites were found to be contingent upon the DNA sequence beyond the immediate neighboring DNA bases. Such was the case for 5' NTTT, 5' TTAT, 5' TCCT, 5' ATCT, 5' TCAT, 5' TACT, 5' TCAC and their complementary sequences.

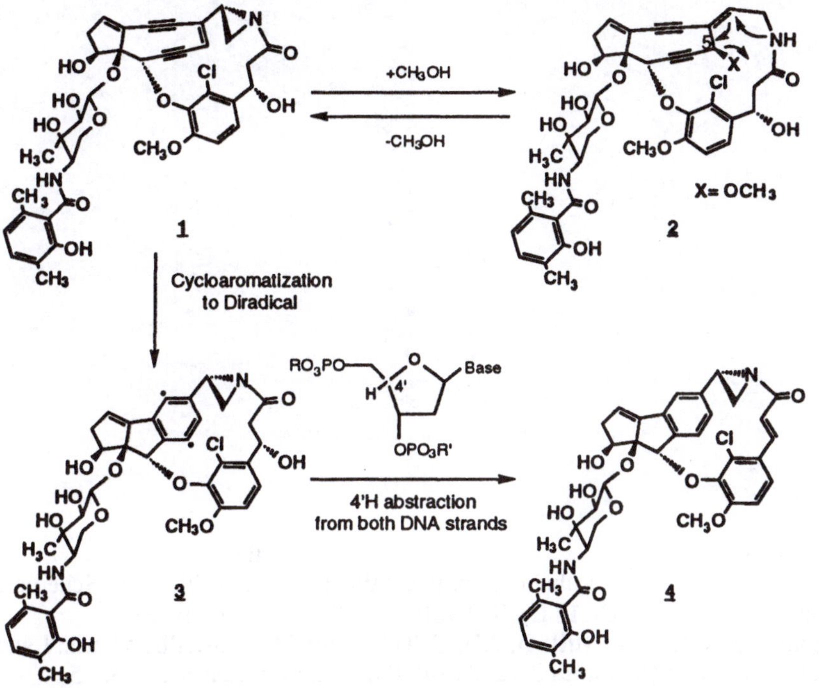

Scheme 2. Proposed mechanism of action of maduropeptin chromophore 1 and the methanol artifact 2.

These observations clearly indicate the importance of the DNA microstructure at the cleavage sites and suggest that certain stretches of DNA are more "accomodating" than others. Conclusive evidence as to the chromophore cleavage chemistry awaits more detailed studies.

As observed with kedarcidin, preincubation with netropsin alters the chromophore cleavage pattern suggesting that the chromophore acts in the minor groove of DNA (Figure 3). Also, by excluding oxygen from the reaction mixture, cleavage is dramatically inhibited. Strand-scission is not affected when the reaction is carried out in presence of excess superoxide dismutase or catalase. These observations along with the specificity of the cleavage argue for a mechanism involving a non-diffusible, carbon-centered radical **3** , as depicted in scheme 2.

It is interesting to note that the 9-membered enediyne chromophores MDP, NCS and C-1027 cleave DS DNA with a two-nucleotide stagger [6, 14, 17]. In contrast, the DS cuts in the case of the 10-membered enediynes, calicheamicin and esperamicin, occur with a three-nucleotide stagger [1].

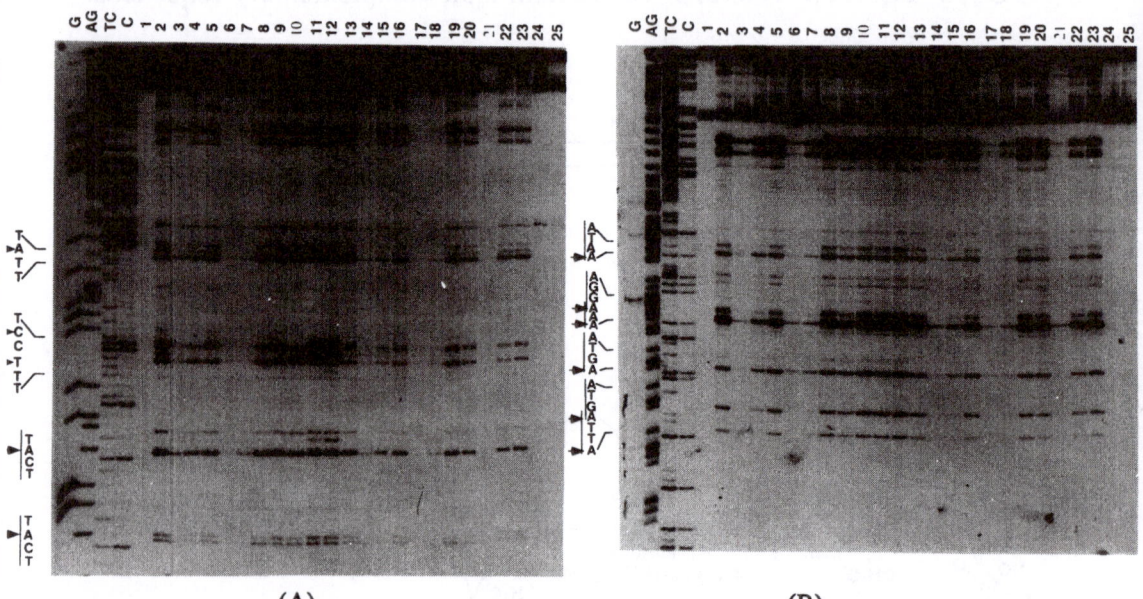

(A)   (B)

Fig.3. Autoradiograms of the reaction of maduropeptin and the methanol artifact **2** in Scheme 2 with (A) the 5' end-labeled fragment pUC18 EcoRI PvuII and (B) the 3' end-labeled complementary fragment pUC18 PstI PvuII. G, AG, C, TC are the Maxam-Gilbert lanes. Lane 1 is control lane. In both (A) and (B) Lanes 2-12, 24 MDP at 50μg/ml, Lanes 13-23, 25 chromophore at 100μg/ml. Lanes 2,13 no further treatment; lanes 3-5 and 14-16 in presence of 10, 1 and 0.1mM $CaCl_2$ respectively; lanes 6, 17 in presence of 10mM $MnCl_2$; lanes 7, 18 in presence of 10mM $MnCl_2$; lanes 8, 19 in presence of 10mM $ZnCl_2$; lanes 9, 20 in presence of 10mM KCl; lanes 10-12, 21-23 in presence of 14, 1.4 and 0.14 mM βSH respectively. Lanes 24, 25 pretreated with netropsin.

As shown with kedarcidin, DNA cleavage is inhibited in the presence of $MgCl_2$ and $CaCl_2$ (Figure 3). NMR studies on **2** (Scheme 2) in presence of $CaCl_2$ suggest the chelation of a calcium cation in the β-hydroxy amide portion of the atoms that bridge the core rings (Figure 4). Such chelation would decrease the nucleophilicity of the amide nitrogen in **2** thus hindering aziridine formation and the concurrent loss of methanol via the intramolecular ring contraction step to form parent enediyne **1** (Scheme 1) and ultimately the DNA-damaging intermediate **3** as well; under these conditions, DNA cleavage cannot occur. The

NMR data also shows the localization of Ca^{2+} near the terminus of the sugar side chain (Figure 4). This observation suggests the involvement of the ortho hydroxy benzamide moiety in DNA binding since in the chelated form, the chromophore cannot associate with DNA and thus this may be the reason DNA cleavage does not occur upon addition of $CaCl_2$. As with kedarcidin, these siderophore-like chelation sites obviously play a role in the interaction of the chromophore with DNA and add to the unique chemical features of these chromophores [6, 15, 16].

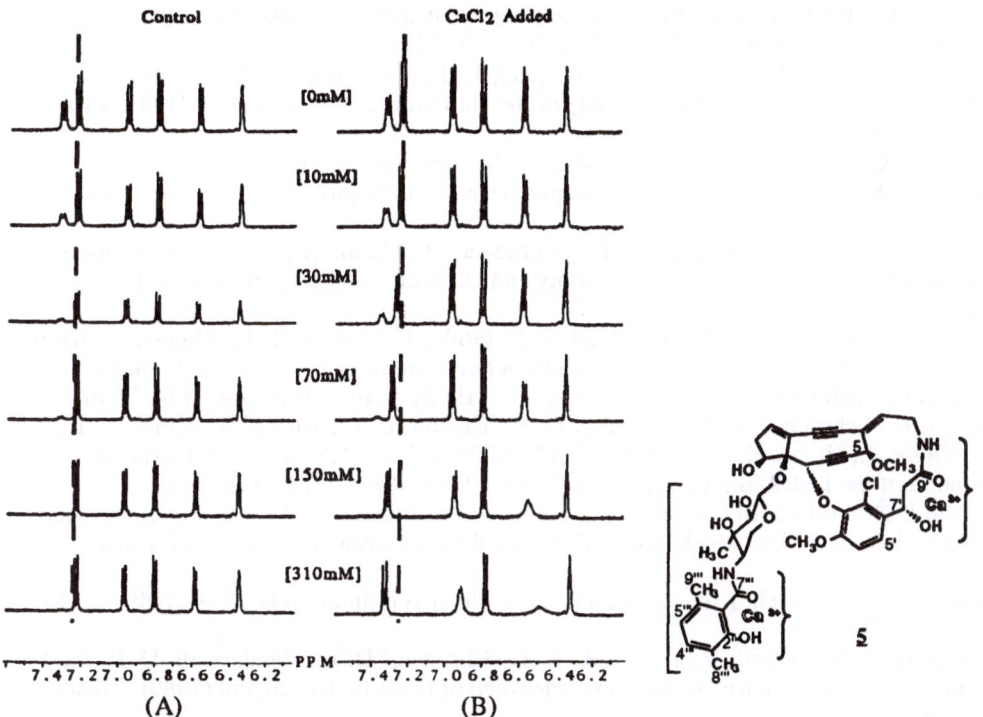

Figure 4. Expanded NMR spectra of the maduropeptin methanol-artifact with increasing amounts of (A) D_2O (control) (B) $CaCl_2$ in D_2O.

Acknowledgments. The author wishes to thank W. Solomon for his invaluable technical assistance, J. Leet and D. Schroeder for providing the chromoproteins and the chromophores and K. Colson for performing the NMR studies.

62

References

1. Doyle, T. W., Borders, D. B. (1994) Enediyne Antitumor Antibiotics, in *Enediyne Antibiotics as Antitumor Agents*, Doyle, T. W., Borders, D. B., Eds., Marcel Dekker Inc., New York, 1994, pp. 1-15 and references therein.
2. Schroeder, D. R., Colson, K.L., Klohr, S. E., Zein, N., Langley, D. R., Lee, M.S., Matson, J. A., Doyle, T. W., Isolation, structure determination, and proposed mechanism of action for artifacts of maduropeptin chromophore (1994) *J. Am. Chem. Soc.* 116, 9351-9352.
3. Zein N., Colson, K. L., Leet, J. E., Schroeder, D. S., Solomon, W., Doyle, T. W., Casazza, A. M., Kedarcidin chromophore: an enediyne that cleaves DNA in a sequence-specific manner (1993) *Proc. Natl. Acad. Sci. USA.*, 90, 2822-2826.
4. Zein, N., Casazza, A. M., Doyle, T. W., Leet, J. E., Schroeder, D. R., Solomon, W., Nadler, S. G., Selective proteolytic activity of the antitumor agent Kedarcidin (1993) *Proc. Natl Acad. Sci. U.S.A.*, 90, 8009.
5. Zein, N., Reiss, P., Bernatowicz, M., Bolgar, M., The proteolytic specificity of the natural enediyne-containing chromoproteins is unique to each chromoprotein (1995) *Chemistry & Biology*, Vol. 2, 7.
6. Zein, N., Solomon, W., Colson, K. L., Schroeder, D., Maduropeptin: a novel antitumor chromoprotein with selective protease activity and DNA cleaving properties (1995) *Biochemistry, in press.*
7. Leet, J. E., Schroeder, D. R., Hofstead, S. J., Golik, J., Colson, K. L., Huang, S., Klohr, S. E., Doyle, T.W., Matson, J. A., Kedarcidin, a new chromoprotein antitumor antibiotic: Structure elucidation of Kedarcidin chromophore (1992) *J. Am. Chem. Soc.* ,114, 7946.
8. Leet, J. E., Schroeder, D. R., Langley, D. R., Colson, K. L., Huang, S., Klohr, S. E., Lee, M. S., Golik, J., Hofstead, S. J., Doyle, T.W., Matson, J. A., Chemistry and structure elucidation of the kedarcidin chromophore (1993) *J. Am. Chem. Soc.* ,115, 8432.
9. Giloni, L., Takeshita, M., Johnson, F., Iden, C. & Grollman, A. P., Bleomycin-induced strand-scission of DNA. Mechanism of deoxyribose cleavage (1981) *J. Biol. Chem.* 256, 8608.
10. Stubbe, J. & Kozarich, J. W., Mechanisms of Bleomycin-induced DNA degradation (1987) *Chem. Rev.* 87, 1107.
11. Kozarich, J. W., Worth Jr., L., Frank, B. L., Christner, D. F., Vanderwall, D. E., Stubbe, J., Sequence-specific isotope effects on the cleavage of DNA by Bleomycin (1989) *Science* 245, 1396.
12. Rabow, L. E., Stubbe, J. A., Kozarich, J. W., Identification and quantitation of the lesion accompanying base release in Bleomycin-mediated DNA degradation (1990) *J. Am. Chem. Soc.* 112, 3196.
12. Zein, N., McGahren, J. M., Morton, G. O., Ashcroft, J. & Ellestad, G. A., Exclusive abstraction of nonexchangeable hydrogens from DNA by calicheamicin (1989) *J.Am. Chem. Soc.* 111, 6888.
13. De Voss, J. J., Townsend, C. A., Ding, W. D., Morton, G. O., Ellestad, G. A., Zein, N., Tabor, A. B., Schreiber, S. L., Site-specific atom transfer from DNA to a bound ligand defines the geometry of a DNA- Calicheamicin γ_1^I complex (1990) *J.Am. Chem. Soc.* 112, 9669.
14. Goldberg, I. H., Kappen, L. S., Neocarzinostatin: Chemical and Biological Basis of Oxidative DNA Damage, in Enediyne Antitumor Antibiotics, in *Enediyne Antibiotics as Antitumor Agents*, Doyle (1994) T. W., Borders, D. B., Eds., Marcel Dekker Inc., New York, pp.327-362 and references therein.

15. Peterson, T, Neilands J. B., Revised structure of a catecholamide spermidine siderophore (1979) *Tetrahedron Lett.* 50, 4805.

16. Telford, J. R., Leary, J. A., Tunstad, L. M. G., Byers, B. R., Raymond, K. N., Amonabactin: Characterization of a series of siderophores from Aeromonas hydrophila (1994) *J. Am. Chem. Soc.* 116, 4499.

17. Xu, Y-J, Zhen, Y-S & Goldberg, I. H., C1027 chromophore, a potent new enediyne antitumor antibiotic, induces sequence-specific double-strand DNA cleavage (1994) *Biochemistry 33*, 5947.

DYNEMICIN A, A NOVEL ENEDIYNE DNA CUTTER

YUKIO SUGIURA & TETSUYA KUSAKABE
Institite for Chemical Research
Kyoto University
Uji, Kyoto 611, Japan

ABSTRACT. Dynemicin A is a hybrid antitumor antibiotic containing anthraquinone and enediyne cores, which contribute to binding and cleavage of DNA duplex. The DNA strand scission of dynemicin A is significantly enhanced by the addition of reductants or by the irradiation of visible light. This unique antibiotic mainly abstracts C-1' hydrogen of DNA deoxyribose. The sequence-specificity of the DNA cleavage is relatively low, but dynemicin A preferentially breaks conformationally flexible regions of DNA such as B-Z junction, bulge, and nick sites. In adition, the drug cleaves specifically at the 3'-shifted position by one base opposite the gap such as 5'-Pu_Pu/3'-PyPuPy sequences. The reaction products of methyl thioglycolate with dynemicin A have been isolated by HPLC technique and identified spectroscopically. On the basis of some experimental results, a two-step mechanism for the double-stranded DNA scission by dynemicin A has also been proposed.

1. Introduction

The enediyne families of antibiotics have become of great interest. This series includes calicheamicin, esperamicin, dynemicin, neocarzinostatin chromophore, C-1027 chromophore, and kedarcidin chromophore. Their novel chemical structures, potent antitumor activities, and fascinating action modes have elicited extensive research activities in chemistry, biology, and medicine.[1,2]

Dynemicin A is characterized as a unique hybrid antitumor antibiotic containing typical two chemotypes, enediyne and anthraquinone.[3] It is believed that dynemicin A exerts its biological effect by causing DNA strand breaks. The DNA cleaving activity of the drug is significantly enhanced by reducing agents such as NADPH and thiol,[4] or by visible light irradiation.[5] Under these experimental conditions, the anthraquinone moiety is reduced to undergo opening of the epoxide, causing a significant conformational rearrangement in the drug molecule. This conformational change leads to the Bergman cyclization of the enediyne unit to form a benzenoid diradical. The reactive radical species can abstract hydrogen atoms from the deoxyribose backbone of DNA duplex. On the basis of molecular mechanism of dynemicin A, a series of enediyne model compounds of dynemicin A type equipped with triggering and modulating devices have also been designed and synthesized.[6-8]

We report here some characteristics of DNA cleavage, conformation-selective DNA strand breaks, and cooperative double-stranded DNA break by dynemicin A. These

B. Meunier (ed.), DNA and RNA Cleavers and Chemotherapy of Cancer and Viral Diseases, 65–73.
© *1996 Kluwer Academic Publishers. Printed in the Netherlands.*

studies elucidate the first model for a "dynamic" targeting of "dynemicin A" at the level of molecular biology.

2. Activation Systems of Dynemicin A and Reaction Products of Dynemicin A with Methyl Thioglycolate

The reaction of methyl thioglycolate with dynemicin A gave two products. These reaction products were isolated by HPLC purification and identified spectroscopically (see Table 1).[9] The major product, dynemicin H, was determined to be a C-8 hydrogen analogue of dynemicins L and N in which the enediyne core is aromatized. The minor product, dynemicin S, is an adduct of methyl thioglycolate at the C-8 position. By using NADPH instead of methyl thioglycolate, the reaction with dynemicin A also gave the same major product (dynemicin H). Futhermore, the photoproduct of dynemicin A was chromatographically identical with the major reaction product (dynemicin H) of the thiol-activated dynemicin A.[5] Dynemicin A also effectively broke DNA strands under alkaline pH condition. The alkali-product of dynemicin A was chromatographically identified with dynemicin N (C-8=OH), suggesting a DNA cleavage mechanism similar to the reductant- and light-induced activation systems of dynemicin A.[10] The nucleotide-specific cleavage of dynemicin A induced by addition of methyl thioglycolate remarkably resembled to that induced by addition of NADPH. On the other hand, the two reaction products, dynemicin H and dynemicin S, showed no DNA cleaving abilities. The formation of dynemicins H and S provides a rationale for the reductive and nucleophilic activations of dynemicin A. Namely, in the first step, dynemicin A is reducively activated to the hydroquinone, which rearranges via epoxide opening to the quinonemethide. This conformational change brings the two acetylene bonds closer together and leads to the Bergman cyclization of the enediyne unit to form a benzenoid diradical. The reactive radical species, when positioned in the minor groove of DNA, can abstract hydrogen atoms from the deoxyribose backbone of DNA and yield the inactivated dynemicin H. The elegance of the action mechanism of dynemicin A is shown in Figure 1.

The product analysis of the drug-cleaved DNA fragments by gel-electrophoresis revealed that dynemicin A abstracts the C-1' hydrogen of DNA deoxyribose, and that the damaged DNA leads to strand breaks with the formation of 5'- and 3'-phosphate termini.[11] The lesions of C-4' hydrogen also occurred at 3'-side of GC base pairs such as 5'-CT and 5'-GA, leading to 5'-phosphate and 3'-phosphoglycolate termini or 4'-hydroxylated abasic sites. It is of interest that the C-1' hydrogen abstraction dominates in the dynemicin A-induced DNA strand scission, and this is distinct from the preferential C-5' hydrogen abstraction of calicheamicin and neocarzinostatin. Therefore, the present result predicts the C-1' hydrogen abstraction from one of the diradical (C27 and C24) of dynemicin A.

3. DNA Cleavage by Dynemicin A for B-Z Conformational Junction, Bulge, and Nick Regions

We previously showed that dynemicin A binds to the minor groove of DNA and produces sequence-preferential cleavage of DNA at the base immediately to the 3'-side of purines.[4] However, its sequence-selectivity is relatively low when compared with other enediyne antibiotics such as esperamicin [12] and calicheamicin.[13] Several DNA cleavers

TABLE 1. ^{1}H NMR spectral data of dynemicins H and S ^{a}

proton	dynemicin H	dynemicin S
4-CH$_3$	0.95 (3 H, d, J = 7 Hz)	0.96 (3 H, d, J = 7 Hz)
4-H	3.23 (1 H, q. J = 7 Hz)	3.27 (1 H, q. J = 7 Hz)
8-H	3.50 (1 H, d, J = 3 Hz)	
-S-CH$_2$-		3.46 (2 H, dd, J = 10. 5 Hz)
-COOCH$_3$		3.61 (3 H, s)
6-OCH$_3$	3.84 (3 H, s)	3.82 (3 H, s)
7-H	3.96 (1 H. d, J = 3 Hz)	4.10 (1 H, s)
2-H	4.78 (1 H. d. J = 5 Hz)	4.72 (1 H, d, J = 5 Hz)
3-OH	5.46^b (1 H. s)	5.39^h (1 H, s)
16-H. 17-H, 24-H	7.13 ⎫	6.98 ⎫
25-H. 26-H, 27-H	7.56 ⎭ (6 H. m)	7.54 ⎭ (6 H, m)
10-H	7.63 (1 H, s)	8.43 (1 H, s)
5-COOH, 11-OH	11.3^b ⎫	11.3^h ⎫
15-OH. 18-OH	14.0^b ⎭ (4 H, br)	14.0^l ⎭ (4 H, br)

^{a}400 MHz in DMSO-d_6 ^{b} The signal disappeared by addition of D$_2$O.

dynemicin A   dynemicin H   dynemicin S

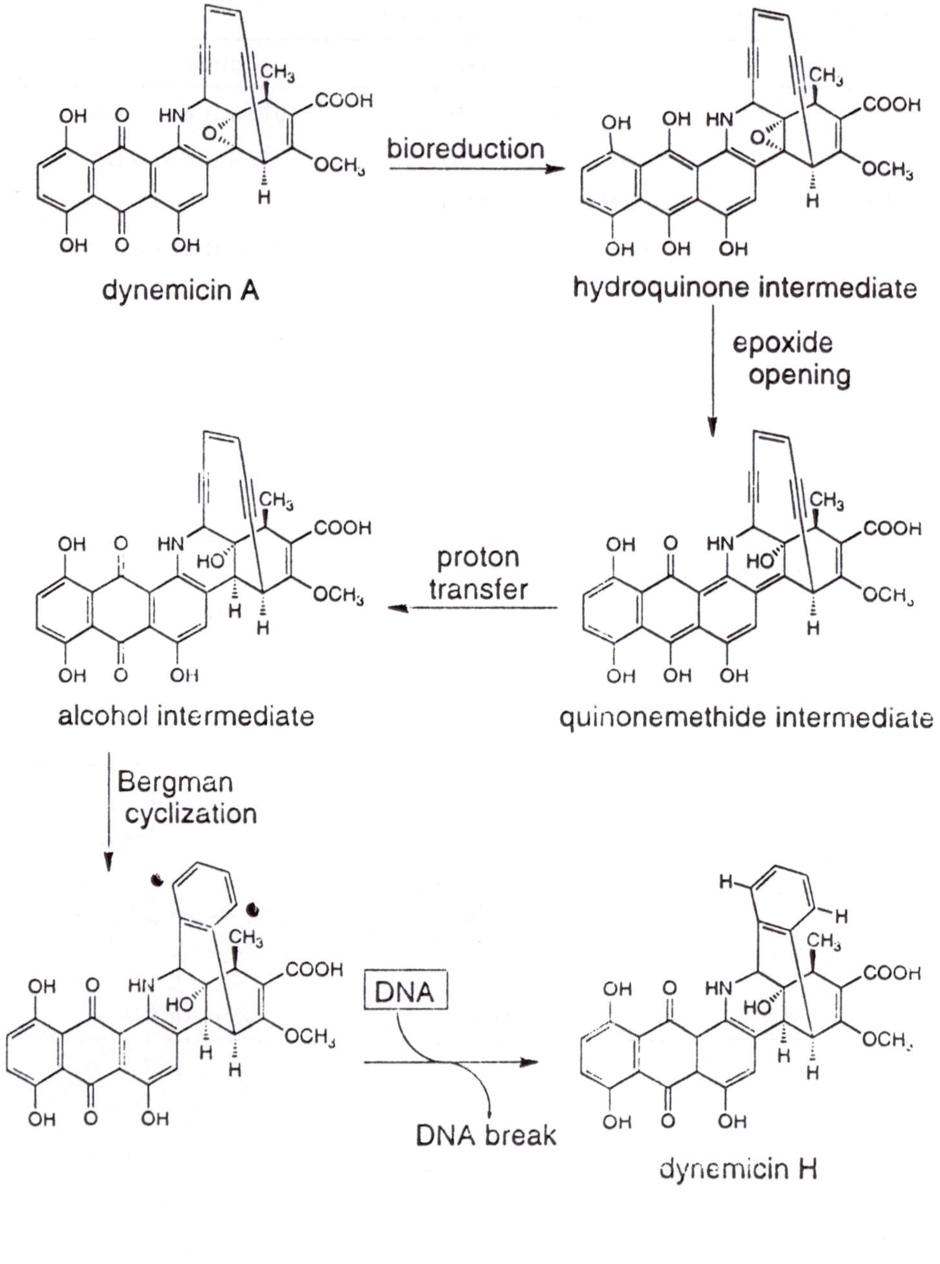

Figure 1. Activation mechanism of dynemicin A

that have low selectivity for a particular base or base-pair step exhibit specificities for certain kinds of DNA higher-structure.

Of special interest is the fact that dynemicin A dramatically enhanced its DNA cleavage in a B-Z conformational junction region.[14] Under high salt condition of pBR Z16 and Z24 DNAs, the cleavage intensity by dynemicin A was evidently increased at the B-Z junction, especially at the A-30 site. In contrast, no significant hyperreactities were detected in the case of pBR Z10 DNA. Indeed, Z transition of pBR Z24 DNA inserted $(dC-dG)_{12}$ oligodeoxynucleotide was confirmed by susceptibility to Os modification. In order to examine the conformational dependence of DNA cleavage by dynemicin A, we synthesized the designed oligodeoxynucleotide substrates. The specific strand breaks occurred at nucleotide residues near the bulge and nick sites as shown in Figure 2.[15] Insertion of the bulged thymidine into the host oligodeoxynucleotide clearly strengthened the cleavage at A-39 and G-40 on the opposite strand. The removal of the bulged thymidine significantly reduced the cleavage at these sites. The result indicates that the introduction of the unpaired thymidine facilitates the DNA cleavage at the 3'-staggered positions on the opposite strand. We also tested the cleavage selectivity of dynemicin A for oligodeoxynucleotide containing a single-nick structure. Dynemicin A gave specific damages at residues around C-15, namely, at 3'-shifted positions from the nick on the opposite strand (Figure 2). Similar cleavages did not occur in the same sequence without any nick. Therefore, this cleavage specificity is due to local flexibility derived from the nick rather than the sequence of the target DNA. Indeed, the introduction of the nick enhanced OsO_4 reactivity at residues C-14, C-15, and T-16, suggesting conformational flexibility around the nick.

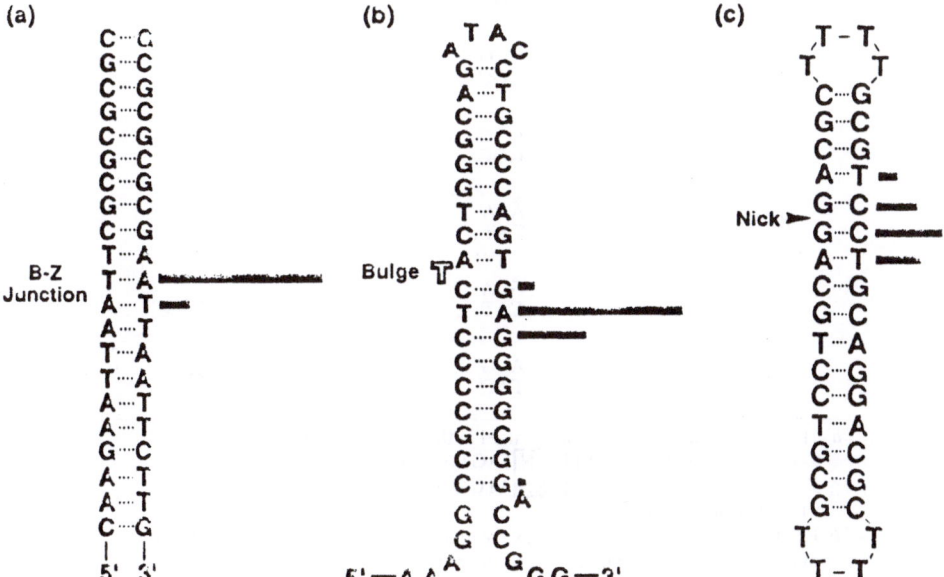

Figure 2. Conformation-specific cleavage of dynemicin A for B-Z junction (a), bulge (b), and nick (c) sites. The bar represents the relative intensity of dynemicin A-induced cleavage determined on the basis of densitometry.

TABLE 2. Dynemicin A-induced cleavage activities for various gap-containing DNAs

entry	N N [a] NNN	relative intensities (%) [b]
II	G G CGC	100
IV	G G CAC	15
V	G G CCC	11
VI	G G CTC	14
VII	A G TGC	62
VIII	A A TGT	32
IX	G A CGT	22
X	G C CGG	34
XI	C G GGC	25
XII	C C GGG	9
XIII	A A TAT	80
XIV	G A CAT	25
XV	A G TAC	68

a The gap-containing DNA duplexes used in this study are as follows:
5'- CCTTTGTAACTGTGAT N N TAGTGCTGTAGATA -3'
3'-   AAACATTGACACTA NNN ATCACGACATCTATCC-5'
where N is the site of cleavage.

b Relative DNA cleavage intensities were obtained from densitometric estimation of gel autoradiogram, and the errors were ± 3 %.

The DNA binding of dynemicin A is expected to be the intercalation of its anthraquinone core into DNA duplex. The DNA unwinding behavior of dynemicin antibiotic and the absorption spectrum of the dynemicin-DNA complex are strongly indicative of its intercalative binding with DNA.[15] Herein, the steric hindrance of its enediyne moiety which is nearly perpendicular to the anthraquinone chromophore, requires a sufficient stereo-space for the stable binding of dynemicin A to DNA. DNAs involving bulge, nick, and B-Z junction may satisfy such a situatoin. Probably, DNA local flexibility is able to create an open space in the minor groove, allows facile intercalation of dynemicin A, and then increases the chances of its DNA damaging event.

The present results evidently show the hyperreactivity of dynemicin A for conformationally flexible regions in duplex DNA. The local flexibility of DNA structure has been believed to be an important factor in protein binding, mutagenesis, and replication of DNA. Therefore, the conformation-selective DNA cleavage by dynemicin A might be related to its potent biological activity.

4. Cooperative Double-Stranded Breaks by Two Dynemicin A Molecules

It is known that enediyne antitumor antibiotics such as neocarzinostatin,[16] calicheamicin,[17] and C-1027[18] induce double-stranded cleavage of DNA. However, the previous computer modeling study predicted that dynemicin A does not cause the DNA double-stranded breaks concomitantly but instead produces single-stranded breaks.[19] Indeed, specific cleavages were restricted to the single-stranded sites of the three conformational regions as mentioned above. On the other hand, our study using nicked DNA substrate proposed that dynemicin A can cause double-sranded breaks cooperatively.[15] Therefore, it is of interest whether the double-stranded cleavage mode elicited by dynemicin A is a two-step or one-step process. Although the detection of the double-stranded cleavage sites was tried by using native gel, we could hardly identify these sites because of remarkable weakness. Therefore, a double-stranded DNA cleavage mechanism by dynemicin A was investigated through sequence-dependent strand breakage of a series of DNA duplexes containing a single nucleotide gap. We found that (1) dynemicin A breaks specifically at the 3'-shifted position by one base opposite the gap, (2) the strong cleavage is detected at 5'-Pu_Pu/3'-PyPuPy sequences, in particular at 5'-G_G/3'-CGC and 5'-A_A/3'-TAT sequences (see Table 2), and (3) dynemicin H gives only a small inhibition effect (20%) on the cleavage of gapped duplex by dynemicin A. In addition, the long half-life (118 min, in the presence of DNA) of aromatization of dynemicin A provided enough time for the second cleavage. These results strongly indicate a two-step mechanism for the double-stranded DNA scission of dynemicin A. Namely, this double-stranded break is caused by two drug molecules, each of which cuts one DNA strand (Figure 3). The first step is the generation of a gapped site by the first dynemicin A molecule. The local flexible conformation around the gap leads to a cleavage reaction of the opposite strand by the second drug molecule. As a result of the two close-spaced single-stranded breaks, an apparent double-stranded breakage ultimately occurs. Since double-stranded breaks are more difficult to repair than single-stranded breaks, the cooperative double-stranded cleavage by dynemicin A may be important for an understanding of its potent cytotoxicity. Our study reveals some characteristics of the

72

conformation-dependent DNA cleavages by dynemicin A, and have a bright prospect of certain applications to cancer chemotherapy.

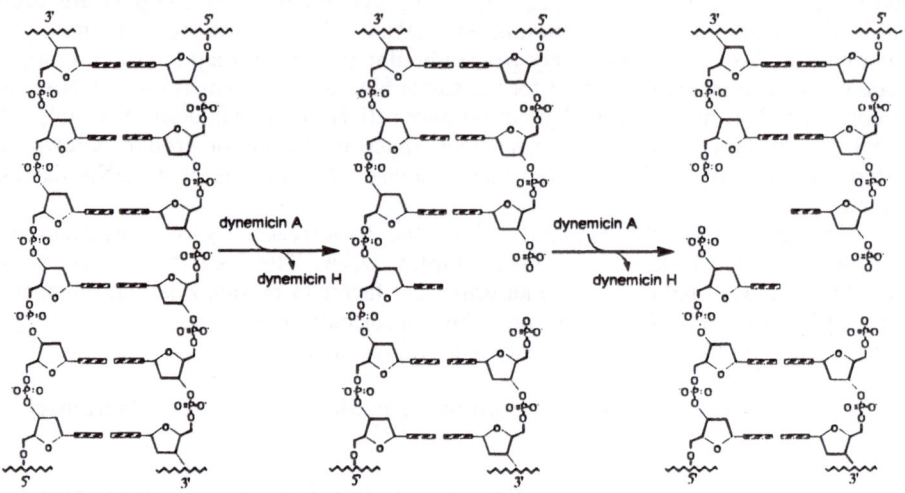

Figure 3. Cooperative double-stranded breaks by two dynemicin A molecules

5.Acknowledgments

We are greatful to our co-workers, collaborators, and colleagues cited in the individual references for their inspired contributions to this research. This study was supported in part by a Grant-in-Aid for Scientific Research on Priority Areas from the Ministry of Education, Science, and Culture, Japan.

6. References

(1) Nicolaou,K.C. and Dai,W.-M. (1991) *Angew.Chem.Int.Ed.Engl.* **30**, 1387-1416.
(2) Nicolaou,K.C., Smith,A.L., and Yue,E.W. (1993) *Proc.Natl.Acad.Sci.USA* **90**, 5881-5888.
(3) Konishi,M., Ohkuma,H., Tsuno,T., Oki,T., VanDuyne,G.D., and Clardy, J. (1990) *J.Am.Chem.Soc.* **112**, 3715-3716.
(4) Sugiura,Y., Shiraki,T., Konishi,M., and Oki,T. (1990) *Proc.Natl.Acad.Sci.USA* **87**, 3831-3835.
(5) Shiraki,T. and Sugiura,Y. (1990). *Biochemistry* **29**, 9795-9798.
(6) Nicolaou,K.C., Dai,W.-M., Tsay,S.-C., Estevez,V.A., and Wrasidlo,W. (1992) *Science* **256**, 1172-1178.
(7) Nicolaou,K.C., Dai,W.-M., Hong,Y.P., Tsay,S.-C., Baldridge, K.K., and Siegel,J.S. (1993) *J.Am.Chem.Soc.* **115**, 7944-7953.
(8) Myers,A.G., Fraley,M.E., Tom,N.J., Cohen,S.B., and Madar,D.J. (1995) *Chem.Biol.* **2**, 33-43.
(9) Sugiura,Y., Arakawa,T., Uesugi.,M., Shiraki,T., Ohkuma,H., and Konishi,M. (1991) *Biochemistry* **30**, 2989-2992.

(10) Arakawa,T., Kusakabe,T., Kuwahara,J., Otsuka,M., and Sugiura,Y. (1993) *Biochem.Biophys.Res.Commun.* **190**, 362-370.

(11) Shiraki,T., Uesugi,M., and Sugiura,Y., (1992) *Biochem.Biophy.Res.Commun.* **188**, 584-589.

(12) Sugiura,Y., Uesawa,Y., Takahashi,Y., Kuwahara,J., Golik,J., and Doyle,T.W. (1989) *Proc.Natl.Acad.Sci.USA* **86**, 7672-7676.

(13) Zein,N., Sinha,A.M., McGahren,W.J., and Ellestad,G.A. (1988) *Science* **240**, 1198-1201.

(14) Ichikawa,A., Kuboya,T., Aoyama,T., and Sugiura,Y. (1992) *Biochemistry* **31**, 6784-6787.

(15) Kusakabe,T., Maekawa,K., Ichikawa,A., Uesugi,M., and Sugiura,Y. (1993) *Biochemistry* **32**, 11669-11675.

(16) Dedon,P.C. and Goldberg,I.H. (1992) *Chem.Res.Toxicol.* **5**, 311-332.

(17) Dedon,P.C. Salzberg,A.A., and Xu,J. (1993) *Biochemistry* **32**, 3617-3622.

(18) Sugiura,Y. and Matsumoto,T. (1993) *Biochemistry* **32**, 5548-5553.

(19) Langley,D.R., Doyle,T.W., and Beveridge,D.L. (1991) *J.Am.Chem.Soc.* **113**, 4395-4403.

(20) Kusakabe,T., Uesugi,M., and Sugiura,Y. (1995) *Biochemistry* **34**, 9944-9950.

Bleomycin: the Paradigm of DNA Cleavers Based on Metal Complexes

POLYNUCLEOTIDE CLEAVAGE AND THE EXPRESSION OF ANTITUMOR ACTIVITY BY BLEOMYCIN

SIDNEY M. HECHT
Departments of Chemistry and Biology
University of Virginia
Charlottesville, Virginia 22901

Abstract

The antitumor agent bleomycin mediates the degradation of both DNA and RNA. In the presence of Fe^{2+} and O_2, DNA and RNA are both degraded oxidatively in a sequence-selective fashion. All major cellular forms of RNA can be degraded, including messenger RNA, ribosomal RNA, transfer RNA and the RNA strand of an RNA-DNA heteroduplex. While the degradation of DNA and RNA each have unique characteristics, the accumulated evidence makes it clear that the processes by which they undergo oxidative degradation are fundamentally analogous. Further, the oxidative destruction of both polynucleotides by bleomycin can obtain within intact cells. In addition to the oxidative destruction of DNA and RNA by bleomycin, both of which probably contribute to the antitumor effects noted for the drug, we have also shown that metal-free bleomycin can mediate RNA strand scission by a hydrolytic mechanism.

1.   Introduction

The bleomycins are a family of glycopeptide-derived antitumor antibiotics; their structures are exemplified by bleomycin A_2 (Figure 1) which is the major constituent of the clinically used mixture of bleomycins denoted blenoxane [1-5]. While several biochemical effects of bleomycin were noted in early experiments employing cell free systems [6-9], the ability of bleomycin to mediate DNA damage quickly became the focus of most studies concerned with the mechanism of action of bleomycin.

In cell free systems, bleomycin has been shown to mediate sequence-selective damage to DNA [4,5,10-12]. The degradation of DNA is oxidative in nature, and requires the presence of one of several redox-active metal ions such as Fe, Cu and Mn [13-16], as well as oxygen or an oxygen surrogate [13,17-20].

Bleomycin has also been shown to degrade chromosomal DNA when incubated in the presence of cultured mammalian cells [21-23]. In a study that measured the effects of a number of bleomycin congeners on cultured human cells, Berry *et al.* [24] demonstrated a reasonable correlation between the ability of a given BLM congener to inflict damage on chromosomal DNA and the observed extent of cell growth inhibition.

While there is little question that DNA constitutes an important therapeutic locus for expression of antitumor activity by bleomycin, it seems not unlikely that other cellular loci also contribute to the observed effects of the drug. The latter might include lipid membranes, which have been shown to undergo BLM-mediated

B. Meunier (ed.), DNA and RNA Cleavers and Chemotherapy of Cancer and Viral Diseases, 77–89.

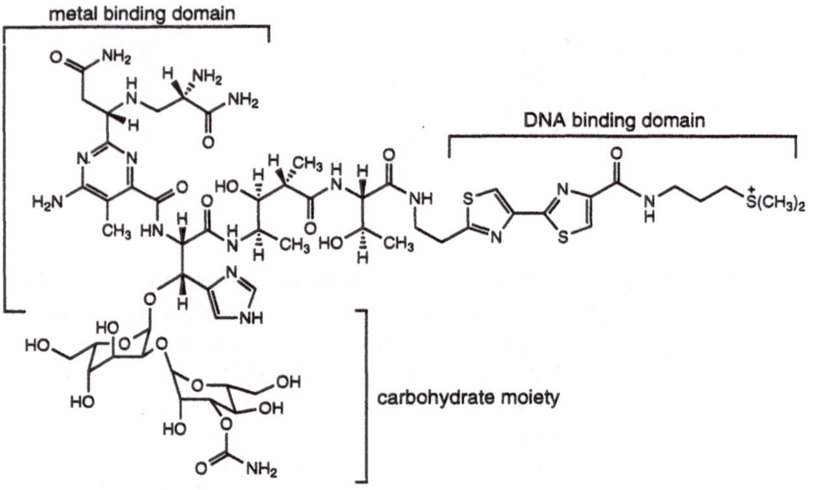

Figure 1. Structure of bleomycin A$_2$ with the functional domains indicated.

peroxidation [25-28], and RNA, whose degradation by BLM in cell free systems is now well documented [29-34].

RNA represents a particularly attractive therapeutic target for BLM since many cellular RNA's are present in the cell cytoplasm where they are more readily accessible to exogenous agents than DNA, the latter of which resides in the cell nucleus. Unlike DNA, which is repaired readily following BLM-mediated damage, there are few characterized mechanisms for RNA repair. Finally, the substantial variety of secondary and tertiary structures assumed by RNA's constitute a rich collection of structurally unique targets amenable to selective binding and degradation.

2. RNA Cleavage by Bleomycin

Transfer RNA's and tRNA precursor transcripts were the first RNA species shown to undergo cleavage when treated with Fe•BLM [29-34]. The cleavage of *B. subtilis* tRNA^{His} precursor was found to proceed with facility [30]. In common with BLM-mediated DNA cleavage, tRNA^{His} degradation was promoted by Fe(II)•BLM but not by Fe(III)•BLM, was potentiated in the presence of ascorbate and required O$_2$, consistent with an oxidative mechanism for RNA degradation. However, in contrast to DNA strand scission Fe(II)•BLM effected scission of the tRNA^{His} precursor at a single major site in this 118-nucleotide substrate.

As shown in Figure 2, this strong cleavage site was at uridine$_{35}$. While this site is part of a G•pyr sequence, which is typical for DNA cleavage, it is the only one of several such sequences that underwent cleavage in this substrate. It is also located at the junction between a single and double-stranded region of the tRNA precursor, assuming that the cloverleaf secondary structure actually obtains for this species. The cleavage patterns noted for several other tRNA's and tRNA precursors seemed equally as unique. Many were not a substrate for cleavage by Fe•BLM under any condition tested; others were cleaved at limited numbers of sites [30,34]. While no firm pattern emerged among the sites of cleavage, a disproportionate number of the sites were at the putative

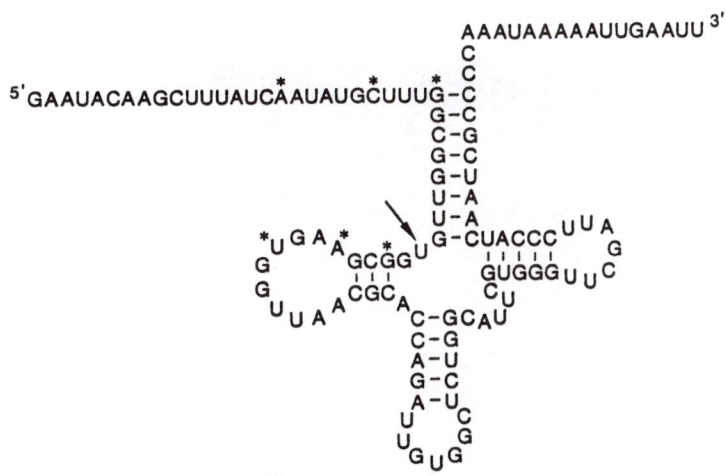

Figure 2. Structure of *B. subtilis* tRNA^{His} precursor showing the major (arrow) and minor (asterisks) sites of cleavage by Fe(II)•BLM A$_2$.

junctions between single and double-stranded regions of the RNA's or involved the pyrimidine nucleotides in G•pyr sequences.

Another unusual feature of RNA cleavage by Fe(II)•BLM was its susceptibility to inhibition by Mg^{2+}. As illustrated in Figure 3 for tRNA^{His} precursor, admixture of Mg^{2+} at concentrations in excess of 125 μM significantly diminished BLM-mediated cleavage. This property was found to vary significantly from one RNA substrate to another, and even at individual cleavage sites within a given RNA substrate [34]. While some RNA substrates were cleaved readily by Fe•BLM in the presence of physiological concentrations of Mg^{2+}, others were not arguing that RNA cleavage by BLM in intact biological systems might be expected to be highly selective. In contrast, the ability of B-form DNA to act as a substrate for Fe•BLM was unaffected by Mg^{2+}, even at rather high concentrations.

Of particular interest in this context was the ability of Fe(II)•BLM A$_2$ to mediate the cleavage of yeast 5S ribosomal RNA, a species whose secondary structure can be specified based on phylogenetic evidence, and conformational mapping studies. As shown in Figure 4, Fe•BLM cleaved this 122-nucleotide substrate at three sites. All three involved the uridine nucleotides in GUA sequences; all were located in double-stranded regions which have one-nucleotide bulges 1 or 2 nucleotides to the 3'-side of the site of cleavage. Interestingly, cleavage at two of these three sites was little affected by 1 mM Mg^{2+}, and was still readily apparent even in the presence of 5 mM Mg^{2+} [34].

In addition to tRNA's and yeast 5S rRNA, Fe(II)•BLM has also been shown to effect the cleavage of two mRNA's. A 270-nucleotide RNA that encoded the N-terminal region of HIV-1 reverse transcriptase was cleaved at four major sites [30]. Dix *et al.* [35] have utilized Fe(II)•BLM to characterize structural changes within the iron regulatory element (IRE) of bullfrog ferritin mRNA; the three sites of cleavage in two stem loop structural variants were all located at the junction between single and double-stranded regions in a stem-loop structure within the IRE.

3. Cleavage of an DNA-RNA Heteroduplex

In addition to the three types of RNA's discussed above, RNA's are also present in cells as components of DNA-RNA heteroduplexes. Such species are formed during both

80

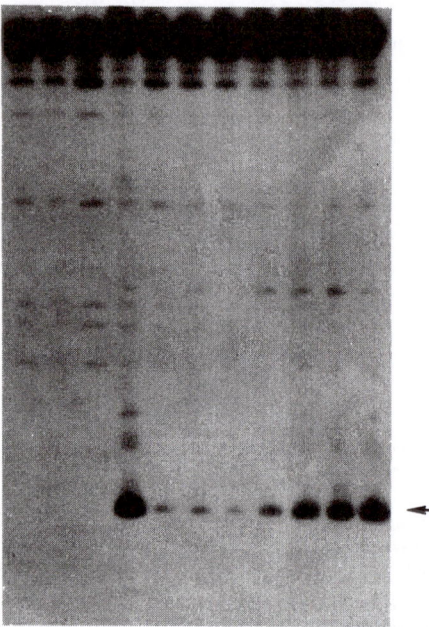

Figure 3. Polyacrylamide gel illustrating the effect of Mg^{2+} on Fe(II)•BLM-mediated cleavage of tRNA^{His} precursor. All lanes contained ~ 17 μM (final nucleotide concentration) 5'-^{32}P-tRNA^{His} precursor in 5 mM Na phosphate buffer, pH 7.5. Lane 1, RNA + 0.5 mM Mg^{2+}; lane 2, 250 μM Fe^{2+} + 0.5 mM Mg^{2+}; lane 3, 250 μM BLM A_2 + 0.5 mM Mg^{2+}; lane 4, 250 μM Fe(II)•BLM A_2. Lanes 5-11 contained 250 μM Fe(II)•BLM A_2 + 0.5, 0.4, 0.3, 0.25, 0.125, 0.1 and 0.05 mM Mg^{2+}, respectively. Mg^{2+} was added directly to the buffered RNA prior to the addition of Fe(II)•BLM A_2. The arrow denotes the major cleavage site.

Figure 4. Sites in yeast 5S ribosomal RNA cleaved by Fe(II)•BLM A_2.

forward and reverse transcription and constitute critical biochemical intermediates. In order to test the susceptibility of such species to cleavage by Fe•BLM, a DNA-RNA heteroduplex was prepared from *E. coli* 5S rRNA by reverse transcription using a suitable primer. The heteroduplex, labeled uniquely either on the DNA or RNA strand of the formed heteroduplex was incubated with varying concentrations of Fe(II)•BLM A_2 [36]. As summarized in Figure 5, cleavage was limited to four sites near the 5'-end of the RNA strand of the heteroduplex and three sites close to the 5'-end of the DNA strand. All three sites on the DNA strand involved G•pyr sequences; in fact a DNA duplex having one strand identical with the first 36 nucleotides of the heteroduplex was cleaved

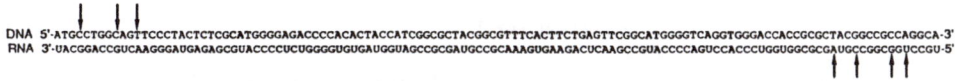

Figure 5. Sites cleaved by Fe(II)•BLM A$_2$ in a DNA-RNA heteroduplex derived from *E. coli* 5S ribosomal RNA, and in a 36-base pair DNA duplex structurally related to one end of the heteroduplex.

at 5 sites, three of which were identical to the sites of cleavage of the heteroduplex. In contrast, none of the four sites of cleavage on the RNA strand involved a G•pyr sequence.

The BLM-mediated cleavage of both strands of the heteroduplex exhibited some sensitivity to added Mg^{2+}, although both did undergo cleavage in the presence of physiological concentrations of Mg^{2+}. Interestingly, cleavage of the DNA strand was somewhat more readily inhibited by Mg^{2+}.

4. Cleavage of a tRNA Transcript and its Corresponding tDNA

While the initial observations concerning Fe(II)•BLM-mediated RNA cleavage seemed to reflect pronounced differences between the degradation of RNA and DNA by bleomycin, the more extensive survey summarized above indicated that many RNA's actually were cleaved at G•pyr sequences, in common with DNA. The observation that a number of RNA cleavage sites occurred at the junction between single and double-stranded regions does not parallel observations made for DNA, but could simply reflect the fact that most DNA's employed as substrates for bleomycin have been B-form DNA. In fact Williams and Goldberg [37] have reported that DNA oligonucleotides containing bulges were cleaved by bleomycin in proximity to those bulges even though the sequences involved would not ordinarily have been conducive to cleavage by Fe•BLM.

In order to obtain a DNA that differed from an authentic RNA substrate for BLM only in the constituent mononucleotides, we prepared a DNA molecule having the same nucleotide sequence as the *B. subtilis* tRNA^{His} precursor shown to be cleaved by Fe(II)•BLM predominantly at U$_{35}$ (Figure 1). Although there was really no direct evidence for the conformation of either of these species, pairs of tRNA's and tDNA's have been studied previously in assays capable of recognizing conformation and have shown remarkable similarities [38,39]. The tDNA related to tRNA^{His} precursor was 5'-^{32}P end labeled at the same specific activity as the tRNA^{His} precursor and incubated with Fe(II)•BLM A$_2$ at several different concentrations of the drug. Remarkably, at low concentrations of Fe(II)•BLM A$_2$, both substrates were cleaved predominantly at position-35 (Figure 6). The most straightforward interpretation of this observation is that the tRNA and tDNA actually do have similar structures, and it is the actual tertiary structure that results in efficient cleavage at the same position.

At higher concentrations of Fe(II)•BLM, the tDNA was cleaved at several additional sites; the full-length substrate was completely consumed [40]. In contrast tRNA^{His} cleavage never proceeded to completion, even in the presence of a 100-fold greater concentration of Fe(II)•BLM. In the belief that this might be due to differences in RNA and DNA binding by Fe(II)•BLM, each of the radiolabeled substrates was treated

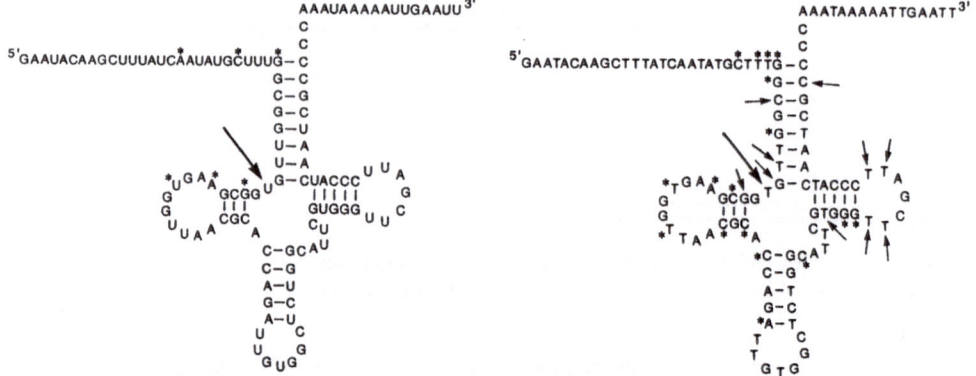

Figure 6. Sites of cleavage of *B. subtilis* tRNA^{His} precursor and the corresponding tDNA by Fe(II)•BLM A$_2$. The major sites of cleavage at U$_{35}$ (T$_{35}$) are denoted by large arrows, other significant sites of tDNA cleavage by small arrows, and minor sites of cleavage by asterisks. Both structures are shown arbitrarily folded into cloverleaf secondary structures.

with Fe(II)•BLM in the presence of much larger amounts of unlabeled tRNA or tDNA
        As shown in Figure 7, there were clear differences in the ability of the unlabeled species to inhibit cleavage of both radiolabeled substrates. Unexpectedly, the unlabeled tRNA proved to be more effective at inhibiting substrate cleavage, indicating that Fe•BLM binds to RNA more effectively than to DNA [40]. Thus the lesser cleavage of the tRNA^{His} precursor by Fe•BLM cannot be due to less efficient binding.
        Given the foregoing observations, it is logical to conclude that the lesser cleavage of RNA must result from RNA binding in some fashion that is not conducive to substrate degradation, or which produces damage not leading to strand scission.

5.   Chemistry of Bleomycin-Mediated RNA Cleavage

The chemical transformations mediated by Fe(II)•BLM + O$_2$ when DNA is the substrate have been studied using DNA oligonucleotides designed to be cleaved at a small number of sites [4,5]. Typical of these efforts was the use of the self-complementary octanucleotide d(CGCTAGCG), which undergoes modification at cytidine$_3$ and cytidine$_7$, both of which are contained within GC sequences. As illustrated in Figure 8, treatment of this substrate with Fe(II)•BLM under aerobic conditions afforded two sets of products, both of which are believed to derive from an initially formed C-4' deoxyribose radical [4,5,41]. One set of products, which results in direct strand scission, involves scission of the C-3' - C-4' bond in the deoxyribose moiety and affords *trans*-cytosine propenal and a dinucleotide terminating in a phosphoroglycolate moiety at the 3'-end [4,5,42-44]. The other affords an alkali-labile product with concomitant release of free cytosine; strand scission results following additional treatment with reagents such as alkali, alkylamines or hydrazine [45-47]. Because strand scission obtains near the end of the oligonucleotide, it is possible to anticipate and prepare the cleavage products as relatively small molecules. These authentic standards can be used to facilitate the identification of products actually resulting from Fe•BLM treatment of the oligonucleotide substrate; they can also be used as reference standards to permit quantification of the amounts of products formed [42,43].
        The finding that both sets of products in Figure 8 derive from a C-4' deoxyribose radical implies that Fe•BLM initiates the destruction of DNA by abstracting

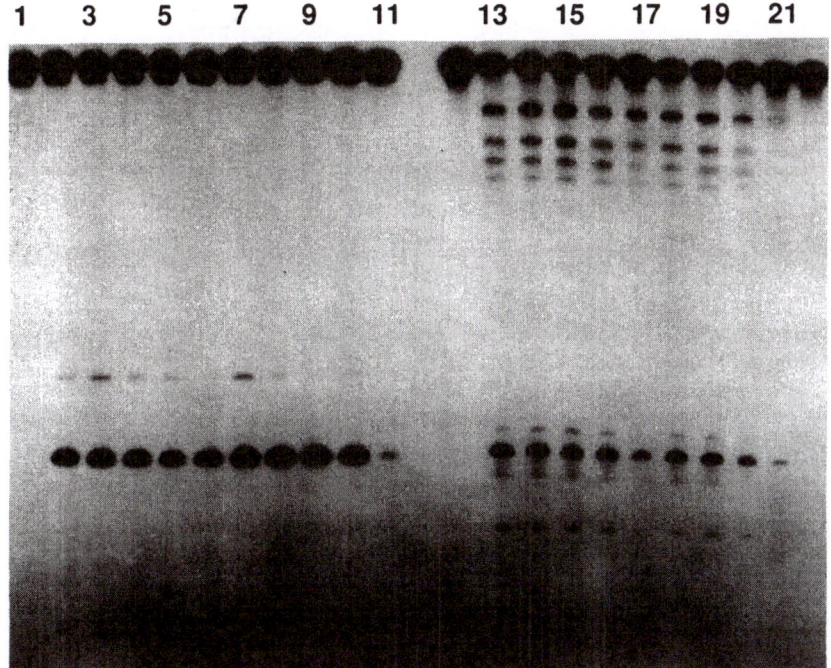

Figure 7. The effects of unlabeled tRNA and tDNA on Fe(II)•BLM-mediated cleavage of 5'-end labeled tRNA[His] and tDNA[His] precursors. Reactions were run using radiolabeled tRNA and tDNA substrates at 7-8 µM nucleotide concentrations. Lane 1, radiolabeled tRNA[His] precursor alone; lanes 2-6, radiolabeled tRNA[His] + 25 µM Fe(II)•BLM A_2 + 0, 4, 8, 16 and 80 µM unlabeled tDNA[His], respectively; lanes 7-11, radiolabeled tRNA[His] + 25 µM Fe(II)•BLM A_2 + 0, 4, 8, 16 and 80 µM unlabeled tRNA[His], respectively. Lane 12, radiolabeled tDNA[His] alone; lanes 13-17, radiolabeled tDNA[His] + 1.25 µM Fe(II)•BLM A_2 + 0, 4, 8, 16 and 80 µM unlabeled tDNA[His], respectively; lanes 18-22, radiolabeled tDNA[His] + 1.25 µM Fe(II)•BLM A_2 + 0, 4, 8, 16 and 80 µM unlabeled tRNA[His], respectively.

a H atom from C-4' of deoxyribose. This H, of course, resides in the minor groove of DNA.

Until the recent finding that the DNA-RNA heteroduplex shown in Figure 5 was cleaved near the 5'-end of the RNA strand [36], there had been no example of an RNA substrate whose chemistry could be explored using a strategy analogous to the one reflected in Figure 8. Therefore, most of what is known about the chemistry of RNA degradation is actually based on the use of a chimeric octanucleotide containing *ribo*-cytidine at position-3 of an oligomer otherwise comprised of deoxyribonucleotides. As shown in Figure 9, this chimeric oligonucleotide, when treated with Fe(II)•BLM, afforded CpGpCH₂COOH suggesting strongly that degradation involved abstraction of C-4' H from *ribo*-cytidine₃ [30,48]. No (hydroxylated) base propenal was observed. However, this species might well be expected to be unstable; in fact, a substantial amount of free cytosine was obtained. That the same transformation actually obtains for an authentic RNA substrate was suggested by the mobility of 3'- and 5'-^{32}P end labeled fragments resulting from treatment of yeast 5S rRNA with Fe(II)•BLM A_2. At present, it is not clear whether RNA degradation also involves the formation of an alkali-labile

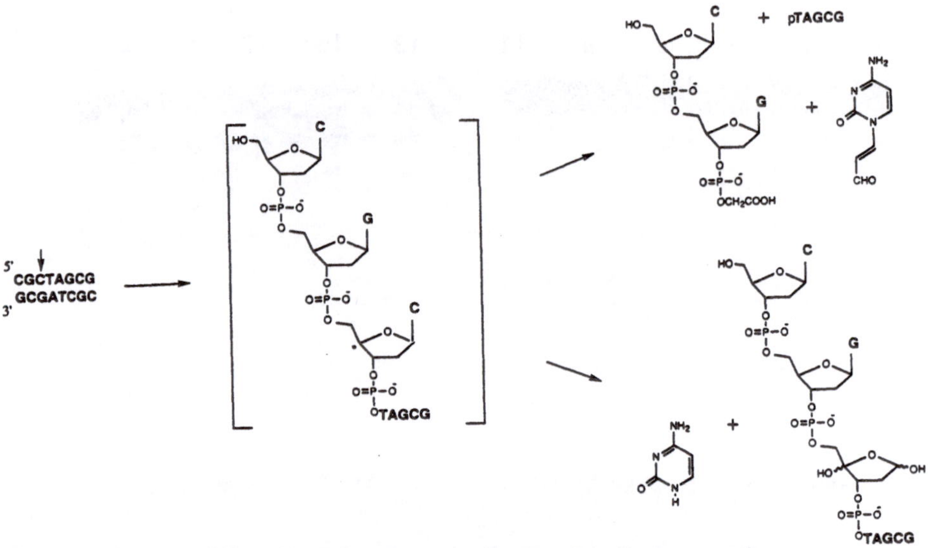

Figure 8. Products resulting from DNA oligonucleotide degradation by Fe(II)•BLM.

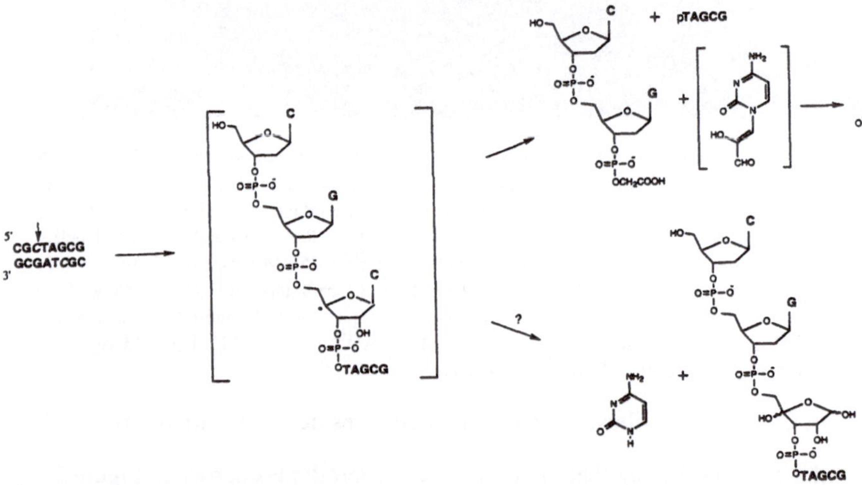

Figure 9. Products resulting from degradation of chimeric oligonucleotides by Fe(II)•BLM.

product. This is technically difficult to determine, since every phosphate ester bond in RNA is intrinsically alkali labile. Thus the ostensibly incomplete cleavage of RNA substrates described above in the context of *B. subtilis* tRNA[His] precursor may be due to the formation of an alkali-labile product.

As discussed previously [48,49], the chemistry of RNA cleavage by Fe(II)•BLM may also reflect the fact that RNA generally exists as an A-form duplex. The minor groove of this type of duplex is wide and shallow. The C-4' H of deoxyribose is not situated prominently in the minor groove of RNA, which could explain the lesser reactivity of RNA even though it is bound effectively by Fe•BLM. Interestingly, C-1' H is present within the minor groove. Although this H has not been found to be abstracted from DNA oligonucleotides by Fe(II)•BLM [50], by the use of the chimeric nucleotide shown in Figure 9 we have obtained evidence that it may well be involved in RNA degradation. As described previously [48], as a result treatment of the octanucleotide successively with Fe(II)•BLM and then diaminobenzene, a dinucleotide quinoxaline derivative was formed. Products resulting from H abstraction at C-1' constituted only 10% of all products in this case. However, a chimeric oligonucleotide containing *ara*-cytidine at position-3 afforded the same products but in a different proportion; 58% of all products resulted from abstraction of H-1'. Since the latter oligonucleotide is also known [51,52] to have a wide minor groove at position-3, it seems possible that some sites in RNA may afford products a larger proportion of which are derived from abstraction of C-1' H. The putative intermediate resulting from C-1' H abstraction does not afford strand scission directly [48], consistent with observations made for RNA cleavage (*vide supra*).

6.   RNA Hydrolysis Mediated by Metal Free Bleomycin

To date, all of the characterized transformations of DNA and RNA by bleomycin have involved oxidative chemistry in which a metallobleomycin employs oxygen or an oxygen surrogate to effect the oxidation of the oligo- or polynucleotide substrate. Recently, we found that metal-free BLM is also capable of mediating the *hydrolysis* of an RNA, yeast tRNA[Phe] [53]. As shown in Figure 10 and Table 1, the major sites of

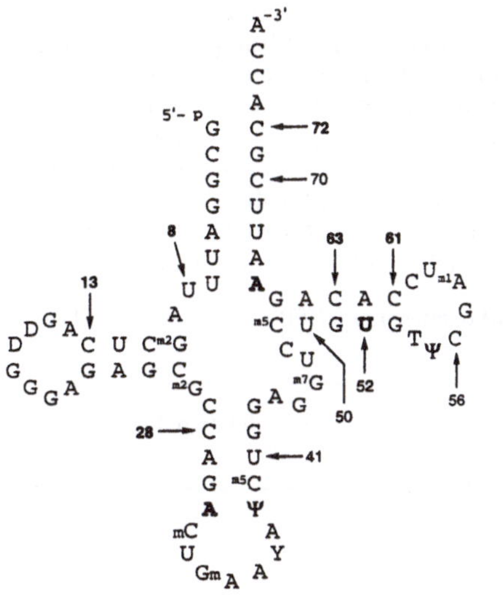

Figure 10.   Sites of hydrolysis of yeast tRNA[Phe] by metal-free BLM A_2. Major sites are in bold numbers; bold letters denote sites of cleavage by Fe(II)•BLM.

hydrolysis involved all (unmodified) pyr•A sites in the tRNA, while minor sites were limited to pyr•G sites. Strand scission occurred between the pyrimidine and purine nucleotide in each case; the products had OH groups at the 5'-termini and 2',3'-cyclic phosphates at the 3'-termini.

In spite of the fact that this type of transformation had not been noted during two decades of investigating the degradation of DNA and RNA by bleomycin, hydrolysis actually proceeds with a facility not much different than that of oxidative cleavage.

TABLE 1. Sites of hydrolysis of tRNA^{Phe} mediated by metal free bleomycin A$_2$

Major sites	Sequence context	Minor sites	Sequence context
U$_8$	5'-UA-3'	U$_{41}$	5'-GU-3'
C$_{13}$	5'-CA-3'	U$_{50}$	5'-UG-3'
C$_{28}$	5'-CA-3'	U$_{52}$	5'-UG-3'
C$_{61}$	5'-CA-3'	U$_{56}$	5'-CG-3'
C$_{63}$	5'-CA-3'	˙U$_{70}$	5'-CG-3'
C$_{72}$	5'-CA-3'		

7.   Does RNA Cleavage by Fe•Bleomycin Obtain *In Vivo*?

While the foregoing studies have established definitively that bleomycin can mediate RNA damage in cell free systems, they do not deal with the ability of BLM to function as an RNA damaging agent *in vivo*. Given the propensity of bleomycin to bind metals avidly, it seems unlikely that the hydrolytic cleavage of RNA noted for metal-free BLM can obtain *in vivo*. Definition of the ability of metallobleomycins to mediate the oxidative transformation of RNA cleavage *in vivo* has not been reported to date. However, it is instructive to consider an experiment carried out recently in the Hecht laboratory involving the synthesis of dihydrofolate reductase in a cell free eukaryotic protein biosynthesizing system. As shown in Table 2, treatment of rabbit reticulocyte lysate with Fe(II)•BLM at either of two concentrations had only a limited effect on the

TABLE 2. Effect of Fe(II)•BLM treatment of DHFR mRNA on the production of full length dihydrofolate reductase

Pretreatment	Dihydrofolate reductase synthesized (%)
None	100
Rabbit reticulocyte lysate, 50 μM Fe(II)•BLM	98
Rabbitt reticulocyte lysate, 250 μM Fe(II)•BLM	88
DHFR mRNA, 50 μM Fe(II)•BLM	55
DHFR mRNA, 250 μM Fe(II)•BLM	13
Complete translation system, 50 μM Fe(II)•BLM	100
Complete translation system	67

ability of the lysate to mediate the synthesis of dihydrofolate reductase (DHFR). Not surprisingly, pretreatment of DHFR mRNA with the same two concentrations of Fe(II)•BLM diminished subsequent DHFR synthesis substantially and in rough proportion to the amount of bleomycin employed. When Fe(II)•BLM was used to pretreat a mixture containing the entire translation system (i.e., containing mRNA and all the factors needed for protein synthesis), the damage mediated by Fe(II)•BLM was much less than that inflicted upon the mRNA alone [54]. Although the predictive value of this experiment for the potential of Fe•BLM as an RNA damaging agent *in vivo* is obviously limited, it does suggest that cellular factors will further increase the selectivity of Fe•BLM-mediated RNA damage beyond that observed in cell free systems.

8.   Acknowledgment

This work was supported at the University of Virginia by Research Grant CA53913 from the National Cancer Institute.

9.   References

1.   Umezawa, H. (1979) Advances in bleomycin studies, in S.M. Hecht (ed.), *Bleomycin: Chemical, Biochemical and Biological Aspects,* Springer-Verlag, New York, pp.24-36.

2.   Carter, S.K., Crooke, S.T., and Umezawa, H. eds. (1978) *Bleomycin: Current Status and New Developments,* Academic Press, New York.

3.   Sikic, B.I., Rozencweig, M., and Carter, S.K. eds. (1985) *Bleomycin Chemotherapy,* Academic Press, New York.

4.   Stubbe, J. and Kozarich, J.W. (1987) Mechanisms of bleomycin-induced DNA degradation, *Chem. Rev.* **87**, 1107-1136.

5.   Natrajan, A. and Hecht, S.M. (1993) Bleomycin: mechanism of polynucleotide recognition and oxidative degradation, in Neidle, S. and Waring, M. (eds.), *Molecular Aspects of Anticancer Drug-DNA Interactions,* MacMillan, London, pp. 197-242.

6.   Tanaka, N., Yamaguchi, H., and Umezawa, H. (1963) Mechanism of action of phleomycin. I. Selective inhibition of the DNA synthesis in *E. coli* and HeLa cells, *J. Antibiot.* **16A**, 86-91.

7.   Falaschi, A. and Kornberg, A. (1964) Phleomycin, an inhibitor of DNA polymerase, *Fed. Proc.* **23**, 940-945.

8.   Mueller, W.E. and Zahn, R.K. (1976) Effect of bleomycin on DNA, RNA, protein, chromatin and on cell transformation by oncogenic RNA viruses, *Prog. Biochem. Pharmacol.* **11**, 28-47.

9.   Ohno, T., Miyaki, M., Taguchi, T., and Ohashi, M. (1976) Actions of bleomycin on DNA ligase and polymerases, *Prog. Biochem. Pharmacol.* **11**, 48-58.

10.   D'Andrea, A.D. and Haseltine, W.A. (1978) Sequence specific cleavage of DNA by the antitumor antibiotics neocarzinostatin and bleomycin, *Proc. Natl. Acad. Sci. U.S.A.* **75**, 3608-3612.

11.   Takeshita, M., Grollman, A.P., Ohtsubo, E., and Ohtsubo, H. (1978) Interaction of bleomycin with DNA, *Proc. Natl. Acad. Sci. U.S.A.* **75**, 5983-5987.

12.   Mirabelli, C.K., Huang, C.-H., and Crooke, S.T. (1983) Role of deoxyribonucleic acid topology in altering the site/sequence specificity of cleavage of deoxyribonucleic acid by bleomycin and talisomycin, *Biochemistry* **22**, 300-306.

13.   Sausville, E.A., Stein, R.W., Peisach, J., and Horwitz, S.B. (1978) Properties and products of the degradation of DNA by bleomycin and iron(II), *Biochemistry* **17**, 2746-2754.

14.   Ehrenfeld, G.M., Shipley, J.B., Heimbrook, D.C., Sugiyama, H., Long, E.C., van Boom, J.H., van der Marel, G.A., Oppenheimer, N.J., and Hecht, S.M. (1987) Copper-dependent cleavage of DNA by bleomycin, *Biochemistry* **26**, 931-942.

15.   Ehrenfeld, G.M., Murugesan, N., and Hecht S.M. (1984) Activation of oxygen and mediation of DNA degradation by manganese-bleomycin, *Inorg. Chem.* **23**, 1496-1498.

16. Burger, R.M., Freedman, J.H., Horwitz, S.B., and Peisach, J. (1984) DNA degradation by manganese (II)-bleomycin plus peroxide, *Inorg. Chem.* **23**, 2215-2217.

17. Ishida, R. and Takahashi, T. (1975) Increased DNA chain breakage by combined action of bleomycin and superoxide radical, *Biochem. Biophys. Res. Commun.* **66**, 1432-1438.

18. Sausville, E.A., Peisach, J., and Horwitz, S.B. (1978) Effect of chelating agents and metal ions on the degradation of DNA by bleomycin, *Biochemistry* **17**, 2740-2746.

19. Kuramochi, H., Takahashi, K., Takita, T., and Umezawa, H. (1981) An active intermediate formed in the reaction of bleomycin-Fe(II) complex with oxygen, *J. Antibiot.* **34**, 576-582.

20. Natrajan, A., Hecht, S.M., van der Marel, G.A., and van Boom, J.H. (1990) A study of oxygen versus hydrogen peroxide-supported activation of iron•bleomycin, *J. Am. Chem. Soc.* **112**, 3997-4002.

21. Barranco, S.C. and Humphrey, R.M. (1971) Effects of bleomycin on survival and cell progression in Chinese hamster cells, *Cancer Res.* **31**, 1218-1223.

22. Hittelman, W.N. and Rao, P.N. (1974) Bleomycin-induced damage in prematurely condensed chromosomes and its relation to cell cycle progression in CHO [chinese hamster ovary] cells, *Cancer Res.* **34**, 3433-3439.

23. Barlogie, B., Drewinko, B., Schumann, J., and Freireich, E.J. (1976) Pulse cytophotometric analysis of cell cycle perturbation with bleomycin *in vitro, Cancer Res.* **36**, 1182-1187.

24. Berry D.E., Chang, L.-H., and Hecht, S.M. (1985) DNA damage and growth inhibition in cultured human cells by bleomycin congeners, *Biochemistry* **24**, 3207-3214.

25. Gutteridge, J.M.C. and Fu, X.-C. (1981) Enahancement of bleomycin-iron free radical damage to DNA by antioxidants and their inhibition of lipid peroxidation, *FEBS Lett.* **123**, 71-74.

26. Ekimoto, H., Takahashi, K., Matsuda, A., Takita, T., and Umezawa, H. (1985) Lipid peroxidation by bleomycin-iron complexes *in vitro, J. Antibiot.* **38**, 1077-1082.

27. Nagata, R., Morimoto, S., and Saito, I. (1990) Iron-peplomycin catalyzed oxygenation of linoleic acid, *Tetrahedron Lett.* **31**, 4485-4488.

28. Kikuchi, H. and Tetsuka, T. (1992) On the mechanism of lipoxygenase-like action of bleomycin-iron complexes, *J. Antibiot.* **45**, 548-555.

29. Magliozzo, R.S., Peisach, J., and Ciriolo, M.R. (1989) Transfer RNA is cleaved by activated bleomycin, *Mol. Pharmacol.* **35**, 428-432.

30. Carter, B.J., de Vroom, E., Long, E.C., van der Marel, G.A., van Boom, J.H., and Hecht, S.M. (1990) Site-specific cleavage of RNA by Fe(II)•bleomycin, *Proc. Natl. Acad. Sci. U.S.A.* **87**, 9373-9377.

31. Carter, B.J., Reddy, K.S., and Hecht, S.M. (1991) Polynucleotide recognition and strand scission by Fe•bleomycin, *Tetrahedron* **47**, 2463-2474.

32. Carter, B.J., Holmes, C.E., Van Atta, R.B., Dange V., and Hecht, S.M. (1991) Metal-catalyzed polynucleotide strand scission, *Nucleosides Nucleotides* **10**, 215-227.

33. Hüttenhofer, A., Hudson, S., Noller, H.F., and Mascharak, P.K. (1992) Cleavage of tRNA by Fe(II)-bleomycin, *J. Biol. Chem.* **267**, 24471-24475.

34. Holmes, C.E., Carter, B.J., and Hecht, S.M. (1993) Characterization of iron(II)•bleomycin-mediated RNA strand scission, *Biochemistry* **32**, 4293-4307.

35. Dix, D.J., Lin, P.-N., McKenzie, A.R., Walden, W.E., and Theil, E.C. (1993) The influence of the base-paired flanking region on structure and function of the ferritin mRNA iron regulatory element, *J. Mol. Biol.* **231**, 230-240.

36. Morgan, M.A. and Hecht, S.M. (1994) Iron (II) bleomycin-mediated degradation of a DNA-RNA heteroduplex, *Biochemistry* **33**, 10280-10293.

37. Williams, L.D. and Goldberg, I.H. (1988) Selective strand scission by intercalating drugs at DNA bulges, *Biochemistry* **27**, 3004-3011.

38. Perreault, J.P., Pon, R.T., Jiang, M., Usmav, N., Pika, J., Ogilivie, K K , and Cedergren, R. (1989) The synthesis and functional-evaluation of RNA and DNA polymers having the sequence of *Escherichia coli* tRNA[fMet], *Eur. J. Biochem.* **186**, 87-93.

39. Khan, A.S. and Roe, B.A. (1988) Aminoacylation of synthetic DNAs corresponding to *Escherichia coli* phenylalanine and lysine tRNAs, *Science* **241**, 74-79.

40. Holmes, C.E. and Hecht, S.M. (1993) Fe•bleomycin cleaves a transfer RNA precursor and its "transfer DNA" analog at the same major site, *J. Biol. Chem.* **268**, 25909-25913.

41. Wu, J.C., Kozarich, J.W., and Stubbe, J. (1985) Mechanism of bleomycin: evidence for a rate-determining 4'-hydrogen abstraction from poly(dA-dU) associated with the formation of both free base and base propenal, *Biochemistry* **24**, 7562-7568.

42. Sugiyama, H., Kilkuskie, R.E., Hecht, S.M., van der Marel, G.A., and van Boom, J.H. (1985) An efficient site-specific DNA target for bleomycin, *J. Am. Chem. Soc.* **107**, 7765-7767.

43. Sugiyama, H., Kilkuskie, R.E., Chang, L.-H., Ma, L.-T., Hecht, S.M, van der Marel, G.A., and van Boom, J.H. (1986) DNA strand scission by bleomycin: catalytic cleavage and strand selectivity, *J. Am. Chem. Soc.* **108**, 3852-3854.

44. Van Atta, R.B., Long, E.C., Hecht, S.M., van der Marel, G.A., and van Boom, J.H. (1989) Electrochemical activation of oxygenated iron-bleomycin, *J. Am. Chem. Soc.* **111**, 2722-2724.

45. Sugiyama, H., Xu, C., Murugesan, N., and Hecht, S.M. (1985) Structure of the alkali-labile product formed during Fe(II)•bleomycin-mediated DNA strand scission, *J. Am. Chem. Soc.* **107**, 4104-4105.

46. Rabow, L.E., Stubbe, J., Kozarich, J.W., and Gerlt, J.A. (1986) Identification of the alkali-labile product accompanying cytosine release during bleomycin-mediated degradation of d(CGCGCG), *J. Am. Chem. Soc.* **108**, 7130-7131.

47. Sugiyama, H., Xu, C., Murugesan, N., Hecht, S.M., van der Marel, G.A., and van Boom, J.H. (1988) Chemistry of the alkali-labile lesion formed from iron(II)-bleomycin and d(CGCTTTAAAGCG), *Biochemistry* **27**, 58-67.

48. Duff, R.J., de Vroom, E., Geluk, A., Hecht, S.M., van der Marel, G.A., and van Boom, J.H. (1993) Evidence for C-1' hydrogen abstraction from modified oligonucleotides by Fe•bleomycin, *J. Am. Chem. Soc.* **115**, 3350-3351.

49. Hecht, S.M. (1994) RNA degradation by bleomycin, a naturally occurring bioconjugate, *Bioconjugate Chem.* **5**, 513-526.

50. Absalon, J.J., Krishnamoorthy, C.R., McGall, G., Kozarich, J.W., and Stubbe, J. (1992) Bleomycin mediated degradation of DNA-RNA hybrids does not involve C-1' chemistry, *Nucleic Acids Res.* **20**, 4179-4185.

51. Pieters, J.M.L., de Vroom, E., van der Marel, G.A., van Boom, J.H., Koning, T.M.G., Kaptein, R., and Altona, C. (1990) Hairpin structures in DNA containing arabinofuranosylcytosine. A combination of nuclear magnetic resonance and molecular dynamics, *Biochemistry* **29**, 788-799.

52. Gao, Y.-G., van der Marel, G.A., van Boom, J.H., Wang, A.H.-J. (1991) Molecular structure of a DNA decamer containing an anticancer nucleoside arabinosylcytosine: conformational perturbation by arabinosylcytosine in B-DNA, *Bichemistry* **30**, 9922-9931.

53. Keck, M.V. and Hecht, S.M. (1995) Sequence-specific hydrolysis of yeast tRNA[Phe] mediated by metal-free bleomycin, *Biochemistry* **33**, xxxx.

54. Karginov, V. and Hecht, S.M., unpublished results.

BLEOMYCIN REACTION PATHWAYS: KINETIC APPROACHES

RICHARD M. BURGER* and KARL DRLICA
Public Health Research Institute
455 First Avenue
New York, NY 10016, USA
and
*Redox Pharmaceutical Corporation
100 Forest Drive
Greenvale, NY 11548, USA

KEYWORDS / ABSTRACT: bleomycin / oxygen / iron / DNA / pathway / kinetics / in vitro

The pathways of bleomycin-mediated oxygen activation and DNA degradation are being dissected by a series of experiments designed to determine the order in which known and inferred chemical changes occur. The rationale of these experiments and various illustrative examples are discussed.

1. Introduction

In the nearly forty years since bleomycin-type antibiotics were first described [1], bleomycin [2, 3] and its congenors [4] have served as leading anti-tumor agents [5] and have become one of the best-understood families of DNA-cleaving antibiotics. Bleomycin-treated cells contain broken DNA [6] and survivors are mutagenized [7, 8], supporting the hypothesis that bleomycin disruption of DNA is significant *in vivo* [9]. Precisely how bleomycin uses oxygen to degrade DNA is understood in considerable detail [10-12], mainly from model studies conducted *in vitro* [13, 14]. The model system is simple: bleomycin and certain combinations of iron (Fe(II) or Fe(III)) and oxygen (dioxygen, superoxide, or peroxide) react with DNA to jointly cause DNA cleavage and/or other degradation. Two sets of DNA products form [15]. In one, broken DNA is terminated by 5'-phosphate [16] and 3'-phosphoglycolate moieties [17] with the release of a derivative of nucleic base and deoxyribose carbons 1-3 [15] characterized as 3-(pyrimidin-1-yl)-2-propenals and 3-(purin-9-yl)-2-propenals, collectively called 'base propenal' [17] (Scheme 1). The other set of products [15] includes alkali-labile DNA [18] in which deoxyribose is converted to a 4'-keto,1'-aldehyde product with loss of the nucleic base [19, 20]. Deducing how reactants become products involves determining stoichiometries, structures, and timing of events. Here we focus on insights arising

91

B. Meunier (ed.), DNA and RNA Cleavers and Chemotherapy of Cancer and Viral Diseases, 91–106.
© *1996 Kluwer Academic Publishers. Printed in the Netherlands.*

from questions about order and timing. Such questions will be broadly defined as matters of 'kinetics.' The gaps that remain in our understanding of bleomycin action are potentially quite interesting, since they involve mechanistic questions common to other oxidative DNA-cleavers as well as the oxygenases and peroxidases that bleomycin appears to model.

2. Activated Bleomycin

The reaction of Fe(II)·bleomycin with O_2 that initiates DNA degradation was initially thought to be rapid, since EDTA, which was known to block the binding of Fe(II) to bleomycin, failed to inhibit the degradation of DNA regardless of how soon it was added after mixing aerobic Fe(II) and bleomycin [21]. However, the lack of partial inhibition made the EDTA experiments ambiguous, since partial inhibition at an early time point was needed to show that the reaction had been stopped. Thus, lack of inhibition could have reflected either a very rapid reaction having escaped inhibition or an ineffective inhibitor permitting the reaction to reach completion. While EDTA is effective in sequestering free Fe(II), its effectiveness with tightly bound Fe(II) was questionable. Since sequestering a reactant is an effective way to prevent or reverse a reaction at equilibrium, but ineffective for stopping reactions that are irreversible or slowly reversible, once such a reaction is started, one must trap, destroy, or in some way change the intermediate to stop it. In the case of bleomycin, stopping the reaction by acidification revealed that it progressed slowly, approaching completion with $t_{1/2} = 8$ sec at room temperature [22] and 2 min at 4 °C [23]. It was later shown that reversal of iron-bleomycin binding by EDTA was indeed very slow [24].

Knowing that the reaction of bleomycin with DNA was slow led to examination of the reactions of O_2 + Fe(II)·bleomycin $\rightarrow$ Fe(III)·bleomycin [25]. When O_2 was added to anaerobic Fe(II)·bleomycin and the reaction was monitored by stop-flow optical spectrometry, two spectral transitions were resolved. The first occurred rapidly, at a rate proportional to O_2 concentration. The second was slower and independent of O_2. A third transition, which was not detected optically, was revealed when a mixture of O_2 and Fe(II)·bleomycin was examined by stop-freeze EPR spectroscopy. Again two transitions were resolved, the earlier one being concurrent with the second optical transition. These reactions reflected the formation and dissolution of a new spectral species resembling, but differing from that already attributed to the end product, Fe(III)·bleomycin. During the third reaction the new species was replaced quantitatively by Fe(III)·bleomycin with $t_{1/2} \sim 2$ min at 4 °C [23, 26]. The coincidence of this third transition with the attack on DNA was striking. The new bleomycin species was the kinetically competent agent of DNA attack, a point verified by demonstrating a 1:1 stoichiometry of DNA lesions with the new bleomycin species. This new, unstable bleomycin intermediate was called activated bleomycin [23].

We now have a reasonable understanding of activated bleomycin as a ferric peroxide. Magnetically labeled Fe and O nuclei were used to demonstrate the presence of ligated Fe-O [23], the Fe(III) valence state was demonstrated by Mössbauer spectroscopy

[27], and the Fe-O oxidation state was inferred to be +5 [23, 27]. When the latter statement was verified by redox titrations, it became clear that activated bleomycin was capable of both one- and two-electron oxidations [28]. This may account for its two different modes of DNA degradation (discussed below). Mass spectroscopy revealed that two oxygen atoms were present [29], a kinetic isotope effect indicated a rate-limiting O-O bond cleavage occurred during activated bleomycin consumption [30] (discussed below), and [17]O-ENDOR spectroscopy indicated that only one oxygen was directly bonded to Fe(III). Thus, activated bleomycin is an end-on ferric peroxide complex [31].

Peroxidases and oxygenases such as cytochrome P-450 show similar ligation of Fe and O as well as similar reaction pathways [32]. However, different intermediates are stable, leading to the proposal [29, 33] that the iron ligation peculiar to bleomycin has poised its peroxide complex in a region of kinetic stability not accessible to dioxygenated intermediates of enzymes such as cytochrome P-450.

The shorter-lived precursor of activated bleomycin is now understood to be a one-electron oxidized analogue of activated bleomycin, but its reactions have not yet been fully characterized. Stop-freeze techniques demonstrated that it is EPR-silent [25], and Mössbauer spectra, obtained when the species was trapped by a 50:1 excess of DNA nucleotides per drug (discussed below), showed it to be an Fe(III)-superoxide complex [27]. As this complex is converted to activated bleomycin, one half is converted to Fe(III)·bleomycin, consistent with oxidation of half the bound superoxide to O_2 [26], which is unreactive with Fe(III)·bleomycin. Thus an ensemble of kinetic experiments led us to the species of bleomycin most closely associated with DNA attack and also uncovered some of the complexities associated with formation of this species.

3. Bleomycin Activation Pathways

The O_2 pathway to activated bleomycin described above is one of several. It was unambiguous because it involved single-turnover reactions from start to finish. The increasing stability of successive bleomycin intermediates made it easy to accumulate observable quantities of each. However, the pathway is not as simple as it first seemed: the conversion of $\cdot O_2^- \cdot$ Fe(III)·bleomycin to activated bleomycin is a one-electron reduction of the former by a disproportionation reaction, in which two equivalents of $\cdot O_2^- \cdot$ Fe(III)·bleomycin generate one equivalent each of activated bleomycin and Fe(III)·bleomycin.

When reducing agents are available, Fe(III) may replace Fe(II) to elicit DNA cleavage by bleomycin [14]. Here the reaction is cyclic, as Fe(II)·bleomycin is regenerated from Fe(III)·bleomycin. In another cyclic reaction, addition of peroxides to Fe(III)·bleomycin produces a steady-state level of a substance having an EPR spectrum like that of activated bleomycin. The next question was whether the peroxide effect represented a new pathway to activated bleomycin or just a reductive recycling of Fe(III)·bleomycin to Fe(II)·bleomycin that could react again with O_2. The first approach was to use an inhibitor to block the Fe(II)-O_2 reaction pathway and thus test the idea that Fe(III) ·bleomycin was simply reduced to Fe(II) ·bleomycin by peroxide.

The O_2 analogue ethyl isocyanide, which binds to Fe(II)·bleomycin and competes with O_2 [22], showed no effect on DNA degradation when Fe(III)·bleomycin was reacted with peroxide [23]. Thus, a second, peroxide pathway to activated bleomycin seemed likely. When the reaction was carried out with peroxide in the absence of O_2, the activated bleomycin EPR spectrum appeared.

Showing that the peroxide-generated EPR spectrum was indeed that of activated bleomycin was based on a procedure that converted the cyclic process to a single cycle. This was accomplished by adding an inhibitor of the Fe(III)·bleomycin + peroxide reaction. In this case concentrated DNA was a suitable inhibitor, since DNA strongly inhibits this reaction at high ratios of DNA to bleomycin as well as serving as a bleomycin target [23]. After preincubation of peroxide and Fe(III)·bleomycin had converted most of the Fe(III)·bleomycin to activated bleomycin, DNA was added to block the pathway forming activated bleomycin from the remaining peroxide and the accumulating Fe(III)·bleomycin product. Reaction aliquots were removed and assayed for DNA degradation and for conversion of the EPR spectral species. These two processes displayed kinetics identical to those with activated bleomycin formed from Fe(II)·bleomycin + O_2. Thus the observed reaction kinetics were consistent with the EPR spectral identity. This apparent identity of activated bleomycin formed with either peroxide or O_2 has been supported by other comparisons, including that of their Mössbauer spectra [27].

Yet another oxygen species, superoxide, can participate in formation of activated bleomycin, but its exact role is ambiguous. Superoxide may be envisioned to react either with Fe(II)·bleomycin to form $\cdot O_2^- \cdot$ Fe(III)·bleomycin, with Fe(III)·bleomycin to form activated bleomycin directly, or simply as a reductant either of Fe(III)·bleomycin (to yield Fe(II)·bleomycin) or of $\cdot O_2^- \cdot$ Fe(III)·bleomycin (to form activated bleomycin;

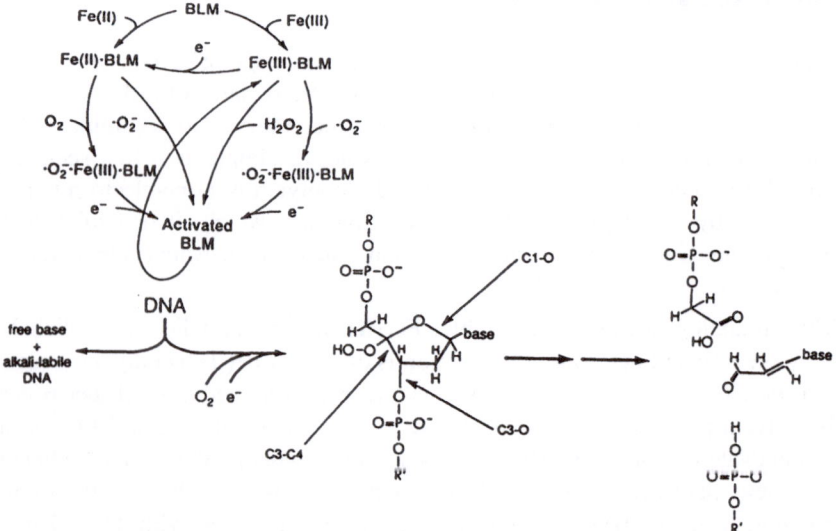

Scheme 1. Pathways of activated bleomycin formation and DNA degradation, abridged.

see Scheme 1). Evidence is available for the last reaction: an excess of DNA inhibits the disproportionation of $\cdot O_2^-$ · Fe(III)·bleomycin to yield activated bleomycin [34], but exogenous superoxide circumvents this and permits activated bleomycin to form [35]. It is not clear whether superoxide displaces O_2 from the $\cdot O_2^-$ · Fe(III)·bleomycin resonance hybrid O_2 · Fe(II)·bleomycin to form activated bleomycin, or simply functions as a reductant. The former displacement reaction is plausible, should O_2 prove dissociable, for superoxide supplied as KO_2 [36], but not its dismutation product mixture, H_2O_2 + O_2, reacts with Fe(II)·bleomycin to initiate DNA cleavage. (Peroxide cannot be the cause of this cleavage because it initiates a suicide-type reaction (see below) with Fe(II)·bleomycin, even in the presence of DNA and oxygen; unpublished result.) The other putative reactions of superoxide with Fe(III)·bleomycin have not been examined experimentally.

4. The Central Reactions

The conversion of activated bleomycin to Fe(III)·bleomycin and the concurrent attack on DNA occur rapidly. For the present purpose we call this ensemble of events the 'central reactions'. Included in this category are the decay of activated bleomycin and the early reactions of DNA, including those responsible for partitioning of the initial lesion into alternative resolution pathways. These reactions are central in two senses: temporally they bridge the more accessible pathways of activated bleomycin formation and the resolution of DNA damage, and intellectually they constitute the kernel of what remains to be understood about bleomycin mechanism.

Since the central reactions seemed difficult to dissect kinetically, the system was simplified by removing DNA and O_2 so activated bleomycin could be studied alone. Omitting DNA was straightforward, but eliminating oxygen required understanding activation. The stop-flow experiments [25] used to define the kinetics of activated bleomycin formation from Fe(II)·bleomycin + O_2 made it clear that their single-cycle reaction was quick and avid enough to consume all the oxygen in a suitably constituted, closed mixture before a significant proportion of activated bleomycin itself reacted. Under these anaerobic conditions activated bleomycin decayed to the air-stable Fe(III)· bleomycin at the same rate as under air [37]. Thus, oxygen seemed to have no role in the decay reaction(s). At that point it became clear that only some of the DNA degradation products required O_2 for formation [23]. Thus there had to be two DNA degradation pathways. The mechanism of product partition is currently the least understood of the bleomycin-DNA reactions.

In addition to its decay to the recyclable Fe(III)·bleomycin, activated bleomycin also decays in a way that produces a complex of Fe(III) with a covalently modified bleomycin (with an EPR spectrum very similar to that of Fe(III)·bleomycin [23, 38]) that has lost the ability attack DNA in subsequent cycles of oxygen activation [22]. This reaction has been termed activated bleomycin suicide. While DNA can prevent this reaction [22], its protective capacity is lost concurrently with activated bleomycin decay [23]. Thus activated bleomycin is kinetically competent in its suicide. Several suicide products have been identified [29], suggesting that the chemistry is complex, but relatively little is known of the responsible reactions other than the lack of an O_2 requirement [37]. Since the products of anaerobic suicide have not been examined, the possible involvement of oxygen in some suicide reactions remains to be evaluated. Understanding the chemistry of the suicide lesion(s) could clarify or verify our understanding of the species responsible for the central reactions.

It is important to stress that activated bleomycin need not be the species that directly attacks DNA. Indeed, the kinetics of activated bleomycin consumption are slow and simple: in phosphate buffer the presence or absence of DNA has no effect on the first-order decay rate [23]. If the rate-limiting reactions were bimolecular, they would be second-order instead of first-order. The observed rate behavior suggests that the reaction of activated bleomycin is a monomolecular conversion to a short-lived proximate attacking species rather than a direct attack on DNA. An intermediate that is consumed more readily than it is formed could be vanishingly rare at steady state and thus escape detection.

A tool for revealing hidden intermediates is the measurement of kinetic isotope effects. We measured the ^{18}O isotope effect on the stability of activated bleomycin in the absence of DNA to address the possibility that cleavage of bound peroxide was the rate-limiting reaction in activated bleomycin decay. Such a reaction might form a monooxygenated 2-electron oxidant bleomycin species analogous to peroxidase compound I. The decay of residual DNA-cleaving activity was faster with $^{16}O_2$ than with $^{18}O_2$ by a factor of 1.054 ±0.008, a magnitude exceeding that predicted for any Fe-O-O or Fe-O reaction other than peroxide O-O cleavage [30]. If, as indicated, the rate-limiting reaction of activated bleomycin decay is a peroxide cleavage, then activated bleomycin must be a peroxide. DNA is not required for, and has no effect on, the rate of this reaction; it seems likely that it is the monooxygenated product of activated bleomycin that reacts directly with DNA.

For a monooxygenated species to yield the final product, Fe(III)·bleomycin, it may oxidize either DNA or some other reductant. However, for spontaneous decay to occur, it must react with an oxygenated bleomycin species. In Fe(II)·bleomycin activation and spontaneous decay, the stoichiometry of Fe(II) consumed to H_2O produced is 2:1 [39]; thus, not all the activated bleomycin oxygen is reduced to water. We recently learned something about the participants in the spontaneous decay reaction by examining a putative product, regenerated O_2. Expecting initially to find that two equivalents of a monooxygenated product could react:

$$2 \text{ bleomycin} \cdot \text{Fe} = O^V \rightarrow 2 \text{ bleomycin} \cdot \text{Fe}^{III} + O_2,$$

we prepared activated bleomycin from a mixture of $^{16}O_2$ and $^{18}O_2$ and submitted the reaction products to mass spectroscopy to detect $^{16}O - ^{18}O$. Finding only the original mixture and no recombinant O_2, we concluded that once separated, activated bleomycin oxygen atoms never recombine (unpublished result, R. M. Burger, B. Luz, and M. Bender). Thus the reaction resembles that of catalase:

$$\text{bleomycin} \cdot \text{Fe} = O^V + \text{bleomycin} \cdot \text{Fe} \cdot {}^-O - OH^V \rightarrow 2 \text{ bleomycin} \cdot \text{Fe}^{III} + O_2 + OH^-,$$

with activated bleomycin acting as the reductant. The apparent first-order kinetics of activated bleomycin decay with DNA absent, which are identical to those of DNA attack, are consistent with activated bleomycin peroxide cleavage being the rate-limiting process over the observed portion of a more complex reaction course.

What happens when DNA is available is more complicated. The early steps are rapid and not easily accessible. The rate-limiting step in DNA attack was shown, by means of a H4′ kinetic isotope effect on both sets of DNA products [40], to involve the H4′ nucleus, which is removed from deoxyribose [41]. Since the stoichiometric efficiency of bleomycin-initiated DNA degradation is high [23], this isotope effect, together with the high stoichiometric efficiency of activated bleomycin, implies that the attacking species is long-lived relative to its mobility on DNA. The ability of activated bleomycin reaction mixtures to perform either one- or two-electron oxidations [28] suggests that the partition of the DNA degradation pathway depends on whether the initial bleomycin-mediated oxidation removes one or two electrons from the deoxyribose C4′ [42-45].

In the base-releasing, alkali-labilizing pathway of DNA degradation, the two-electron oxidation product, a C4′-cation, is thought to react with water to form a 4′-hydroxylated deoxyribose intermediate [42]. This hemiketal collapses to form a deoxyribose 4′-keto,1′-aldehyde derivative with loss of the nucleic base glycosidic bond [41]. These reactions, which are expected to require no additional oxygen, account for water oxygen incorporation in the 4′ carbonyl [42] and for the oxidation state of the DNA product. It is unclear whether the oxidation occurs in one step or by a one-electron oxidation of a 4′ deoxyribose radical.

In one-electron oxidation, hydrogen atom abstraction would give a C4′-centered radical that is thought to form an adduct with endogenous O_2 [10, 46-48]. This is followed by a series of reactions (considered below) yielding DNA cleavage. The radical is not long-lived, for O_2 must be made available promptly for DNA cleavage to occur [37]. Since the one atom incorporated of this O_2 appears in the 3′-terminal phosphoglycolate [49], a Criegee-type rearrangement was proposed in which an oxygen would be inserted between C3′ and C4′ as a distinctive feature of the consequent DNA cleavage pathway (Scheme 2 [50]). That a C4′ radical actually forms is suggested by products of a radical-trapping analogue of the target oligonucleotide d(CCGG) [44]. Still, it is likely that the product-partitioning step occurs prior to the reaction of the DNA lesion with O_2. The fate of the C4′ radical in the absence of O_2 is unclear, since the alternative, base-releasing pathway discussed above is not enhanced [37] and since no other product has been identified.

Soon after forming the initial lesion, bleomycin and DNA may react a second time. The reaction, proposed [45, 51] to account for a high incidence [18] of closely-spaced double-strand cleavages, is as follows. When O_2 adds to the C4′ radical, the resulting peroxyl radical must be reduced by one electron to form the DNA peroxide precursor of cleavage. The proximate active bleomycin species, having just made a one-electron oxidation and thus achieved the oxidation level of peroxidase compound II, might furnish the necessary electron to the DNA peroxyl radical and thereby revert to its two-electron oxidant status [45]. The oxygen just added to DNA would thus function to re-oxidize the bleomycin species to its state just prior to DNA-oxidation, making this an oxygen-driven chain reaction: O_2 would execute one-electron oxidations of both the DNA and bleomycin products immediately after their formation. If the reoxidized bleomycin species is the unstable proximate active species, it is likely to react again before it can travel far [45], and thus it would account for the closely-spaced double-strand lesions that constitute 10-20% of cleavage events [18]. The plausibility of the model is reinforced by the finding that double lesions may include one alkali-labile DNA lesion, but they never include a pair of them: aerobic (one-electron) DNA degradation is required for this phenomenon [45].

Another sort of double lesion has been seen that may have the same origin. Examination of the size distribution of same-strand cleavage events revealed an anomalous >2-fold enhancement of same-strand double cleavages spaced 3 nucleotides apart, even at low levels of total cleavage [52]. A similar distribution was also found after treatment with alkali to reveal lesions made by the base-releasing pathway. However, when DNA is degraded by an anaerobic preparation of activated bleomycin, the size distribution after alkali-treatment lacks the anomaly [52]. Thus, single-strand lesion pairs, like opposite-strand pairs, must include at least one frank, oxygen-dependent cleavage. However limited the mobility of the proximate attacking species, it may still suffice for the formation of both opposing and same-strand lesions.

The putative reoxidation of a bleomycin·Fe = O^{IV} product by ·O – O-C4′-DNA as a chain reaction is a welcome proposal because it argues for the existence of a bleomycin intermediate that is an obvious origin for the electron required to convert a peroxyl radical to the putative 4′-deoxyribose peroxide shown in Scheme 1. How then is the chain terminated? One possibility is the decay of active bleomycin while dissociated from DNA. However, decay while dissociated from DNA would be expected to result in substantial bleomycin suicide, which is not seen. The 2-electron oxidation of DNA, which leads to 4′-hydroxylation and its consequences, seems a more likely reaction of chain termination. One prediction of this hypothesis is that the yield of hydroxylation should be independent of the base propenal yield. This has been observed [37]. Another prediction is that each attacking bleomycin should generate one hydroxylated site, provided that adequate DNA is available. This prediction differs from the current stoichiometric estimates. Comparing the observed stoichiometries of base propenal formation to hydroxylation (free base release) [15] and of base propenal formation to activated bleomycin [23] gave an estimate of 0.5 hydroxylations per activated bleomycin instead of the expected 1.0. To better test this chain-termination hypothesis, the ratio of deoxyribose hydroxylation to activated bleomycin should be reexamined, particularly under a range of conditions known to alter the yield of base propenal. Another prediction of the chain hypothesis is that conditions known to enhance the yield of base propenal should also enhance the proportion of closely spaced lesions. A large enhancement of base propenal yield was seen in incubation mixtures containing ^2H_2O [22]. Mixing experiments with solutions of DNA and of activated bleomycin prepared in different solvents indicate that the ^2H_2O effect is exerted at the time of bleomycin attack and not during drug activation or DNA cleavage resolution. We are currently testing the prediction of increased closely-spaced lesions in this system.

The possibility that activated bleomycin reacts to form a proximate DNA-attacking species is consistent with the discovery of closely-spaced DNA lesions: their close spacing would reflect the short lifetime expected of such a species [45]. The availability of a high concentration of the proximate attacking species at a well defined time would be very valuable in facilitating the kinetic examination of subsequent reactions otherwise too rapid to resolve. However, difficulties are expected in observing such a short-lived species even if it could be formed directly by means of single-oxygen donors such as iodosylbenzene or persulfate salts [53, 54]. The alternative approach is to infer the formation of such a fleeting species by the formation of bleomycin-type DNA products in the reaction mixture. To do this, the short-lived proximate attacking species would probably need be generated *in situ* on DNA, assuming that steric access of the oxygen donor to bleomycin was preserved. This could be difficult. Nevertheless, DNA cleavage has been seen with bleomycin, Fe(III), and $KHSO_5$ as an oxygen donor [53, 54]. It remains to be verified that this attack on DNA is due to the same mechanism as seen with activated bleomycin by showing that the DNA products and bleomycin suicide products are identical to those obtained with activated bleomycin. Iodosylbenzene presents a cautionary example: its effects, which appeared to include induction of bleomycin suicide, were found to be independent of metal complexation and were eventually attributed not to oxygen but to hypervalent iodine [55].

5. O_2-Independent DNA Degradation

Kinetic methodologies have contributed relatively little to illucidating the events of the base-releasing pathway. However, the finding that DNA still retained the capacity to release the majority of its free base yield even after it had released most of its base propenal led to an appreciation of the independence of the two pathways [15]. This was reinforced by the discovery that only cleavage was oxygen-dependent [23]. The proposal of 4'-hydroxylation and consequent ketone formation coupled to base release was based on the finding that H3' release is associated only with base release [41]. This H3' release is expected to be even slower than base release, but it has not been measured. The base-releasing intermediate [56], as well as the final, alkali-labile product [57], have been structurally characterized. The events of lesion development subsequent to partition seem well documented.

6. O_2-Mediated DNA Degradation

The events of DNA cleavage, like those of bleomycin activation, have been elucidated by kinetic means. As mentioned above, the separate requirement of O_2 for activation and cleavage was distinguished by means of their well-resolved kinetic loci [23]. This separation was later exploited to show that only the second oxygen appears in a DNA product [49]. That the oxygen (one atom) was found in the phosphoglycolate moiety supported the idea of oxygen addition at a C4' lesion. The only bleomycin pathway leading to DNA cleavage is that yielding base propenal, since there is no cleavage in excess of base propenal formation [52] and since formation of base propenal, which includes the deoxyribose C3′ atom, must break DNA. Since the events of DNA cleavage are spread over a considerable span of time and occur as a series of single-turnover reactions, the events could be observed sequentially. In this case the intermediates were difficult to trap, so real-time methods of monitoring, such as spectroscopy, were used to order the cleavage reactions [58, 59].

As with bleomycin activation, inability to halt DNA cleavage in mid-course made it seem that cleavage might be a rapid reaction. However, visual and instrumental observation of changes in reaction mixture viscosity made it clear that cleavage took place over a scale of seconds at room temperature and minutes at 4 °C [58]. Greater ability to detect single-strand cleavage was achieved by a fluorimetric assay of plasmid relaxation, and for cleavage of synthetic octanucleotides [59] we used the hyperchromic shift that accompanies cleavage-initiated denaturation. All these measurements showed that cleavage occurs with $t_{1/2}$ ~ 4 min. Knowing just when cleavage occurs made it possible to test a mechanistic proposal in which cleavage was concerted with the loss of a H2′, a loss that had been well characterized structurally and stoichiometrically [60] but not kinetically. By using a rapid alcohol-precipitation method to remove DNA from the mixture in mid-reaction, it was shown that the H2′ atom appeared in water with $t_{1/2}$ ~ 1.8 min [58], which led to a revision of the proposed degradation pathway [50].

Earlier proposals for the DNA cleavage mechanism suggested that all products formed together [10, 46-48], but kinetic measurements indicated that DNA cleavage was not the last event in the cleavage pathway [58]. To form all the end products of DNA cleavage, three carbon bonds [15, 17] must break (Scheme 1), but fewer may suffice for cleavage (Scheme 2). The possibility that degradation was stepwise rather than a concerted process was first indicated by the very slow rate ($t_{1/2}$ ~ ½h) at which base propenal ceased to co-precipitate with DNA in alcohol [58]. However, this measurement was troublesome because the kinetics of propenal release were complex and dependent on the individual nucleic base [15]. To challenge the assumption that base propenal formation was signaled by its release from DNA, proton NMR was adapted as a real-time assay of octanucleotide degradation, taking the advent of the propenal aldehyde proton signal as an indication of propenal formation [59]. This made it clear that propenal formation ($t_{1/2}$ ~ 7 min) followed DNA cleavage, although more closely than previously inferred from measurements with large DNA and alcohol precipitation. We resolved the difference between DNA and the octanucleotide as targets by showing that the departure of base propenals from DNA could be hastened by adding propidium iodide as a competing intercalating agent [59]. Thus cleavage precedes base propenal formation, which itself is not coincident with its release when large DNA is being cleaved.

To determine whether the event forming base propenal was coupled to the C3′-phosphate-O cleavage, real-time ^{31}P-NMR was employed. The advent of a phosphomonoester signal concurrent with propenal formation and following DNA cleavage established that this phosphate bond is the last cleaved. Thus strand scission must arise from cleavage of two sugar carbon bonds.

The present bleomycin-mediated DNA cleavage pathway is the latest in a series of models based first on the identification of reaction products and reactants and later

The present bleomycin-mediated DNA cleavage pathway is the latest in a series of models based first on the identification of reaction products and reactants and later modified to accommodate the kinetic resolution of intermediates. The insertion of an O_2 atom between deoxyribose C3' and C4' was not proposed until essential details of other mechanisms proved impossible. Kinetic experiments give access to reaction intermediates, which may be reassuring, surprising, or both with respect to 'paper chemistry'.

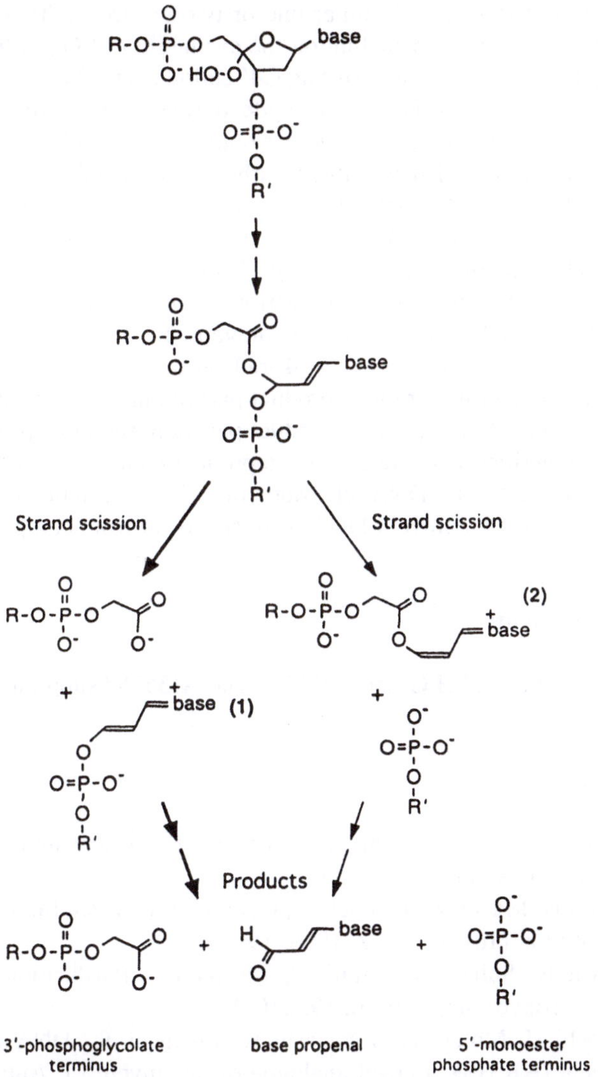

Scheme 2. Proposed intermediates between DNA oxygenation and base propenal formation [50]. These are consistent with the observed kinetics of H2' release, DNA strand scission, and base propenal formation. Only the alternative shown with bold arrows (left) is also consistent with observed kinetics of phosphomonoester formation.

7. Concluding Remarks

Our understanding of bleomycin-mediated DNA degradation pathways has moved from the beginnings and ends toward the center. There are at least three ways to begin and form the relatively stable, kinetically competent intermediate called activated bleomycin. Attack on DNA then occurs with the removal of the deoxynucleotide-4'-proton, together with either one or two electrons. Two sets of degradation products then develop, one depending on the availability of O_2. When the DNA lesion is oxidized by two electrons, the resulting C4'-cation is attacked by water and the lesion develops an alkali-labile 4'-keto,1'-aldehyde derivative of deoxyribose with loss of nucleic base. If the attack on DNA involves one electron and O_2 is present, DNA cleavage ensues, with the loss of deoxyribose carbons and nucleic base in the form of base propenal. This leaves DNA fragments terminated with 5'-phosphate and 3'-phosphoglycolate. While the reactions leading to activated bleomycin and those developing products subsequent to pathway partition have been verified and characterized in some detail, gaps remain among the central reactions. We suspect that the proximate attacking product of activated bleomycin is a bleomycin $\cdot$ Fe = O^V complex of uncertain fate. How it reacts with DNA to produce a C4'-radical or a (less-documented) C4'-cation is also uncertain, as are a role for O_2 in product partitioning and the fate of the C4'-radical in the absence of O_2. Subsequent to product partition, the most prominent mechanistic feature yet to be verified is the deoxyribose product enlarged by a C3'-O-C4' insertion. This putative precursor to DNA cleavage may be detectable by solid-state NMR. Clearly, unraveling the bleomycin-DNA pathways is far from complete.

8. Acknowledgements

ACS grant CH-443 and NIH Grants AI33337 and AI35257 supported the authors.

9. References

1. Maeda K, Kosaka H, Yagishita K, Umezawa H: A new antibiotic, phleomycin. J. Antibiot. (Ser. A) 1956; 9: 82-85.
2. Umezawa H, Maeda K, Takeuchi T, Okami Y: New antibiotics, bleomycin A and B. J. Antibiot. (Tokyo) Ser. A 1966; 19: 200-209.
3. Umezawa H, Suhara Y, Takita T, Maeda K: Purification of bleomycins. J. Antibiot. (Tokyo) Ser. A 1966; 19: 210-215.
4. Kuramochi H, Motegi A, Takahashi K, Takeuchi T: DNA cleavage activity of liblomycin (NK313), a novel analogue of bleomycin. J. Antibiot. (Tokyo) 1988; 41: 1846-1853.
5. Sikic BI, Rosencweig M, Carter SK: *Bleomycin Chemotherapy*. Academic Press, Orlando, 1985.

6. Murray V, Martin RF: The sequence specificity of bleomycin-induced DNA damage in intact cells. J. Biol. Chem. 1985; 260: 10389-10391.

7. Moore CW: Bleomycin-induced mutation and recombination in *Saccharomyces cerevisiae*. Mutation Res. 1978; 58: 41-49.

8. Povirk LF, Goldberg IH: A role for oxidative DNA sugar damage in mutagenesis by neocarzinostatin and bleomycin. Biochimie 1987; 69: 815-823.

9. Suzuki H, Nagai K, Yamaki H, Tanaka N, Umezawa H: Mechanism of action of bleomycin. Studies with the growing culture of bacterial and tumor cells. J. Antibiot. (Tokyo) 1968; 21: 379-386.

10. Stubbe J, Kozarich JW: Mechanisms of bleomycin-induced DNA degradation. Chem. Rev. 1987; 87: 1107-1136.

11. Dedon PC, Goldberg IH: Free-radical mechanisms involved in the formation of sequence-dependent bistranded DNA lesions by the antitumor antibiotics bleomycin, neocarzinostatin, and calcheamicin. Chem. Res. Toxicol. 1992; 5: 311-322.

12. Hecht SM: RNA degradation by bleomycin, a naturally occurring bioconjugate. Bioconj. Chem. 1994; 5: 513-526.

13. Nagai N, Yamaki H, Suzuki H, Tanaka N, Umezawa H: The combined effects of bleomycin and sulfhydryl compounds on the thermal denaturation of DNA. Biochim. Biophys. Acta 1969; 179: 165-171.

14. Sausville EA, Peisach J, Horwitz SB: Effects of chelating agents and metal ions in the degradation of DNA by bleomycin. Biochem. Biophys. Res. Commun. 1976; 73: 814-822.

15. Burger RM, Berkowitz AR, Peisach J, Horwitz SB: Origin of malondialdehyde from DNA degraded by Fe(II)-bleomycin. J. Biol. Chem. 1980; 255: 11832-11838.

16. Sugiyama H, Kilkuskie RE, Hecht SM, van der Marel GA, van Boom JH: An efficient, site-specific target for bleomycin. J. Am. Chem. Soc. 1985; 107: 7765-7767.

17. Giloni L, Takeshita M, Johnson F, Iden C, Grollman AP: Bleomycin-induced strand scission of DNA. Mechanism of deoxyribose cleavage. J. Biol. Chem. 1981; 256: 8608-8615.

18. Povirk LF, Wubker W, Kohnlein W, Hutchinson F: DNA double-strand breaks and alkali-labile bonds produced by bleomycin. Nucleic Acids Res. 1977; 4: 3573-3580.

19. Sugiyama H, Xu C, Murugesan N, Hecht SM: Structure of the alkali-labile product formed during iron(II)-bleomycin-mediated DNA strand scission. J. Am. Chem. Soc. 1985; 107: 4104-4105.

20. Rabow LE, Stubbe J, Kozarich JW, Gerlt JA: Identification of the alkaline-labile product accompanying cytosine release during bleomycin-mediated degradation of d(CGCGCG). J. Am. Chem. Soc. 1986; 108: 7130-7131.

21. Sausville EA, Peisach J, Horwitz SB: Properties and products of the degradation of DNA by bleomycin. Biochemistry 1978; 17: 2746-2754.

22. Burger RM, Peisach J, Blumberg WE, Horwitz SB: Iron-bleomycin interactions with oxygen and oxygen analogues: Effects on spectra and drug activity. J. Biol. Chem. 1979; 254: 10906-10912.

23. Burger RM, Peisach J, Horwitz SB: Activated bleomycin, a transient complex of drug, iron and oxygen that degrades DNA. J. Biol. Chem. 1981; 256: 11636-11644.

24. Aronovitch J, Godinger D, Goldstein S, Czapski G: Substitution reactions of iron-bleomycin-DNA with EDTA, DETAPAC or desferrioxamine. J. Antibiot. (Tokyo) 1987; 40: 1344-1348.

25. Burger RM, Horwitz SB, Peisach J, Wittenberg JB: Oxygenated iron bleomycin. A short-lived intermediate in the reaction of ferrous bleomycin with O_2. J. Biol. Chem. 1979; 254: 12299-12302.

26. Kuramochi H, Takahashi K, Takita T, Umezawa H: An active intermediate formed in the reaction of bleomycin-Fe(II) complex with oxygen. J. Antibiot. (Tokyo) 1981; 34: 578-582.

27. Burger RM, Kent TA, Horwitz SB, Munck E, Peisach J: Mossbauer study of iron bleomycin and its activation intermediates. J. Biol. Chem. 1983; 258: 1559-1563.

28. Burger RM, Blanchard JS, Horwitz SB, Peisach J: The redox state of activated bleomycin. J. Biol. Chem. 1985; 260: 15406-15409.

29. Sam JW, Tang X, Peisach J: Electrospray mass spectrometry of iron bleomycin: Demonstration that activated bleomycin is a ferric peroxide complex. J. Am. Chem. Soc. 1994; 116: 5250-5256.

30. Burger RM, Tian G, Drlica K: Oxygen isotope effect on activated bleomycin stability. J. Am. Chem. Soc. 1995; 117: 1167-1168.

31. Veselov A, Sun H, Sienkiewicz A, Taylor H, Burger RM, Scholes CP: Iron coordination of activated bleomycin probed by Q- and X-band ENDOR: Hyperfine coupling to activated ^{17}O-Oxygen, ^{14}N, and exchangeable ^{1}H. J. Am. Chem. Soc. 1995; 117: 7508-7512.

32. Dawson JW: Probing structure-function relationships in heme-containing oxygenases and peroxidases. Science 1988; 240: 433-439.

33. Loeb KE, Zaleski JM, Westre TE, Guajardo RJ, Mascharak PK, Hedman B, Hodgson KO, Solomon EI: Spectroscopic definition of the geometric and electronic structure of the non-heme iron active site in iron(II) bleomycin: Correlation with oxygen reactivity. J. Am. Chem. Soc. 1995; 117: 4545-4561.

34. Albertini JP, Garnier-Suillerot A, Tosi L: Evidence of a long-lived bleomycin-iron-oxygen intermediate. Biochem. Biophys. Res. Commun. 1982; 104: 557-563.

35. Ciriolo MR, Magliozzo RS, Peisach J: Microsome-stimulated activation of ferrous bleomycin in the presence of DNA. J. Biol. Chem. 1987; 262: 6290-6295.

36. Valentine JS, Curtis AB: A convenient preparation of solutions of superoxide anion and the reaction of superoxide anion with a copper(II) complex. J. Am. Chem. Soc. 1975; 97: 224-226.

37. Burger RM, Peisach J, Horwitz SB: Effects of O_2 on the reactions of activated bleomycin. J. Biol. Chem. 1982; 257: 3372-3375.

38. Nakamura M, Peisach J: Self-inactivation of Fe(II)-bleomycin. J. Antibiot. (Tokyo) 1988; 41: 638-647.

39. Barr JR, Van Atta RB, Natrajan A, Hecht SM, van der Marel GA, van Boom JH: Iron(II)-bleomycin-mediated reduction of O_2 to water: An ^{17}O nuclear magnetic resonance study. J. Am. Chem. Soc. 1990; 112: 4058-4060.

40. Wu JC, Kozarich JW, Stubbe JA: Mechanism of bleomycin: Evidence for a rate-determining 4'-hydrogen abstraction from poly (dA-dU) associated with the formation of both free base and base propenal. Biochemistry 1985; 24: 2562-2568.

41. Wu JC, Kozarich JW, Stubbe JA: The mechanism of free base formation from DNA by bleomycin. J. Biol. Chem. 1983; 258: 4694-4697.

42. Rabow LE, McGall GH, Stubbe J, Kozarich JW: Identification of the source of oxygen in the alkaline-labile product accompanying cytosine release during bleomycin-mediated oxidative degradation of d(CGCGCG). J. Am. Chem. Soc. 1990; 112: 3203-3208.

43. Sugiyama H, Sera T, Dannoue Y, Marumoto R, Saito I: Bleomycin-mediated degradation of aristeromycin-containing DNA. Novel dehydrogenation activity of iron^{II}-bleomycin. J. Am. Chem. Soc. 1991; 113: 2290-2295.

44. Sugiyama H, Ohmori K, Saito I: Oligonucleotides containing a radical trap: The trapping of deoxyribose C4' radical in bleomycin-mediated reactions. J. Am. Chem. Soc. 1994; 116: 10326-10327.

45. Absalon MJ, Wu W, Kozarich JW, Stubbe J: Sequence-specific double-strand cleavage of DNA by Fe-bleomycin. 2. Mechanism and dynamics. Biochemistry 1995; 34: 2076-2086.

46. Takeshita M, Grollman AP: A molecular basis for the interaction of bleomycin with DNA; In SM Hecht (ed): Bleomycin: Chemical, Biochemical, and Biological Aspects. New York, Springer-Verlag, 1979, p. 207-221.

47. Grollman AP, Takeshita M, Pillai KMR, Johnson F: Origin and cytotoxic properties of base propenals derived from DNA. Cancer Res. 1985; 45: 1127-1131.

48. Saito I, Morii T, Matsuura T: Chemistry of 4'-hydroperoxy nucleosides as a model for the intermediate in bleomycin-induced degradation of DNA. J. Org. Chem. 1987; 52: 1008-1012.

49. McGall GH, Rabow LE, Stubbe J, Kozarich JW: Incorporation of 18-O into glycolic acid obtained from the bleomycin-mediated degradation of DNA: Evidence for 4'-radical trapping by 18-O2. J. Am. Chem. Soc. 1987; 109: 2836-2837.

50. McGall GH, Rabow LE, Ashley RG, Wu SH, Kozarich JW, Stubbe J: New insight into the mechanism of base propenal formation during bleomycin-mediated DNA degradation. J. Am. Chem. Soc. 1992; 114: 4958-4967.

51. Steighner RJ, Povirk LF: Bleomycin-induced DNA lesions at mutational hot spots: Implications for mechanism of double-strand cleavage. Proc. Natl. Acad. Sci. USA 1990; 87: 8350-8354.

52. Burger RM, Peisach J, Horwitz SB: Stoichiometry of DNA strand scission and aldehyde formation by bleomycin. J. Biol. Chem. 1982; 257: 8612-8614.
53. Pratviel G, Bernadou J, Meunier B: DNA breaks generated by the bleomycin-iron III complex in the presence of $KHSO_5$, a single oxygen donor. Biochem. Biophys. Res. Commun. 1986; 136: 1013-1020.
54. Pratviel G, Bernadou J, Meunier B: Evidence for high-valent iron-oxo species active in the DNA breaks mediated by iron-bleomycin. Biochem. Pharmacol. 1989; 38: 133-140.
55. Sam JW, Tang X, Magliozzo RS, Peisach J: Electrospray mass spectrometry of iron bleomycin II: Investigation of the reaction of Fe(III)-bleomycin with iodosylbenzine. J. Am. Chem. Soc. 1995; 117: 1012-1018.
56. Sugiyama H, Xu C, Murugesan N, Hecht SM, van der Marel GA, van Boom JH: Chemistry of the alkali-labile lesion formed from iron(II) bleomycin and d(CGCTTTAAAGCG). Biochemistry 1988; 27: 58-67.
57. Rabow LE, Stubbe J, Kozarich JW: Identification and quantitation of the lesion accompanying base release in bleomycin-mediated DNA degradation. J. Am. Chem. Soc. 1990; 112: 3196-3203.
58. Burger RM, Projan SJ, Horwitz SB, Peisach J: The DNA cleavage mechanism of iron bleomycin. Kinetic resolution of strand scission from base propenal release. J. Biol. Chem. 1986; 261: 15955-15959.
59. Burger RM, Drlica K, Birdsall B: The DNA cleavage pathway of iron bleomycin. Strand scission precedes deoxyribose-3-phosphate bond cleavage. J. Biol. Chem. 1994; 269: 25978-25985.
60. Ajmera S, Wu JC, Worth, Jr. L, Rabow LE, Stubbe J, Kozarich JW: DNA degradation by bleomycin: Evidence for 2'R-proton abstraction and for C-O bond cleavage accompanying base propenal formation. Biochemistry 1986; 25: 6586-6592.

SELECTED GENERATION AND CLEAVAGE
OF C4'-MODIFIED OLIGONUCLEOTIDES

B. GIESE
Department of Chemistry, University of Basel
St. Johanns Ring 19, CH-4056 Basel, Switzerland

DNA strand scission induced by Fe^{2+} complexes of bleomycin [1] or methidiumpropyl-EDTA [2] occurs mainly *via* 4'-DNA radical **1**.

The subsequent cleavage reaction of the C3'- and C5'-phosphate bonds generates phosphates **2** and **3** as fragments of lower molecular weight. In addition, phosphoglycolate **4** and base propenal **5** are formed which contain the oxidized fragments of the 4'-nucleotide radical.

B. Meunier (ed.), *DNA and RNA Cleavers and Chemotherapy of Cancer and Viral Diseases*, 107–115.
© 1996 *Kluwer Academic Publishers. Printed in the Netherlands.*

We have now developed methods to synthesize 4'-nucleotide radicals selectively by independent methods. As radical precursors we used alkene **6**, phenylselenide **7**, and ketone **8**.

The first experiments were carried out with dinucleotide **9**. Addition of thiol radicals cleaved the dinucleotide **9** in a fast and quantitative reaction [3].

Intermediate in this reaction is C4'-radical **10** which could give enolether **11** either by homolytic (**10**→**11**) or heterolytic (**10**→**12**) C,O–bond cleavage.

The influence of the solvent polarity and the acidity of the leaving group on the rate of the cleavage [4], as well as photocurrent [5], and trapping experiments [4] demonstrate that radical cations of structure 12 or 18 are formed as intermediates of the reaction.

Thus generation of radical 17 from selenide 13 in methanol leads to products 14–16 [4]. These products are formed by trapping of radical cations 18 and 20 by the nucleophilic solvent methanol.

Also ketone 21 leads to trapping products 14-16 in which the phosphates at C3'- and C5'-positions are cleaved off.

In order to carry out this kind of experiments with oligonucleotides the modified mononucleotides 26 and 31 were synthesized by oxidation of deoxyriboses 22 and 27, respectively. Selenation of aldehyde 23 led to isomers 24a,b. Separation of 24b and introduction of suitable protective groups yielded phenylselenide 26 [6].

In a similar way, ketone **31** was synthesized from aldehyde **28** *via* ketone **29** and subsequent hydroxymethylation (**29**→**30**) [7].

Synthesis of oligomers like **32** which contain a modified mononucleotide as radical precursor can be carried out with a solid phase DNA synthesizer [8]. Irradiation with light of λ>320 nm generated radical **33** which underwent cleavage and gave phosphate **34** and **35** in the absence of O$_2$.

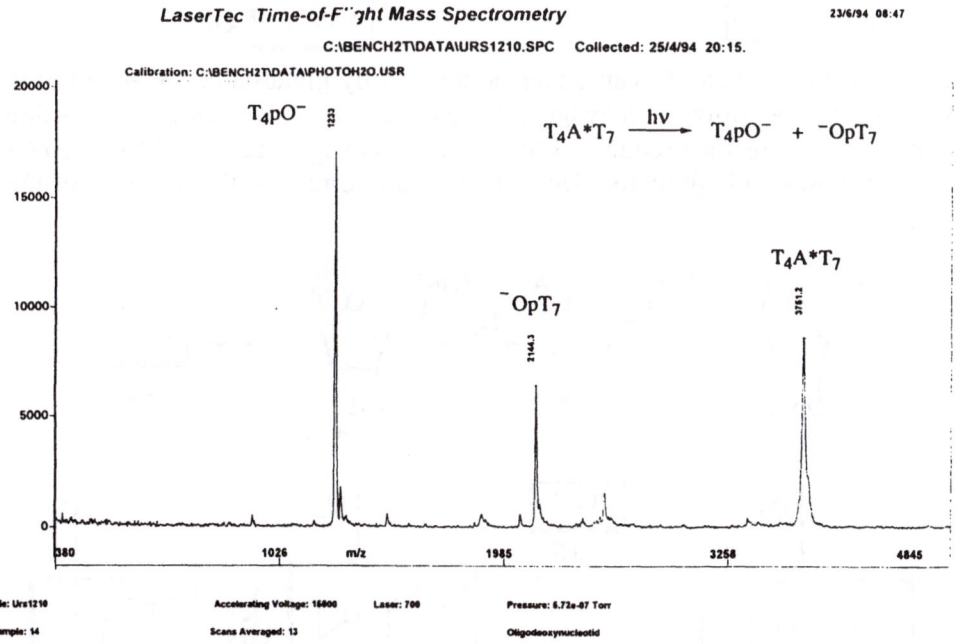

MALDI-MS (Fig. 1) and HPLC were used as analytical tools [8].

Figure 1. Photolysis of modified oligonucleotide $T_4A^*T_7$ (32) analyzed by MALDI-MS. The 3'-phosphate T_4pO^- (34) and the 5'-phosphate $^-OpT_7$ (35) are the main cleavage products.

The strand scission of the 4'-DNA radical occurs *via* heterolytic cleavage of the secondary C3',O bond (33 → 35 + 36), subsequent addition of H_2O (36→37), and cleavage of the primary C5',O bond (37 → 34 + 38). The mechanism could be proved by experiments in the presence of glutathione that trapped radical 33 and 37 and generated 39 and 40 [8].

Thus, 4'-DNA radical 33 can either be trapped by glutathione (33→39) or it undergoes a heterolytic cleavage reaction (33→36). A similar competition situation explains the products of the aerobic cleavage reactions. MALDI-TOF spectra show that hydroperoxides 41 and 42 are formed in the presence of O_2 [9].

Whereas **41** is the hydroperoxide of the intact DNA, hydroperoxide **42** is formed after cleavage of the radical (**33**→**36**). Isotope studies show that in **41** two oxygen atoms arise from O_2 whereas in **42** one more oxygen is introduced by H_2O.

In model studies we could show that glycolate **43** is not only formed from **41** but also from the hydroperoxide **42**. Because of the free OH-group at C3' compound **42** undergoes a Grob fragmentation (**42**→**44**)with subsequent elimination (**44** → **43** + **45**)[8].

Further experiments showed that oligonucleotides with two modified mononucleotides can also be cleaved easily.

114

Thus, aerobic photolysis of oligonucleotide **46**, containing two radical precursors A*, gave three sets of cleavage products. Compounds **47** and **48** each contain 3'-phosphates, 3'-phosphoglycolates, and hydroperoxides whereas **49** consists of a 5'-phosphate as the only compound.

Similar cleavage experiments can be carried out with double stranded oligonucleotides in which one strand contains a modified mononucleotide. The "melting points" of these modified double strands are reduced by about 2°C if 12-mers are used.

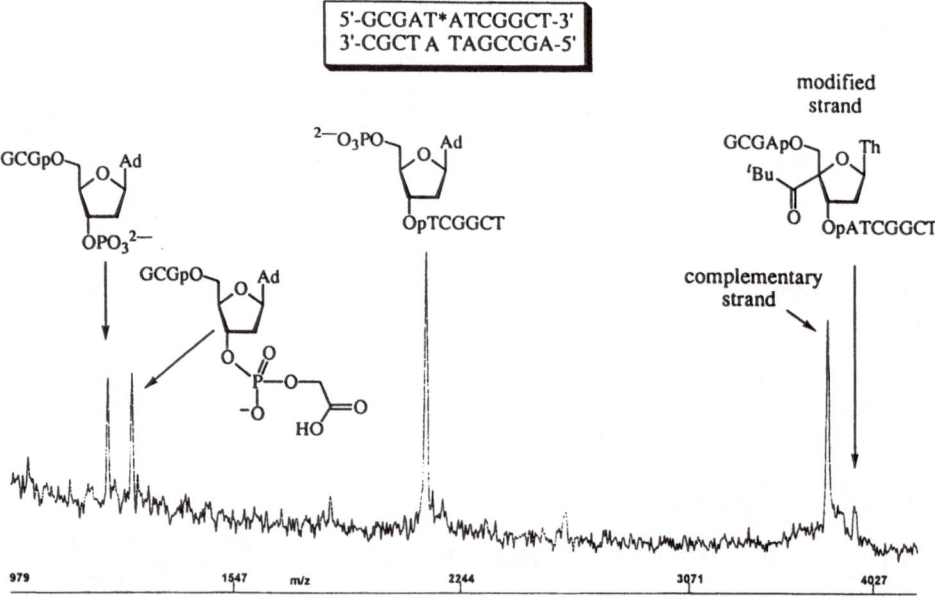

Figure 2. MALDI-MS of photolysis products of a double stranded oligonucleotide with one modified mononucleotide in one strand.

Figure 2 shows that aerobic photolysis of a modified double stranded oligomer yielded 3'-phosphate, 3'-phosphoglycolate, and 5'-phosphate as typical cleavage products. Intermediate hydroperoxides could not be detected. We assume that they are not stable enough under the high salt concentrations used in these experiments. Interestingly, the non-modified complementary strand remains intact in these experiments.

Acknowledgements: This work was supported by the Swiss National Science Foundation.

References

1. Stubbe, J. and Kozarich, J. W. (1987) Mechanisms of bleomycin-induced DNA degradation, *Chem. Rev.* **87**, 1107–1136. Pratviel, G., Bernadou, J. and Meunier, B. (1995) Carbon-hydrogen bonds of DNA sugar units as targets for chemical nucleases and drugs, *Angew. Chem. Int. Ed. Engl.* **34**, 746–770.
2. Hertzberg, R. P. and Dervan, P. B. (1984) Cleavage of DNA with methidiumpropyl-EDTA–iron(II): reaction conditions and product analyses, *Biochemistry* **23**, 3934–3945.
3. Giese, B., Burger, J., Kang, T.W., Kesselheim, C., and Wittmer, T. (1992) Model studies on the radical induced DNA strand cleavage, *J. Am. Chem. Soc.* **114**, 7322–7324.
4. Giese, B., Beyrich-Graf, X., Burger, J., Kesselheim, C., Senn, M. and Schäfer, T. (1993) The mechanism of anaerobic, radical-induced DNA strand scission, *Angew. Chem. Int. Ed. Engl.* **32**, 1742–1743.
5. Giese, B., Erdmann, P., Giraud, L., Göbel, T., Petretta, M., Schäfer, T., and von Raumer, M. (1994) Heterolytic C,O-bond cleavage of 4'-nucleotide radicals, *Tetrahedron Lett.* **35**, 2683–2686.
6. Giese, B., Erdmann, P., Schäfer, T. and Schwitter, U. (1994) Synthesis of 4'-selenated 2-deoxyadenosine derivative: A novel precursor suitable for the preparation of modified oligonucleotides, *Synthesis* 1310–1312.
7. Giese, B., Imwinkelried, P. and Petretta, M. (1994) Synthesis of a modified thymidine and preparation of precursors of oligonucleotide radicals, *Synlett* 1003–1004.
8. Giese, B., Dussy, A., Elie, C., Erdmann, P. and Schwitter, U. (1994) Synthesis and selective radical cleavage of C–4'-modified oligonucleotides, *Angew. Chem. Int. Ed. Engl.* **33**, 1861–1863.
9. Giese, B., Beyrich-Graf, X., Erdmann, P., Petretta, M. and Schwitter, U. (1995) The chemistry of single-stranded 4'-DNA radicals: influence of the radical precursor on anaerobic strand cleavage, *Chem. & Biol.* 2, 367–375. Giese, B., Beyrich-Graf, X., Erdmann, P., Giraud, L., Imwinkelried, P., Müller, S. N., Schwitter, U. (1995) Cleavage of single-stranded 4'-oligonucleotide radicals in the presence of O_2, *J. Am. Chem. Soc.* **117**, 6146–6147.

Acknowledgement: This work was supported by the Swiss National Science Foundation.

References

Site-Specific DNA Cleavage

SITE SPECIFIC OXIDATIVE SCISSION OF NUCLEIC ACIDS AND PROTEINS

D. S. SIGMAN
Molecular Biology Institute
University of California, Los Angeles
Los Angeles, CA 90095-1570

1. INTRODUCTION

The 1,10-phenanthroline-cuprous complex with hydrogen peroxide as a coreactant cleaves DNA and RNA by oxidative attack on the (deoxy) ribose moiety [1]. Recent studies have demonstrated that the oxidative species formed by the reductive cleavage of hydrogen peroxide also degrades polypeptide chains as well [2]. Because of its intrinsically high reactivity, the copper-oxo species formed by the oxidation of cuprous ion by hydrogen peroxide must be generated near the site of its target or it will be quenched nonspecifically. As a result, the kinetic mechanism for reactions involving 1,10-phenanthroline copper generally involve the reversible formation of a noncovalent complex followed by a hydrogen peroxide dependent oxidative step (CN= chemical nuclease).

$$CN + RNA/DNA \rightleftharpoons CN\text{----}RNA/DNA \xrightarrow{H_2O_2} \text{nicked products} \qquad Eq.1$$

If the reaction is intramolecular as in the case secondary structure mapping of membrane proteins to be discussed below, noncovalent complexes are not obligatory to insure proximity of the oxidant to its target [2].

Our most extensive studies of this system have focused on the reactivity of 1,10-phenanthroline-copper with nucleic acids. This coordination complex has proved useful in two contexts [3,4]. As the untargeted tetrahedral 2:1 1,10-phenanthroline-cuprous complex, the chemical nuclease has shown a remarkable array of reaction preferences based on the binding affinity of the hydrophobic tetrahedral cation. The reagent is useful for footprinting DNA- and RNA- ligand interactions especially because it can reveal sequence-specific contacts in complexes isolated by gel retardation assays. Tethered to a ligand, the oxidative reactivity of the 1:1 1,10-phenanthroline-cuprous has provided an approach to investigate the phenomena of molecular recognition and conformation of nucleic acids and proteins [5].

2. REACTION SPECIFICITY OF (OP)$_2$Cu$^+$

The tetrahedral (OP)$_2$Cu$^+$ shows reaction preference for B-DNA [6] for single stranded-loops or bulges in RNA [7]. The specificity for B-DNA is based on favorable binding of the coordination complex to the minor groove [8,9]. The structure of the chelate's binding site for

119

B. Meunier (ed.), DNA and RNA Cleavers and Chemotherapy of Cancer and Viral Diseases, 119–132.
© 1996 Kluwer Academic Publishers. Printed in the Netherlands.

RNA has resisted characterization but the cleavage reaction is sufficiently stringent that it can be used to "audit" postulated secondary structures of RNAs [10,11]. The products of the scission of DNA by $(OP)_2Cu^+$ have been determined and include, free bases, 5' and 3' phosphorylated termini and the oxidized deoxyribose derivative 5-methylene-furanone [12]. In all cases, where the resolution of the sequencing is adequate, 3-phosphoglycolate products, which presumably arise from C-4H attack, have also been detected in yields which range from 1% to 20%. [13]

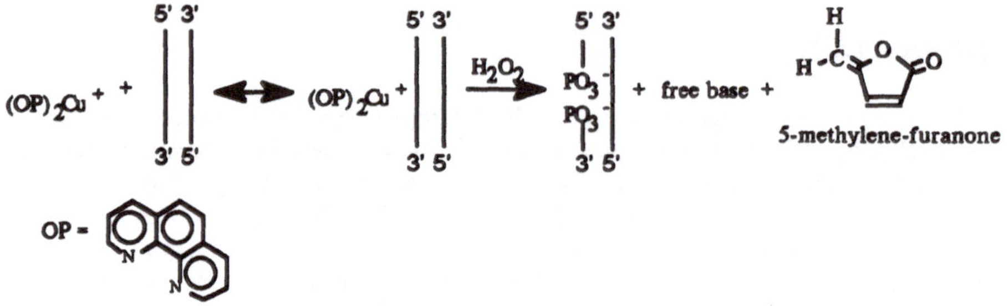

Fig. 1. Chemical Mechanism of DNA Scission

The most intriguing reactivity of the tetrahedral cuprous complexes is their specificity for single stranded DNA formed at the active site of RNA polymerase during catalysis [14]. The open complex at the start of transcription creates a binding site for the tetrahedral cuprous complexes of 1,10-phenanthroline and its derivatives [8]. These complexes bind to the open complex and cleave the template strand of DNA exclusively. The patterns of cleavage are a function of the 1,10-phenanthroline derivative used to form the cuprous chelate. Some derivatives, such as the 5-phenyl-1,10-phenanthroline cleave the open complex very efficiently while others such as the 4,7-diphenyl-1,10-phenanthroline are incapable of open complex scission although they are reactive with B-DNA. This binding site formed by the melting of the DNA at the start of transcription by RNA polymerase therefore exhibits shape selective specificity.

The reactivity of these chelates with the single-stranded DNA of steady state intermediates formed during transcription can be used to follow the downstream progression of RNA synthesis [8,15,16]. When an incomplete complement of nucleotide is added, very short oligonucleotides are synthesized and the cleavage sites of the chelate trail the positions at which the internucleotide bond is synthesized. However, when the RNA strand length increases so that the initially transcribing complex is converted into the elongation complex, the site of hyperreactivity is at or slightly in advance of the site of phosphodiester bond synthesis.

The inference that this cleavage specificity is a result of the creation of a binding site for the tetrahedral chelate is supported by the demonstration that redox-inert isosteres, the tetrahedral chelates 2,9-dimethyl-1,10-phenanthroline (neocuproine) cuprous ion $((NC)_2Cu^+)$ are efficient transcription inhibitors which block the scission of the open complex by 1,10-phenanthroline-copper [17]. In fact, these neocuproine complexes are unique inhibitors of transcription because they bind to the enzyme-substrate complex near the catalytic site.

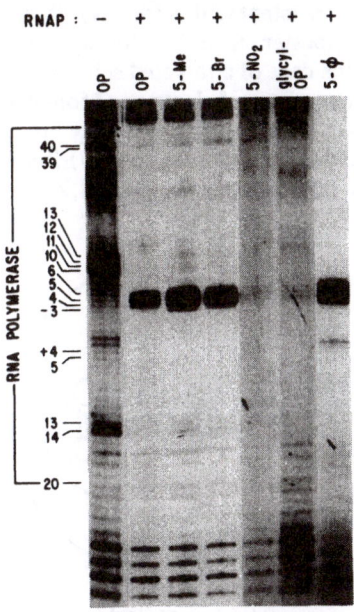

Figure 2. Footprinting the lac UV 5-RNA polymerase Open Complex with Cuprous Complexes of 5'-Substituted 1,10-Phenanthrolines

Other transcription inhibitors (e.g actinomycin) are high affinity DNA ligands that bind at specific sites but destabilize protein binding or antibiotics (e.g. rifampicin) which interact with RNA polymerase and inhibit transcription initiation [18]. Rifampicin binds to the open complex but does not protect the scission of the template strand by $(OP)_2Cu^+$ and its derivatives. The discrete scission sites within the open complex provide additional evidence that the oxidative species formed by the oxidation of cuprous ion by hydrogen peroxide is not freely diffusible but is instead a copper oxene species possibly akin to the ferryl-oxo species formed as an intermediate in cytochrome P-450 mediated reactions [19].

3. TARGETING THE NUCLEASE ACTIVITY OF 1,10-PHENANTHROLINE-COPPER

Early studies of the untargeted chemical nuclease activity of 1,10-phenanthroline-copper indicated that the 1:1 OP-Cu complex was nucleolytically inactive [20]. These conclusions were based on a comparison of the degradative activity in the presence of excess OP relative to copper (the 2:1 complex) to that in the presence of an excess of copper relative to OP (the 1:1 complex). An efficient chemical nuclease activity was only observed when OP was present in excess. Subsequent studies have strongly indicated that these differences in reactivity are due to

122

the productive binding of the coordination complex to DNA or RNA and not to an inherent difference in the chemistry of the two complexes. Oligonucleotides, peptides, proteins, and DNA binding drugs all direct the scission of RNA and DNA [10,21-23]. In every case, scission occurs at all nucleotides consistent with a (deoxy) ribose directed chemistry and the generation of 5' an 3' phosphomonoester termini. Although no attempt has been made to isolate 5-methylene furanone in the case of a targeted scission reaction, there is no evidence of any underlying difference in the chemistry of cleavage. More detailed studies on the mechanism of the targeted scission of the nuclease activity are in progress. They will focus on the initial site of hydrogen abstraction from the (deoxy)ribose using kinetic isotope effects and the atom source of oxygen in the lactone products [24,25].

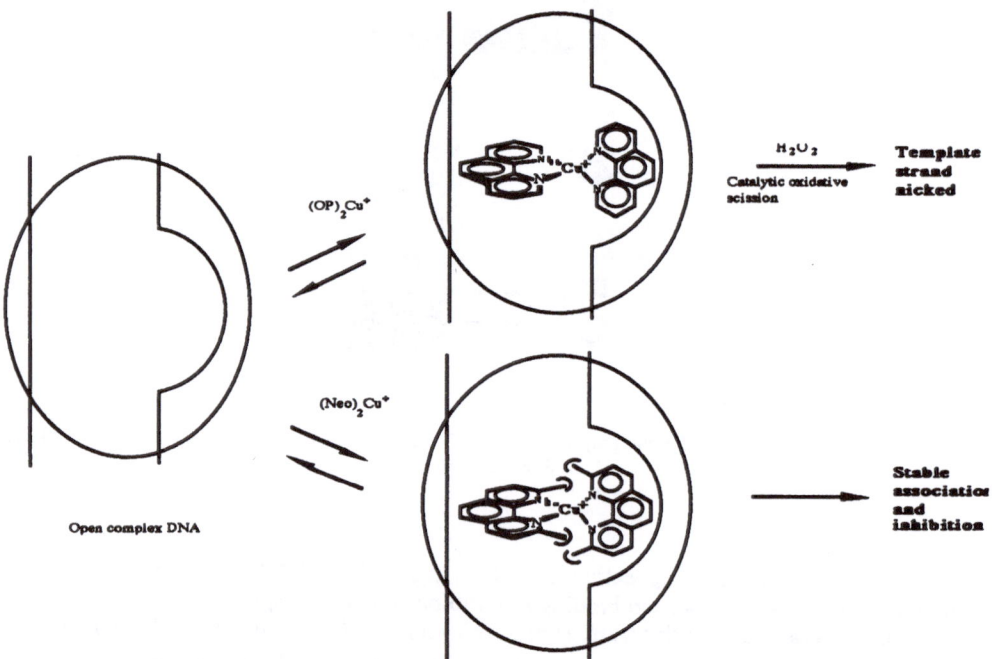

Figure 3. Competitive Binding of Isosteric Tetrahedral Chelates to Open Complex

4. TARGETED SCISSION BY GENE-SPECIFIC INHIBITORS

The targeted scission of DNA by 1,10-phenanthroline-copper linked to ligands has yielded intriguing results in several different contexts. Possibly the most useful may in the design of gene specific transcription inhibitors. The demonstration that the single strand of DNA within the open complex is accessible to both $(OP)_2Cu^+$ and $(NC)_2Cu^+$ suggested that the template strand of DNA should also be accessible to hybridization by complementary oligonucleotides. In effect, complementary oligonucleotides should be mechanism-based inhibitors because they will only bind to sites created during the course of catalysis. This inference has received clear experimental support for bacterial promoters [26] and is currently under intense investigation in eucaryotic systems [27]. 5'-rUGGAA which is complementary to sequence positions -3 to +2 of the single stranded template strand at the start of transcription of the lac UV-5 promoter binds to the open complex with submicromolar affinity and blocks de novo transcription *in vitro* [26]. This pentamer does not inhibit transcription from the trp EDCBA promoter even though the complementary sequence in this case would be 5'-rUGUAA and therefore differs at only 2 of the 5 sequence positions. 5'-rUGUAA inhibits de novo transcription from the trp EDCBA promoter but not from the lac UV-5 promoter. Although the sequence stringency indicates these oligonucleotides bind to their respective open complexes by hybridization to the single stranded template, site specific cleavage of the template strand of a 1,10-phenanthroline-derivatized hybridizing oligonucleotide provides unambiguous evidence for the formation of an antiparallel heteroduplex within the open complex [26]. Recent studies have indicated that the hybridization site within the open complexes binds RNase resistant abiological oligonucleotides but does not interact tightly with deoxyoligonucleotides and peptide-nucleic acids of identical sequence [28]. RNA polymerase contributes to the stability of the duplexes formed in the open complex although the structural features of the hybridizing oligonucleotide that the enzyme will tolerate must be determined empirically.

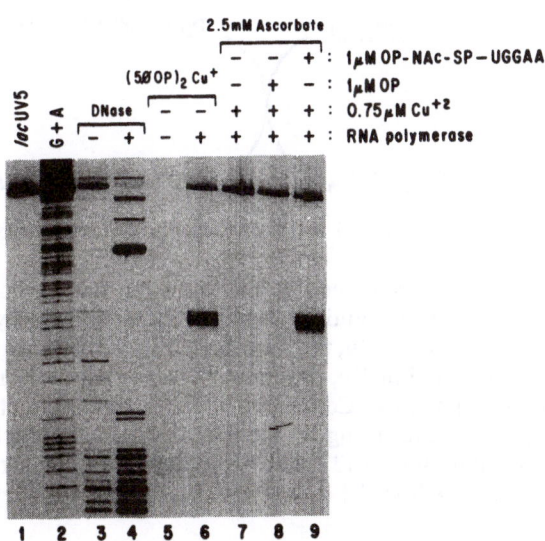

Figure 4. Directed Cleavage of the Open Complex by OP-derivatized-UGAAA

5. TARGETED SCISSION BY DNA BINDING PROTEINS

The conversion of DNA binding proteins into site specific nucleases by chemically tethering 1,10-phenanthroline is proving to a fertile area of experimentation. Following initial studies using iminothiolane as a bridging reagent [29], a two step protocol has been developed in which a nonessential amino acid near the redox sensitive minor groove is converted into a cysteine residue which is then alkylated by 5-iodoacetamido-1,10-phenanthroline, or analogous derivatives, to yield a cleavage competent DNA binding protein-1,10-phenanthroline chimera [30,31].

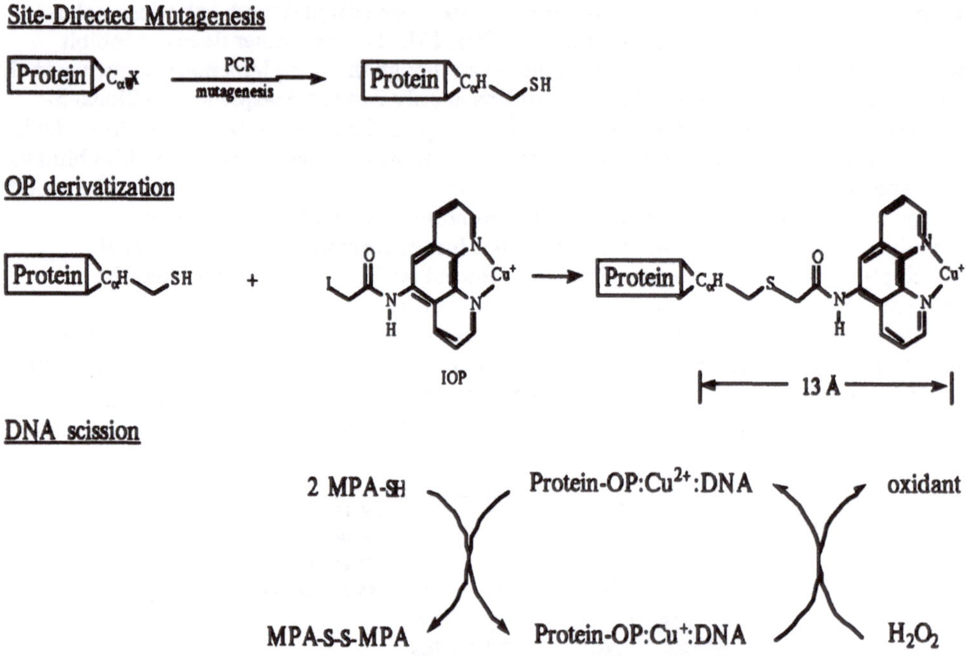

Site-Directed Mutagenesis

OP derivatization

DNA scission

Figure 5. Scheme for the conversion of DNA Binding Proteins Into Site Specific Nucleases

These semisynthetic nucleases can be used in the following ways. First, they can test models of protein binding based on high resolution structures obtained by x-ray crystallography and nuclear magnetic resonance. Secondly, they can serve as DNA cleavage reagents which recognize 10-15 bp sequences. Finally, they have the potential to identify biologically relevant sites of regulatory proteins. Conversion of DNA binding proteins into site-specific nucleases was initially established using the lambda phage cro protein [30] and has now been extended to the E. coli Trp Repressor [32] and Fis proteins [33], the Drosophila engrailed homeodomain [34] and mouse Msx-1 [35] and the cyclic AMP binding protein [36,37].

a) E. coli Trp Repressor

The most efficient scission reagent generated in our laboratory so far was derived from the E. Coli Trp Repressor (Trp R)by alkylation of the trp E49C mutant with 5-iodoacetamido 1,10-phenanthroline [32]. Virtually quantitative cleavage of the three known trp operators was achieved. The precision of the scission permitted the detection of the three distinct subsites within the trp EDCBA, two subsites within the aro H operator and a single site within the trp R operator. In all cases, L-tryptophan, which is essential for repression, was required for site specific cleavage. In fact, in its absence, nonspecific background scission is observed. Tryptophan not only targets the recognition sequence of Trp R for scission but it also prevents apo Trp R E49C-OP from random DNA binding. Alteration of the binding specificity modifies the DNA scission specificity of Trp R. Thus a site specific nuclease was derived from E49C,I79K whose cleavage preferences mirrored the binding of Trp R I79K and not the wild type Trp R [38]. These experiments indicate a family of engineered nuclease can be generated based on the Trp R conformation.

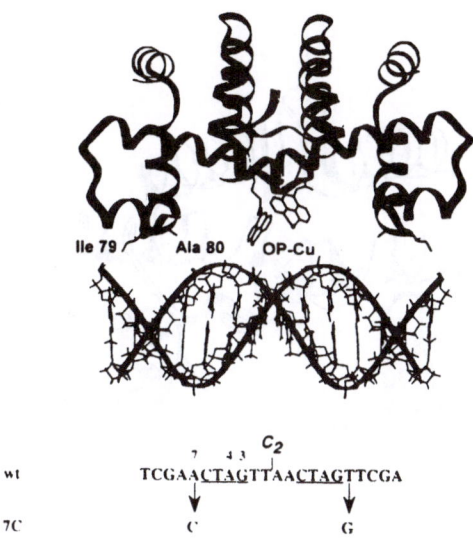

Figure 6. Ribbon diagram of the *Trp R* dimer structure showing the Ile 79 side chains and the 1,10-phenanthroline-copper adduct with Cys 49 aligned with the consensus operator.

b) E. coli Factor for Inversion Stimulation

Our most recent work with the *E. Coli* Fis protein illustrates the ability of these proteins to identify as yet unknown sites within a genome. The Fis protein regulates a diverse set of reactions including recombination, transcription, and replication (for review see Finkel & Johnson, 1992 [39]). Studies on these systems have revealed a large number of specific Fis binding sites, though these are only poorly related at the primary sequence level. Approximately half of the known sites conform to a highly degenerate 15 bp sequence (5' G N N C/T A/T N N A/T N N T/C G/A N N C 3', where N can be any base) [39,40]. Only a fraction of sequences containing this motif bind Fis. These properties preclude identification of specific Fis binding sites based on sequence information alone. In addition to binding specifically at a limited number of sites, Fis binds DNA relatively randomly at lower affinity and forms complexes which can be readily competed by excess nonspecific DNA.

The structure of Fis complexed to one of its recognition sequences has not been determined experimentally. However, model building based on the solved high resolution structure of the protein indicates that the DNA must be bent when a complex is formed [39]. A particularly interesting construct is the Fis N73C-OP mutant. Based on model building, this chimera should be an efficient scission reagent. In total congruence with the model, Fis N73C-OP cleaves all of the 26 known sites tested [33,41]. Cleavage by Fis N73C-OP therefore provides a stringent test for identifying a high affinity functional Fis site. This chimera identifies a Fis site not only on the basis of its internal consensus sequence but also its bendability around the Fis protein.

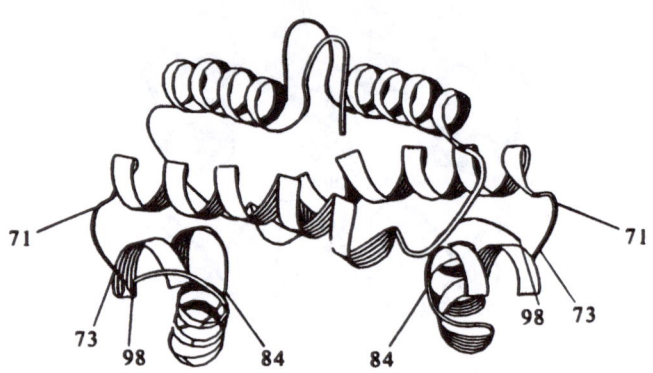

Figure 7. Ribbon diagram of Fis dimer highlighting the side chains of the four amino acids that have been separately mutated to cysteine and alkylated with OP derivatives. The N-terminal A α-helices are on the top and the C-terminal C and D α helices are on the bottom.

Three new sites for Fis-OP have been identified using these chimera. One of these sites is in the *lac* regulatory region. The *lac* operon is activated by the catabolite activator protein (CAP) and repressed by the Lac repressor (LacI). The CAP core binding site is located from -69 to -54 with respect to the major start of transcription and therefore partially overlaps the Fis core binding site which stretches from -60 to -46. Since the Fis binding site in the *lacP* region overlaps the CAP binding site, it seemed possible that Fis would negatively affect CAP activation. However, a negative effect by Fis on *lacP* activity was not observed under inducing conditions. On the other hand, the absence of Fis produced a more completely repressed state. A negative effect of Fis on the efficiency of LacI-mediated repression indicates a different role for Fis in the control of the *lac* operon. It has been suggested that the tetrameric LacI can simultaneously bind to two operators resulting in strong repression through cooperativity mediated by DNA looping [42,43]. A secondary operator, O_3, is centered at -83 and experiments by Borowiec et al. [44] together with the implications of the recent crystal structure of the repressor core tetramer [45] strongly suggest that a single repressor tetramer holds O_1 and O_3 together. Moreover, Oehler et al. [46,47] have shown that removal of O_3 decreases repression efficiency up to 3.5-fold. As diagrammed in Figure 8, bending of the DNA by Fis bound between O_1 and O_3 may inhibit cooperativity by directing the DNA into a path which is not compatible with loop formation between these operators, thus reducing repression efficiency [41].

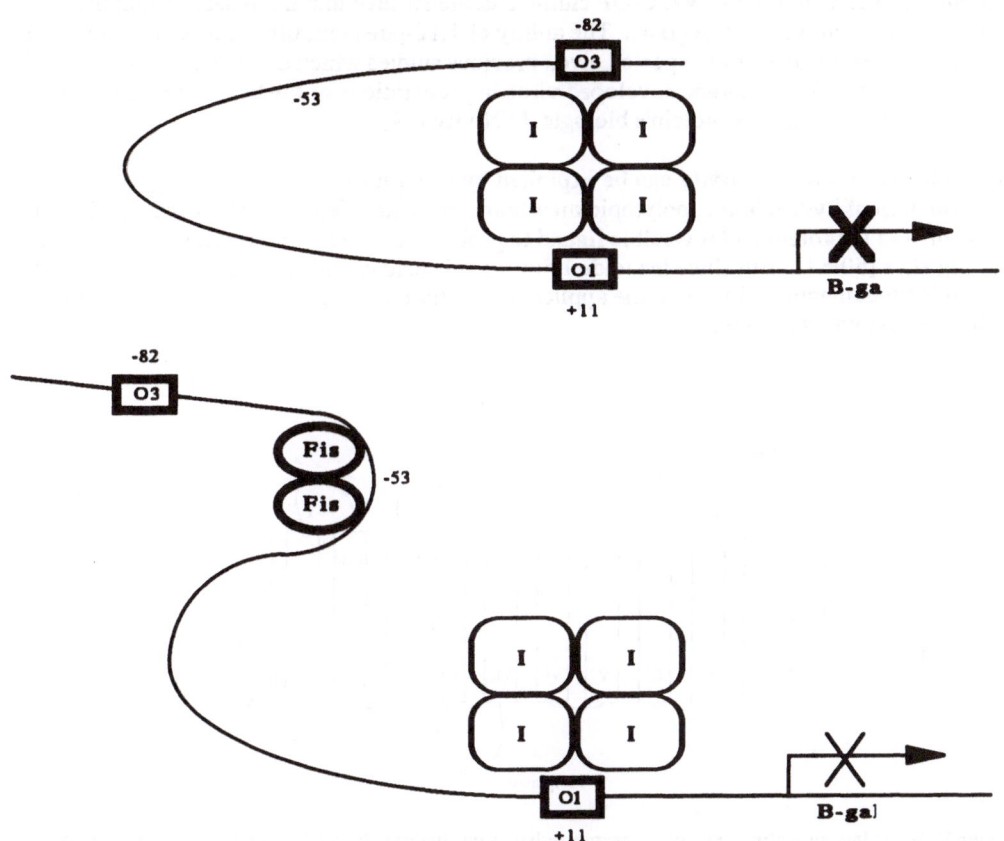

Figure 8. A model of de-repression of the *lac* promoter by Fis. Fis-mediated DNA bending may inhibit looping between the two operators and reduce lac I repression efficiency.

DNA binding protein-OP chimeras provide a new paradigm for the discovery of binding sites for widely dispersed biological regulators. The assay for these sites is direct because specific scission relies on high affinity binding of the protein. In the discovery of new Fis sites on plasmids and in lambda phage, restriction mapping and PCR are sufficient to localize the recognition sites at the sequence level. For larger genomes, ligation-mediated PCR techniques, that have proven useful in genomic footprinting applications [48], should play a major role.

6. CHEMICAL PROTEASE ACTIVITY OF 1,10-PHENANTHROLINE-COPPER

DNA binding protein-OP chimeras vary in their efficiency in the cleavage of nucleic acid because of differences in the distance and orientation of the OP-Cu to the target sequence. Ongoing studies with the Fis N73C-OP chimera demonstrated another reason is that the derivatized proteins can self-degrade. The ability of 1,10-phenanthroline-copper to oxidatively degrade peptide bonds was unexpected given previous studies which demonstrated that OP-Cu could be used to kill membrane enveloped virus in preparations of the blood clotting factor VIII without affecting this protein's biological activity [49].

This oxidative protease activity can be exploited by using it to study the topology and 3-D conformation of hydrophobic, polytopic membrane proteins. Ferrous-EDTA has previously been tethered to protein and reversible ligand to probe nearest neighbor relationships at protein binding sites [50-54]. But since the underlying nucleic acid chemistry of ferrous-EDTA and OP-Cu is fundamentally different, the application of the 1,10-phenanthroline-copper complex in this context was warranted.

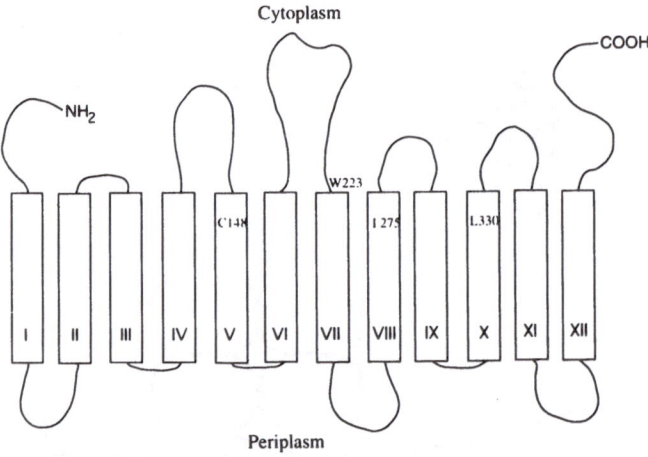

Figure 9. Secondary-structure model of lac permease based on hydropathy analysis and a series of lac permease-alkaline phosphatase (lacY/phoA) fusions.

The initial question that we chose to address was the packing of the 12 transmembrane helical segments of the lac permease, an extensively studied membrane associated protein which regulates lactose import in the presence of a proton gradient [55]. We have studied the cleavage of four lac permease mutants with 1,10-phenanthrolines covalently attached at sequence positions 147, 148, 149 and 330 to determine the nearest neighbors consistent with . previous spectroscopic and chemical modification studies [2]. In the case of lac permease, the chemical cleavage data unambiguously positions helix V relative to helices VII and VIII. This oreintation is supported by the scission of helices VIII and VII by the lac permease with OP-linked to a cysteine at position 148, the exclusive scission of helix VII when OP is linked to position 149 and the absence of any scission when OP is linked to position 147 presumably on the exterior of the permease (see Figs 10 & 11).

The chemical mechanism of the scission has not yet been established although it may be similar to that proposed for the copper-dependent oxidative cleavage of polypeptide catalyzed by amidating enzyme [56]. As long as methods are available to analyze the cleavage products, the chemical protease should be a robust method for determining protein conformation in the presence and absence of ligand and protonmotive force.

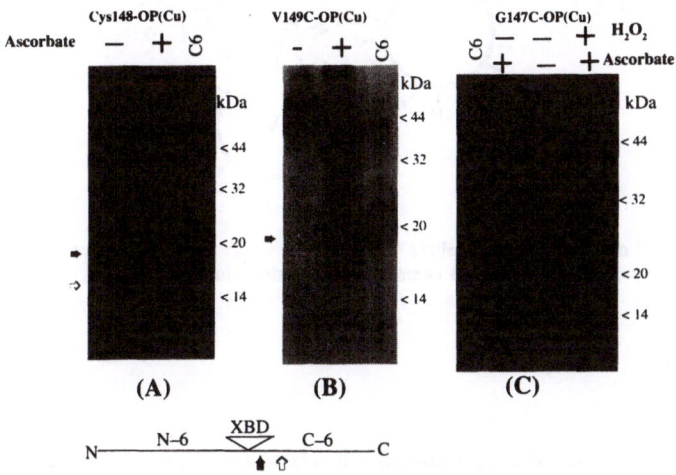

Figure 10. Site-directed proteolysis with single-Cys148-OP (A), V149C-OP (B) or G147C-OP (C) lac permease using purified permeases containing single-Cys residues and a biotin acceptor domain in the middle cytoplasm loop. A and B are linear schematics of lac permease showing the position of the biotin acceptor domain and the sites of cleavage from positions 148 and 149, respectively.

130

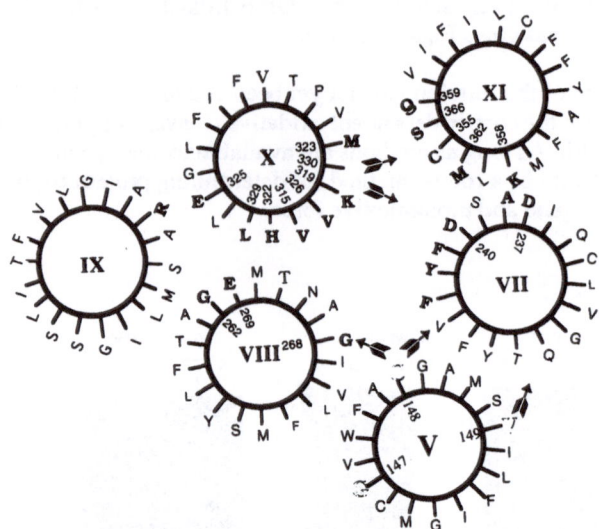

Figure 11. Helical wheel model of putative helices V and VII-XI in lac permease viewed from the periplasmic surface based on cleavage, spectroscopic and chemical modification studies.

7. CONCLUSION

The generation of new functional biomolecules has been the focus current research in catalytic antibodies, ribozymes, *in vitro* selection schemes and site-directed mutagenesis. Derivatizing biomolecules with the 1,10-phenanthroline cuprous complex provides another direction in bioorganic chemistry in the development of reactive molecules with designed specificities. The reactivity of these conjugates has proven remarkably robust and has led to the insight into protein-DNA reactions and a novel method for mapping the conformation of membrane proteins.

8. REFERENCES

1. D. S. Sigman, A. Mazumder and D. M. Perrin (1993) Chemical Nucleases, *Chem. Rev.*, **93**, 2295-2316.
2. J. Wu, D. M. Perrin, D. S. Sigman and H. R. Kaback (1995) Helix packing of lactose permease in Escherichia coli studied by site-directed chemical cleavage, *Proc. Natl. Acad. Sci. USA*, **92**, in press.
3. D. S. Sigman (1990) Chemical Nucleases, *Biochemistry*, **29**, 9097-9105.
4. D. S. Sigman and C.-h. B. Chen (1990) Chemical Nucleases: New Reagents in Molecular Biology, *Ann. Rev. Biochem.*, **59**, 207-236.
5. D. S. Sigman, T. W. Bruice, A. Mazumder and C. L. Sutton (1993) Targeted Chemical Nucleases, *Acc. Chem. Res.*, **26**, 98-104.
6. L. E. Pope and D. S. Sigman (1984) Secondary Structure Specificity of the Nuclease Activity of the 1,10-Phenanthroline-copper Complex, *Proc. Natl. Acad. Sci. USA*, **81**, 3-7.
7. G. J. Murakawa, C.-h. B. Chen, M. D. Kuwabara, D. Nierlich and D. S. Sigman (1989) Scission of RNA by the Chemical Nuclease 1,10-Phenanthroline-Copper Ion. Preference for Single-Stranded Loops, *Nucleic Acids Res.*, **17**, 5361-5369.
8. T. Thederahn, A. Spassky, M. D. Kuwabara and D. S. Sigman (1990) Chemical Nuclease Activity of 5-Phenyl-1,10-Phenanthroline-Copper Ion Detect Intermediates in Transcription Initiation by E. Coli RNA Polymerase, *Biochem. Biophys. Res. Comm.*, **168**, 756-762.

9. J. M. Veal and R. L. Rill (1988) Sequence Specificity of DNA Cleavage by Bis(1,10-phenanthroline)Cu$^+$, *Biochemistry*, **27**, 1822-1827.

10. C.-H. B. Chen and D. S. Sigman (1988) Sequence-Specific Scission of RNA by 1,10-Phenanthroline-Copper Linked to Deoxyoligonucleotides, *J. Am. Chem. Soc.*, **110**, 6570-6572.

11. L. Pearson, C. C.-h. B., R. P. Gaynor and D. S. Sigman (1994) Footprinting RNA-Protein Complexes following Gel Retardation Assay: Application to the R-17-Procoat-RNA and Tat-TAR Interactions, *Nucleic Acids Res.*, **22**, 2255-2263.

12. T. E. Goyne and D. S. Sigman (1987) Nuclease Activity of 1,10-Phenanthroline-copper Ion. Chemistry of Deoxyribose Oxidation, *J. Am. Chem. Soc.*, **109**, 2846-2848.

13. M. Kuwabara, C. Yoon, T. E. Goyne, T. Thederahn and D. S. Sigman (1986) Nuclease Activity of 1,10-Phenanthroline-copper Ion: Reaction with CGCGAATTCGCG and Its Complexes with Netropsin and *Eco*RI, *Biochemistry*, **25**, 7401-7408.

14. A. Spassky and D. S. Sigman (1985) Nuclease Activity of 1,10-Phenanthroline-copper Ion. Conformational Analysis and Footprinting of the *lac* Operon, *Biochemistry*, **24**, 8050-8056.

15. A. Spassky (1986) Following the Progression of the Transcription Bubble, *J. Mol. Biol.*, **188**, 99-103.

16. D. M. Perrin, V. M. Hoan, Y. Xu, A. Mazumder and D. S. Sigman (1995) Inhibitors of E. Coli RNA Polymerase Specific for the Single Stranded DNA of Transcription Intermediates. Tetrahedral Cuprous Chelates of 1,10-Phenanthrolines, *Biochemistry*, **34**, submitted.

17. A. Mazumder, D. M. Perrin, K. H. Watson and D. S. Sigman (1993) A transcription inhibitor specific for unwound DNA in RNA polymerase-promoter open complexes, *Proc. Natl. Acad. Sci. USA*, **90**, 8140-8144.

18. A. Mazumder, D. M. Perrin, D. McMillen and D. S. Sigman (1994) Interactions of Transcription Inhibitors with the Elscherichia] coli RNA polymerase-*LacUV5* Promoter Open-Complex, *Biochemistry*, **33**, 2262-2268.

19. Y. Watanabe and J. T. Groves (1992) *Molecular Mechanism of Oxygen Activation by P-450*, XX, 406-447.

20. L. E. Marshall, D. R. Graham, K. A. Reich and D. S. Sigman (1981) Cleavage of Deoxyribonucleic Acid by the 1,10-Phenanthroline-cuprous Complex. Hydrogen Peroxide Requirement and Primary and Secondary Structure Specificity, *Biochemistry*, **20**, 244-250.

21. C.-h. B. Chen and D. S. Sigman (1986) Nuclease Activity of 1,10-Phenanthroline-copper: Sequence-specific Targeting, *Proc. Natl. Acad. Sci. USA*, **83**, 7147-7151.

22. C.-h. B. Chen, A. Mazumder, J.-F. Constant and D. S. Sigman (1993) Nuclease Activity of 1,10-Phenanthroline-Copper. New Conjugates with Low Molecular Weight Targeting Ligands, *Bioconj. Chem.*, **4**, 69-77.

23. C.-h. B. Chen, M. B. Gorin and D. S. Sigman (1993) Sequence-Specific Scission of DNA by the Chemical Nuclease Activity of 1,10-Phenanthroline-Copper(I) Targeted by RNA, *Proc. Natl. Acad. Sci. USA*, **90**, 4206-4210.

24. J. W. Kozarich, J. L. Worth, B. L. Frank, D. F. Christner, D. E. Vanderwall and J. Stubbe (1989) Sequence-Specific Isotope Effects on the Cleavage of DNA by Bleomycin, *Science*, **245**, 1396-1399.

25. G. Pratviel, J. Bernatou and B. Meunier (1995) Carbon-Hydrogen Bonds of DNA Sugar Units as Targets for Chemical Nucleases and Drugs, *Angew Chem Int Ed Eng I*, **34**, 746-769.

26. D. M. Perrin, A. Mazumder, F. Sadeghi and D. S. Sigman (1994) Hybridization of a complementary riboologonucleotide to the transcription start site of the *lacUV-5*-*Escherichia coli* RNA polymerase open complex. Potential for gene-specific inactivation reagents, *Biochemistry*, **33**, 3848-3854.

27. D. M. Perrin, L. Pearson, A. Mazumder and D. S. Sigman (1994) Inhibition of prokaryotic and eukaryotic transcription by the 2:1 2,9-dimethyl-1,10-phenanthroline-cuprous complex, a ligand specific for open complexes, *Gene*, **149**, 173-178.

28. D. M. Perrin and D. S. Sigman (1995) A new Approach for Mechanism-Base, Gene-Specific Inhibition: Hybridization of Oligonucleotides to Transcription Start Sites, *J. Amer. Chem. Soc*, **117**, submitted.

29. C.-h. B. Chen and D. S. Sigman (1987) Chemical Conversion of a DNA-binding Protein into a Site-specific Nuclease, *Science*, **237**, 1197-1201.

30. T. W. Bruice, J. Wise, D. S. E. Rosser and D. S. Sigman (1991) Conversion of Lambda Phage Cro into an Operator-Specific Nuclease, *J. Am. Chem. Soc.*, **113**, 5446-5447.

31. C. Q. Pan, R. Landgraf and D. S. Sigman (1994) DNA Binding Proteins as Site-Specific Nucleases, *Mol. Micro.*, **12**, 335-342.

32. C. Sutton, A. Mazumder, C.-h. B. Chen and D. S. Sigman (1993) Transforming the *Escherichia coli* Trp Repressor into a Site-Specific Nuclease, *Biochemistry*, **32**, 4225-4230.

33. C. Q. Pan, J. Feng, S. E. Finkel, R. Landgraf, R. Johnson and D. S. Sigman (1994) Structure of the Escherichia coli fis protein-DNA complex probed by protein conjugated with 1,10-phenanthroline copper(I) complex, *Proc. Natl. Acad. Sci. USA*, **91**, 1721-1725.

132

34. C. Q. Pan, R. Landgraf and D. S. Sigman (1995) Drosophila engrailed-1,10-phenanthroline chimeras as probes of homeodomain-DNA complexes, *J. Protein Science*, **4**, in press.
35. Z. Shang, Y. Ebright, N. Iler, S. Pendergrast, Y. Echelard, A. McMahon, R. Ebright and C. Abate (1994) DNA affinity cleaving analysis of homeodomain-DNA interaction: identification of homeodomain consensus sites in genomic DNA, *Proc. Natl. Acad. Sci. USA*, **91**, 118-122.
36. R. H. Ebright, Y. W. Ebright, P. S. Pendergrast and A. Gunasekera (1990) Conversion of a helix-turn-helix motif sequence-specific DNA binding protein into a site-specific DNA cleavage agent, *Proc. Natl. Acad. Sci. USA*, **87**, 2882-2886.
37. P. S. Pendergrast, Y. W. Ebright and R. H. Ebright (1994) High-Specificity DNA Cleavage Agent: Design and Application to Kilobase and Megabase DNA Substrates, *Science*, **1994**, 959-962.
38. J. Pfau, D. N. Arvidson, P. Youderian, L. L. Pearson and D. S. Sigman (1994) A site-specific endonuclease derived from a mutant *Trp R*epressor with altered DNA-binding specificity, *Biochemistry*, **33**, 11391-11403.
39. S. E. Finkel and R. C. Johnson (1992) The Fis protein: it's not just for DNA inversion anymore, *Molec. Microbio.*, **6**, 3257-3265.
40. P. Hubner and W. Arber (1989) *EMBO J.*, **8**, 577-585.
41. C. Q. Pan, R. C. Johnson and D. S. Sigman (1995) Identification of New Fis Binding Sites by DNA Scission with Fis-1,10-Phenanthroline Copper (I) Chimeras, subm itted.
42. M. C. Mossing and M. T. Record, Jr. (1986) Upstream Operators Enhance Repression of the *lac* Promoter, *Science*, **233**, 889-892.
43. H. Kramer, M. Niemoller, M. Amouyal, B. Revet, B. von Wilcken-Bergmann and B. Muller-Hill (1987) *Lac* Repressor Forms Loops with Linear DNA Carrying Two Suitably Spaced *Lac* Operators, *EMBO J.*, **6**, 1481-1491.
44. J. A. Borowiec, L. Zhang, S. Sasse-Dwight and J. D. Gralla (1987) DNA Supercoiling Promotes Formation of a Bent Repression Loop in *lac* DNA, *J. Mol. Biol.*, **196**, 101-111.
45. A. M. Friedman, T. O. Fischmann and T. A. Steitz (1995) Crystal Structure of lac repressor core tetramer and its implications for DNA looping, *Science*, **268**, 1721-1727.
46. S. Oehler, M. Amouyal, P. Kolkhof, B. W. Bergman and B. Muller-Hill (1994) Quality and Position of the Three Lac Operators of E. Coli Define Efficiency of Repression, *EMBO J.*, **13**, 3348-3355.
47. S. Oehler, E. R. Eismann, H. Kramer and B. Muller-Hill (1990) The Three Operators of the lac Operon Cooperate in Repression, *EMBO J.*, **9**, 973-979.
48. G. P. Pfeifer, S. D. Steigerwald, P. R. Mueller, B. Wold and A. D. Riggs (1989) Genomic Sequencing and Methylation Analysis by Ligation Mediated PCR, *Science*, **246**, 810-813.
49. K. J. Lembach, M. B. Dobkin and R. E. Louie (1989) *Inactivation of Viruses with 1,10-Phenanthroline*, in Virus Inactivation in Plasma Productts, J-J. Morgenthaler Ed. p. 97-108.
50. T. M. Rana and C. F. Meares (1990) Specific Cleavage of a Protein by an Attached Iron Chelate, *J. Am. Chem. Soc.*, **112**, 2457-2458.
51. T. M. Rana and C. F. Meares (1991) Transfer of oxygen from an artificial protease to peptide carbon during proteolysis, *Proc. Natl. Acad. Sci. USA*, **88**, 10578-10582.
52. D. Hoyer, H. Cho and P. G. Schultz (1990) A New Strategy for Selective Protein Cleavage, *J. Am. Chem. Soc.*, **112**, 3249-3250.
53. A. Schepartz and B. Cuenoud (1990) Site-Specific Cleavage of the Protein Calmodulin Using a Trifluoperazine-Based Affinity Reagent, *J. Am. Chem. Soc.*, **112**, 3247-3249.
54. I. E. Platis, M. R. Ermacora and R. O. Fox (1993) Oxidative Polypeptide Cleavage Mediated by EDTA-Fe Covalently Linked to Cysteine Residues, *Biochemistry*, **32**, 12761-12767.
55. H. R. Kaback (1989) *Molecular Biology of Active Transport: From Membrane to Molecule to Mechanism*, Series 83, 77-105.
56. R. C. Bateman, Jr., W. W. Youngblood, W. H. Busby, Jr. and J. S. Kizer (1985) Nonenzymatic Peptide α-Amidation: Implications for a Novel Enzyme Mechanism, *J. Biol. Chem.*, **260**, 9088-9091.

SEQUENCE SELECTIVE DNA CLEAVAGE BY PNA-NTA CONJUGATES

J. LOHSE, C. HUI[±], S.H. SÖNNICHSEN[+] and P.E. NIELSEN[*]

Center for Biomolecular Recognition
Department of Medical Biochemistry & Genetics
Laboratory B, The Panum Institute
Blegdamsvej 3, 2200 N Copenhagen
Denmark

± *Present address*: University of Gothenburg, Sweden

+ *Present address*: Clinical Chemical Department, Glostrup Hospital, 2600 Glostrup, Denmark

* Corresponding author

1. Introduction

The DNA mimic, PNA (peptide nucleic acid) (Figure 1) (1-4), binds sequence specifically to homopurine sequences in double stranded DNA by strand invasion (1,5) <u>via</u> an internal $PNA_2 \cdot DNA$ triplex (6) and displacement of the homopyrimidine DNA strand (Figure 2). This recognition is relying on conventional Watson-Crick and Hoogsteen base pairings (T-A·T & C-G·C⁺ triplets) (6,7) and the binding is therefore analogously to the formation of DNA triplexes of this motif strongly pH dependent. However, the introduction of the C⁺-mimic, pseudoisocytosine, into bis-PNAs is able to relieve this pH-dependence (8).

Despite the very high thermal and thermodynamic stability of decamer PNA

133

B. Meunier (ed.), DNA and RNA Cleavers and Chemotherapy of Cancer and Viral Diseases, 133–141.

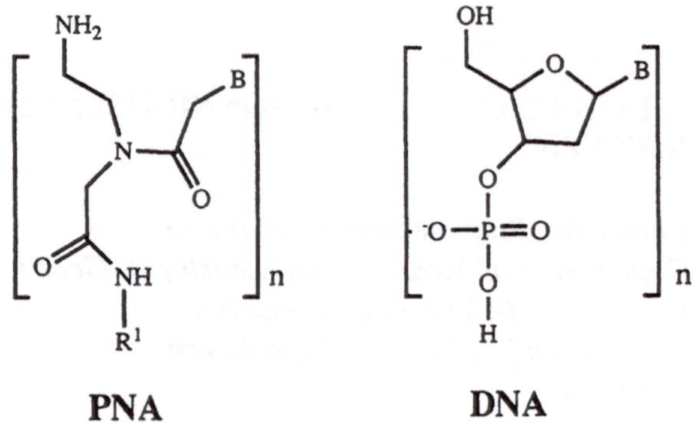

PNA   **DNA**

<u>Figure 1</u>. Chemical structure of PNA.

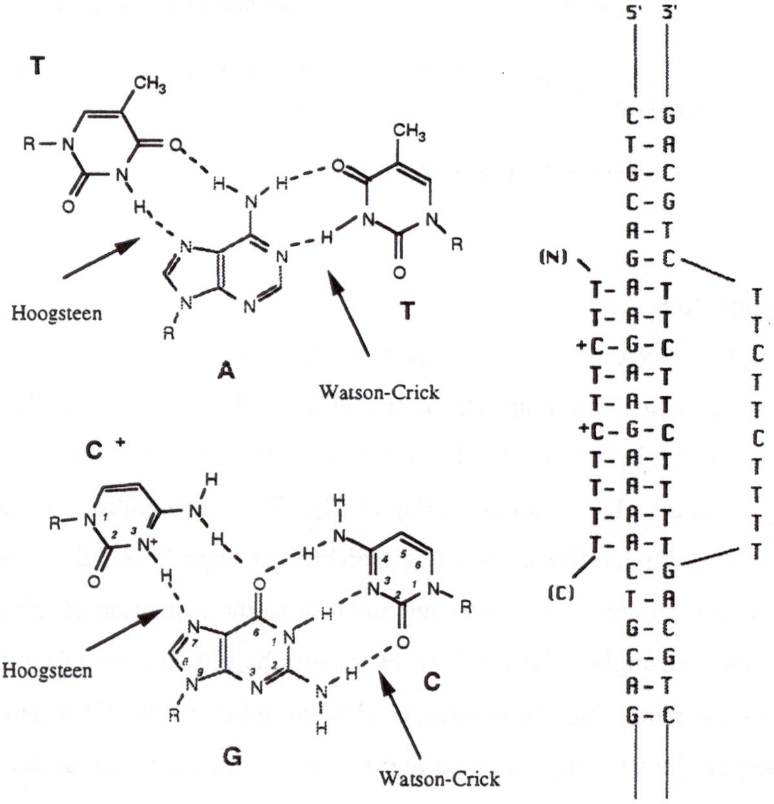

<u>Figure 2</u>. Schematic drawing of a PNA·dsDNA strand displacement complex and the triplexes involved.

complexes with double stranded DNA targets, such PNAs exhibit high sequence discrimination (9-11) because the binding is kinetically controlled (11).

Triplex forming oligonucleotides have successfully been conjugated to EDTA and phenanthroline to produce sequence specific DNA cleaving reagents when complexed to Fe^{2+} or Cu^+, respectively (12). In view of the DNA binding properties of PNA, it was of interest to prepare analogous conjugates of PNA. We now present PNA-NTA conjugates (NTA = nitrilo triacetic acid) and show that these efficiently and sequence selectively cleave double stranded DNA.

2. Materials & Methods

The PNAs were prepared by coupling of the properly protected NTA to the PNA on the solid support using our standard procedure with TFMSA cleavage/deprotection (13). The synthesis of the di-protected NTA will be reported elsewhere. The PNAs were purified by HPLC and their identity was assured by FAB[+] mass spectrometry.

DNA plasmid preparation and DNA restriction fragment labelling and purification were performed as previously described (6). The DNA cleavage rections were done in 10 mM Na-phosphate buffer pH 6.5.

3. Results and Discussion

The DNA cleavage propensity of the PNAs was examined using an ^{32}P-endlabeled DNA restriction fragment containing a target (AAAAGAAGAA) for the PNA. The cleavage products were analyzed by high resolution polyacrylamide gel electrophoresis.

The results presented in Figure 3 show that an NTA-PNA conjugate (Figure 4) does indeed sequence selectively cleave a double stranded DNA target. Furthermore, the cleavage requires a reducing agent (DTT, lane 3). However, the

136

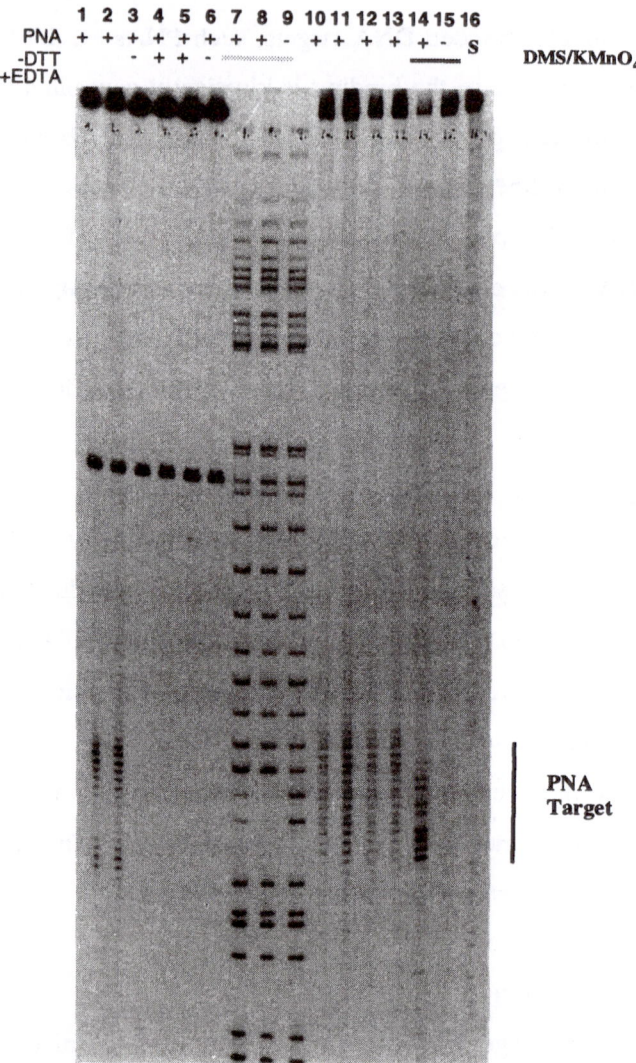

Figure 3. Cleavage of a dsDNA target with an NTA-PNA. The conditions were as described in Materials and Methods and as indicated above each lane. In lanes 1-9, the 3'-^{32}P-endlabelled DNA fragment was used, while the 5'-^{32}P-endlabelled DNA fragment was used in the samples of lanes 10-16. The PNA concentrations were 6 μM, and the incubation time with PNA prior to addition of DDT was 60 min at 37°C. Cleavage was performed for 60 min at 37°C. DMS and KMnO$_4$ probing as well as gel analysis was performed as desribed in (6). Fe^{2+} was added to the samples of lanes 1,4,7,8,13 (0.1 μM), 10 (2 μM) 11, (1 μM) & 12 (0.2 μM). DDT (3mM) was added to the samples of lanes 1,2,4,5 & 10-13, and EDTA (1mM) was addede to those of lanes 4-6.

addition of Fe^{2+} apparently was not necessary (lane 2). On the other hand, the cleavage was completely suppressed in the presence of EDTA (lanes 4-5). Thus, the NTA-PNA conjugate most probably captures sufficient Fe^{2+} from the medium (minute contaminants from buffers and/or glasware).

As a measure of binding of the PNA to the DNA target, we also performed a permanganate and a DMS-protection probing (5) and these results (Figure 3, lanes 7,8 and 14) verified that the PNA did indeed bind to the target by PNA_2-DNA triplex strand displacement.

Surprisingly, using this PNA we observed two regions of cleavage one at either end of the DNA target (Figure 5a). Since, however, PNAs prefer the anti-parallel orientation for binding of the Watson-Crick strand and the parallel orientation for the Hoogsteen strand, and the target would allow for an eight base pair match in the antiparallel orientation of the PNA, these results indicate that a major part of the complexes adopt this configuration thereby having the two PNAs antiparallel. Thus we ascribe the cleavage at the 5'-end of the purine strand to originate from the "Watson-Crick PNA", while the cleavage at the 3'-end is ascribed to the "Hoogsteen PNA" (Figure 5b).

This conclusion is compatible with results obtained using an NTA-bis-PNA (Figure 4) which only cleave the target at the 5'-end (Figure 5a) fully consistent with the predicted preferred binding orientation for such a bis-PNA (8).

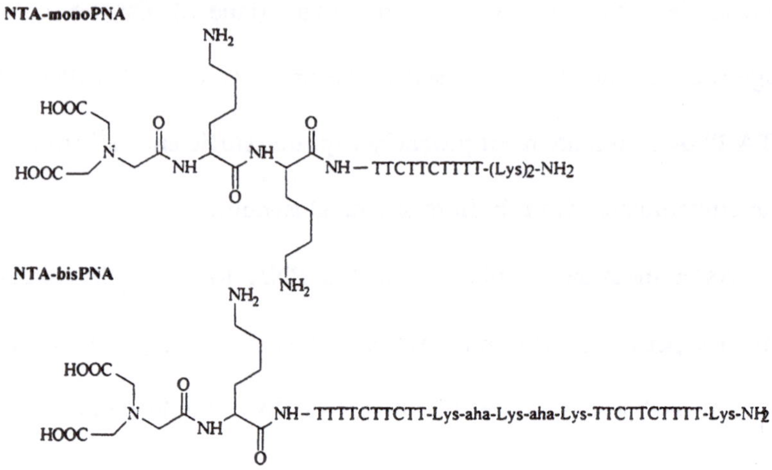

NTA-monoPNA

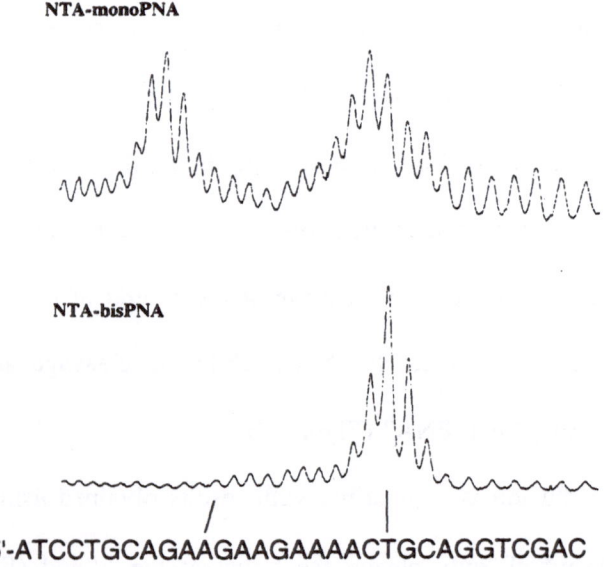

NTA-bisPNA

Figure 4. Structures of the NTA-PNA conjugates employed. Lys = L-lysine, aha = 6-aminohexanoic acid.

NTA-monoPNA

NTA-bisPNA

5'-ATCCTGCAGAAGAAGAAAACTGCAGGTCGAC

Figure 5. (a) Quantitative representation of the cleavage of the purine strand of the dsDNA target by NTA-monoPNA and (b) NTA-bis-PNA (Figure 4).

Mono PNA

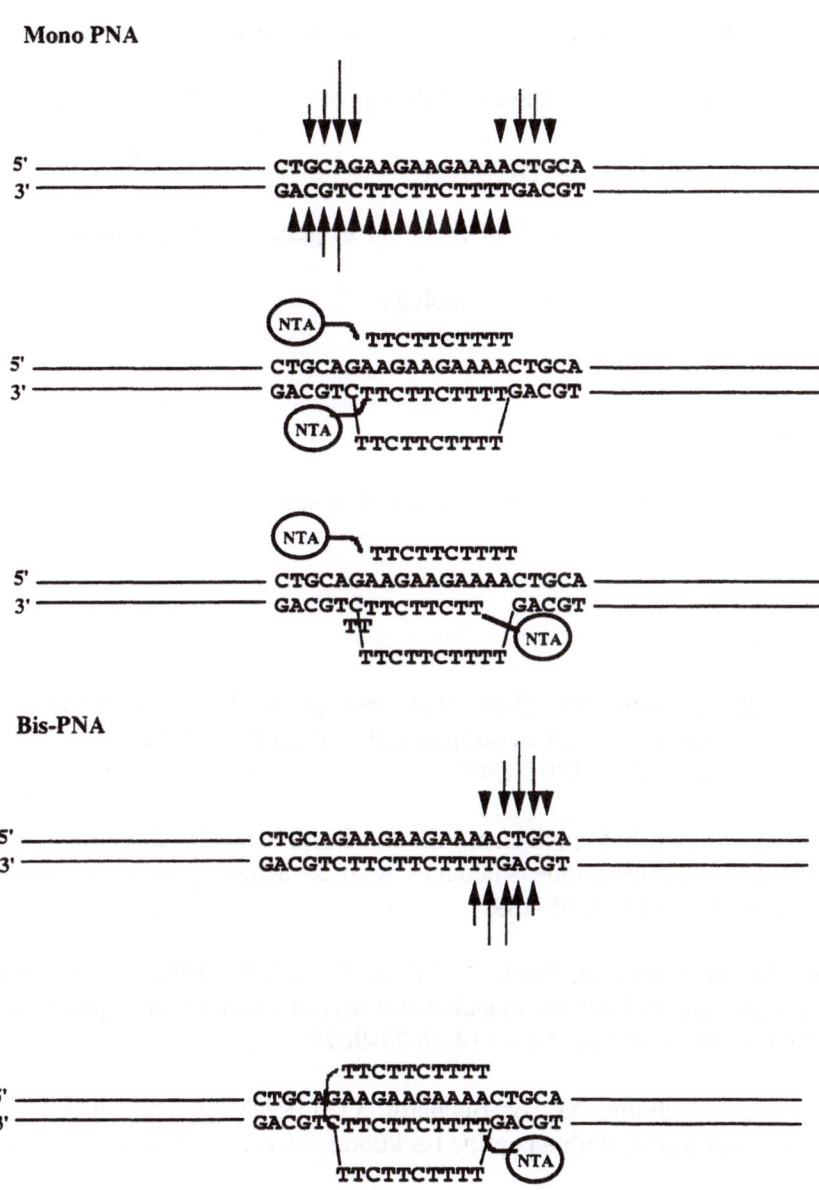

Bis-PNA

<u>Figure 5</u>. (b) Proposed model for the binding and cleavage by the NTA-PNA.

140

In conclusion we have shown that NTA-PNA conjugates in the presence of Fe^{2+} and a reducing agent cleaves dsDNA proximal to the PNA binding site. Thus in analogy to EDTA-or phenanthroline-oligonucleotide conjugates, such NTA-PNA conjugates could find applications in gene mapping and analysis as well as in antisense and anti gene technology.

4. Acknowledgement

This work was supported by the Danish National Research Foundation.

5. References

1. Nielsen, P.E., Egholm, M., Berg, R.H. and Buchardt, O. (1991) Sequence selective recognition of DNA by strand displacement with a thymine-substituted polyamide, *Science* **254**, 1497-1500.

2. Egholm, M., Buchardt, O., Nielsen, P.E. and Berg, R.H. (1992) Peptide nucleic acids (PNA). Oligonucleotide analogues with an achiral peptide backbone, *J. Amer. Chem. Soc.* **114**, 1895-1897.

3. Egholm, M., Buchardt, O., Nielsen, P.E. & Berg, R.H. (1992) Recognition of guanine and adenine in DNA by cytosine and thymine containing peptide nucleic acids (PNA), *J. Amer. Chem. Soc.* **114**, 9677-9678.

4. Nielsen, P.E., Egholm, M., & Buchardt, O. (1994) Peptide Nucleic Acids (PNA). A DNA mimic with a peptide backbone, *Bioconjugate Chemistry* **5**, 3-7.

5. Cherny, D.Y., Belotserkovskii, B.P., Frank-Kamenetskii, M.D., Egholm, M., Buchardt, O., Berg, R.H. and Nielsen, P.E. (1993) DNA unwinding upon strand displacement of binding of PNA to double stranded DNA, *Proc. Natl. Acad. Sci. USA.* **90**, 1667-1670.

6. Nielsen, P.E., Egholm, M. and Ole Buchardt (1994) Evidence for (PNA)₂/DNA triplex structure upon binding of PNA to dsDNA by strand displacement, *J. Mol. Recognition* 7, 165-70.

7. Demidov, V.V., Cherney, D., Kurakin, A.V., Yanilovich, M.V., Malkov, V.A., Frank-Kamenetskii, M.D., Sönnichsen, S.H. and Nielsen, P.E. (1995) Electron Microscopy Mapping of Oligopurine Tracts in Duplex DNA by Peptide Nucleic Acid (PNA) Targeting, *Nucleic Acids Res.* 22, 5218-5222.

8. Egholm, M., Christensen, L., Dueholm, K., Buchardt, O., Coull, J. and Nielsen, P.E. (1995) Efficient pH Independent Sequence Specific DNA Binding by Pseudoisocytosine-containing Bis-PNA, *Nucleic Acids Res.* 23, 217-222.

9. Nielsen, P.E., Egholm. M., Berg, R.H. and Buchardt, O. (1993) Peptide Nucleic Acids (PNA). Potential antisense and anti-gene Agents, *Anti Cancer Drug Design.* 8, 53-63.

10. Nielsen, P.E., Egholm. M., Berg, R.H. and Buchardt, O. (1993) Sequence specific inhibition of restriction enzyme cleavage by PNA, *Nucleic Acids Res.* 21, 197-200.

11. Demidov, V.V., Yavnilovich, M.V., Belotserkovskii, B.P., Frank-Kamenetskii, M.D. and Nielsen, P.E. (1995) Kinetics and mechanism of PNA binding to duplex DNA, *Proc.Natl.Acad.Sci. USA* 92, 2637-2641.

12. Hélène, C. (1993) Sequence-selective recognition and cleavage of double stranded DNA, *Curr. Opinion Biotech.* 4, 29-36.

13. Christensen, L., Fitzpatrick, R., Gildea, B., Petersen, K.H., Hansen, H.F., Koch, T., Egholm, M., Buchardt, O., Nielsen, P.E., Coull, J. and Berg, R.H. (1995) Solid-phase synthesis of peptide nucleic acids (PNA), *J. Peptide Sci.* 3, 175-183.

THE SYNTHESIS AND DNA/RNA HYBRIDIZATION PROPERTIES OF DNG, A PUTATIVE ANTISENSE AGENT WITH AN *ATTRACTIVE* POLYCATIONIC GUANIDINIUM BACKBONE.

ROBERT O. DEMPCY AND THOMAS C. BRUICE

Department of Chemistry, University of California at Santa Barbara

Santa Barbara, California 93106

KEYWORDS/ABSTRACT:　antisense / antigene / DNA / deoxyribonucleic guanidine (DNG) / hybrid duplex / RNA / triple helix / polycationic backbone

Replacement of the phosphodiester linkages of the polyanion DNA with guanidine linkers provides the polycation deoxyribonucleic guanidine (DNG). Polycationic pentameric thymidyl DNG binds with unprecedented affinity to negatively charged ssDNA and ssRNA to provide both double and triple helices. The dramatic stability of these hybrid structures is shown by the denaturation temperatures. For example, the double helix of the pentameric thymidyl DNG and poly(dA) or poly(rA) does not dissociate in boiling water ($\mu = 0.12$) while the T_m for pentameric thymidyl DNA associated with poly(dA) is ca. 13 °C ($\mu = 0.12$). The effect of ionic strength on T_m for DNG:DNA and DNG:RNA hybrids shows an opposite correlation compared to dsDNA or dsRNA and is much more dramatic.

1. Introduction

Putative drugs consisting of oligonucleotide analogs capable of combining with RNA or DNA, thereby arresting cellular processes at the translational or transcriptional level,

B. Meunier (ed.), DNA and RNA Cleavers and Chemotherapy of Cancer and Viral Diseases, 143–162.

are known as antisense and antigene agents.[1]   The backbones of viable antisense/antigene agents do not incorporate phosphodiester linkages because of the susceptibility of this linkage to degradation by cellular nucleases. To be effective, such agents must bind with fidelity to target nucleic acids *via* Watson-Crick and Hoogsteen base pairing.   Since antisense/antigene agents must compete with specific oligonucleotides and proteins for RNA/DNA targets, it is desirable that these agents bind with high affinity to compatible RNA/DNA sequences. The stability of double- and triple-stranded RNA and DNA would increase if the electrostatic repulsion among the polyanionic single strands could be alleviated. This is seen in the enhanced binding of the neutrally charged peptide nucleic acids (PNA) to ssDNA.[2] One might suspect, therefore, that a strand complementary to DNA and connected together by positively charged linkages would act as a particularly effective antisense/antigene agent since the repulsive electrostatic effects in dsDNA would be replaced by attractive electrostatic interactions. On the other hand, the electrostatic bonding between polycationic and polyanionic structures might be quite nonspecific and independent of complementary base pairing. To evaluate the properties of a positively charged DNA analog we synthesized a thymidyl pentamer of deoxyribonucleic guanidine (DNG) in which the phosphodiester linkages $\{-O-(PO_2^-)-O-\}$, found in DNA, are replaced by guanidinium linkages $\{-NH-C(=NH_2^+)-NH-\}$ (**1**, Fig. 1). This DNG model compound binds to complementary DNA and RNA tighter than previously studied antisense agents.

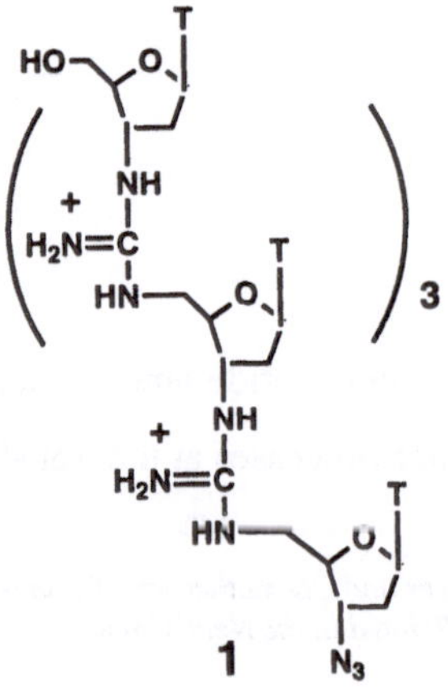

Figure 1.

2. Molecular Modeling

2.1. DNA•DNG COMPLEXES

Model building and computational analysis of DNA•DNG complexes were performed on a Silicon Graphics 4D/320GTX workstation using CHARMm version 22 [adopted-basis Newton-Raphson (ABNR), steepest descents (SD) minimization, DNAH.RTF topology file] and Quanta version 3.3 (Nucleic Acid Builder, Molecular Editor) programs (Molecular Simulations, Burlington, MA). A structure for the thymidyl DNG dimer (Fig. 2) was modeled by using the x-ray crystal structures for the thymidylyl(5'-3')-thymidylate(5') dinucleotide (pTpT) and tri-O-acetyl-ß-D-xylopyranosylazide.[3,4] Replacement of the phosphate of the nucleotide linkage with a guanidinium moiety mandates adjustment of the torsion angle C5'-N5'-C-N3' of the new linkage between the thymidine nucleosides. This angle is near -180° in the model of the DNG dimer and is required for planarity of the guanidine. For comparison, the dihedral counterpart in pTpT, C5'-O5'-P-O3', or (alpha), is between -39° and -73° in

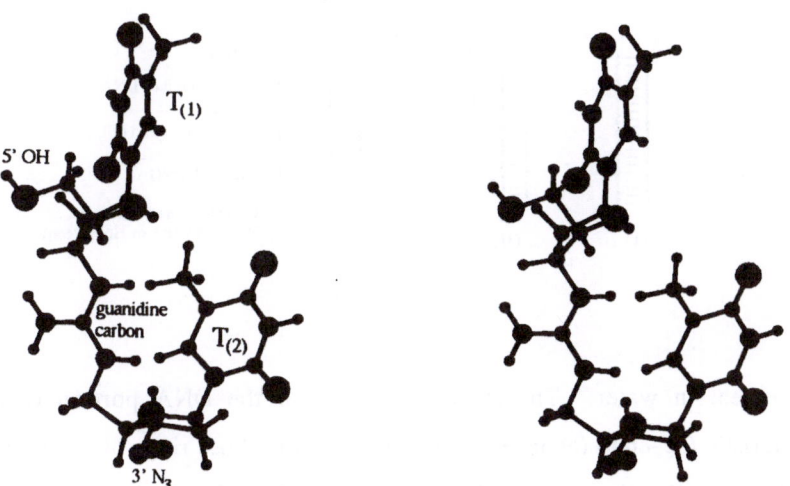

Figure 2. Stereoview of a ball-and-stick model of the thymidyl DNG dimer generated from the x-ray coordinates of thymidyly(5'-3')thymidylate(5') hydrate.

146

typical conformations of DNA and RNA and is -66.3° in the structure of pTpT. The change in the C5'-N5'-C-N3' torsion brought about by energy minimization causes the juxtaposition of the two thymine rings of the DNG dimer to alter, so that the bases are further apart than in the x-ray structure of pTpT. Several options exist for the composition of base-paired oligomeric DNA•DNG. The following are combinations of decameric DNG and DNA strands which we consider at the present time (Scheme 1): *(i)* A hybrid duplex of thymidyl DNG $d(gT)_{10}$ with a d(Ap)10 complement of DNA [$d(gT)_{10}•d(Ap)_{10}$]; *(ii)* a hybrid triplex of double-stranded DNA with a $d(gT)_{10}$ as a Hoogsteen paired strand in the major groove of the DNA [$d(Tp)_{10}•d(Ap)_{10}•d(gT)_{10}$]; and *(iii)* a hybrid triplex $d(Ap)_{10}$ with two strands of $d(gT)_{10}$ [$d(gT)_{10}•d(Ap)_{10}•d(gT)_{10}$].

(i) Fig. 3 shows a model of $d(gT)_{10}·d(Ap)_{10}$ after prolonged energy minimization (2000 steps) in CHARMm. The backbones of the two strands approach one another due to electrostatic interactions between repeating -O-(PO_2^-)-O- and -NH-($C=NH_2^+$)-NH- units. It is interesting that it is the minor groove that contracts. In gas-phase calculations this contraction is from ca. 11.5 Å to ca. 4 Å with retention of Watson-Crick base pairing. These electrostatic interactions should be attenuated to

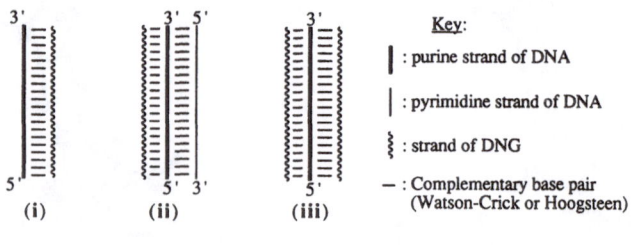

Scheme 1.

some extent in water. The sugar puckering of the DNA portion remains the characteristic C2'-endo (as in B-DNA), whereas the ribose rings of the DNG strands resemble an O4'-endo conformation, with the five-membered rings approaching flat geometries. In the gas-phase calculations a repeating hydrogen bond between N(5')H to O(4') in the same residue is observed. The latter may be responsible for the conversion of the ribose sugar puckering from C(2')-endo toward O(4')-endo.

(ii) Building of $d(Tp)_{10} \cdot d(Ap)_{10} \cdot d(gT)_{10}$ ensued from the structure proposed for the $d(Tp)_{10} \cdot d(Ap)_{10} \cdot d(Tp)_{10}$ triple helix of DNA.[5] The guanidinium ions of the Hoogsteen base-paired $d(gT)_{10}$ strand in the major groove of the decameric double-stranded DNA reside quite near the phosphates of the purine strand, so that $d(Ap)_{10}$ and $d(gT)_{10}$ experience electrostatic attractions between the backbones of one another. Energy minimization resulted in the model presented in Fig. 4. The structure reveals enforced $d(pA)_{10}$ and $d(gT)_{10}$ backbone interactions. The Watson-Crick paired $d(Tp)_{10}$ is remote from the guanidine backbone of $d(gT)_{10}$ and is not involved in electrostatic interactions in the complex to any significant extent.

(iii) We finally consider the possible hybridization of $d(gT)_{10}$ with a purine strand of single-stranded DNA, as well as strand invasion into duplex DNA by $d(gT)_{10}$ A precedent from the recent antisense/antigene literature for antagonistic behavior toward duplex DNA is found in the case of polyamide ("peptide") nucleic acids.[6] In the final complex, a central triple helix is produced leaving an extruded single strand of DNA. Additional electrostatic interactions would be gained, due to the presence of two anionic DNA strands and two cationic DNG strands in the same complex.

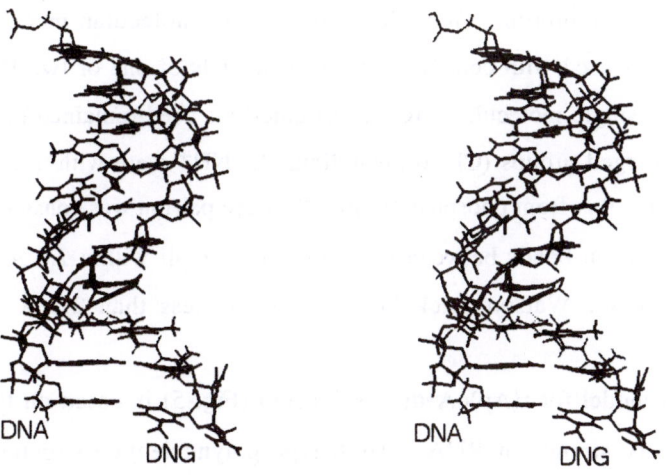

Figure 3. Stereoview of the duplex hybrid of $d(gT)_{10} \cdot d(Ap)_{10}$. The structure was computer generated from the structure of B-[$d(pT)_{10} \cdot d(Ap)_{10}$] by replacing the phosphate linkages with guanidinium linkages and energy minimization via ABNR in CHARMm.

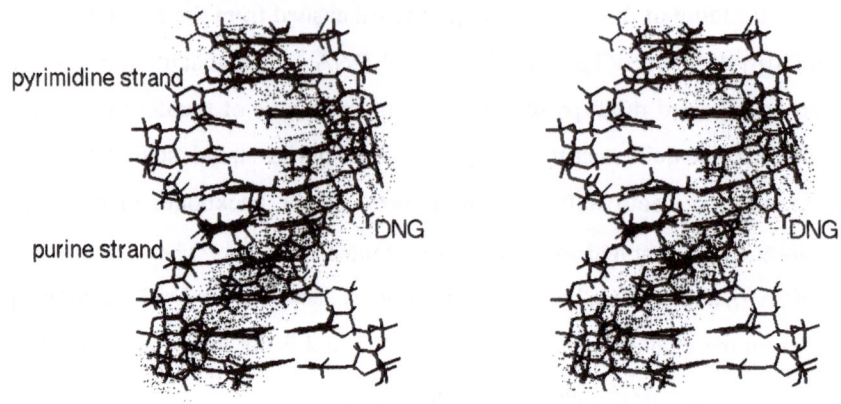

Figure 4. Stereoview of the triple-helical structure of d(Tp)$_{10}$•d(Ap)$_{10}$•d(gT)$_{10}$. The model was created from the coordinates of d(Tp)$_{10}$•d(Ap)$_{10}$•d(Tp)$_{10}$ after Raghunathan *et al.*

2.2. RNA•DNG COMPLEXES

Similar binding arrangements are possible for RNA•DNG complexes. The adopted-basis Newton Raphson minimization algorithm in the molecular mechanics program CHARMm was used with constraints to yield models of one or two DNG molecules annealed to a single molecule of RNA. Counterions used remained proximal to their respective charged groups (Cl$^-$ to guanidinium's -NH$^+$ were within 2.3 Å while Na$^+$ were less than 2.9 Å from phosphate's >PO$_2^-$). Base pairing was considered maintained since the hydrogen bonds between the base pairs are all within 2.0 Å from donor to acceptor atom for Watson-Crick interactions and less than 2.2 Å for Hoogsteen interactions.

A model for r(Ap)$_9$A•d(Tg)$_9$T-azido (Fig. 5) is based on the structure of complementary strands of RNA. An A-type polynucleotide structure is generally maintained with all sugars of both strands in the C3'-endo conformation except for the DNG-terminal sugar which is attached to the azido functionality; this sugar is in the

O4'-endo conformation. In addition, this short and wide structure has an axial rise of only 2.4 Å/nucleotide which also suggests an A-conformation.[7] A comparison of groove widths for the DNG•RNA duplex relative to the RNA•RNA analog shows that the major groove width increases by nearly 2.5 Å while the minor groove contracts by 1.4 Å. This is due to the electrostatic attractions rather than repulsions found in the different nucleic acid backbones.

The model for the triple helical decamer d(Tg)9T-azido•r(Ap)9A•d(Tg)9T-azido presented in Fig. 6 is based on the model of d(pT)$_{10}$•d(pA)$_{10}$•d(gT)$_{10}$.[8] The sugars of the RNA strand are predominately puckered in the C2'-endo or C2'-endo/C4'-endo twist conformations while both strands of DNG are in the C2'-endo conformation, except

RNA

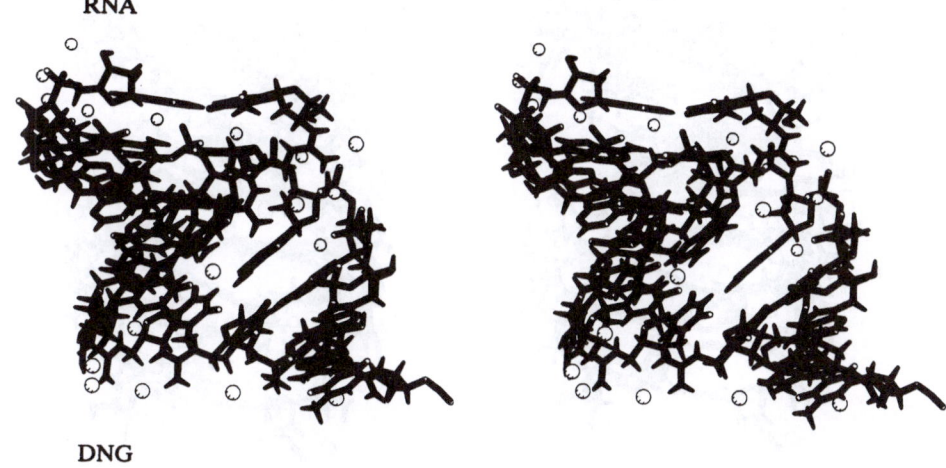

DNG

Figure 5. Stereoview of the duplex hybrid of r(Ap)9A•d(Tg)9T-azido. The model was computer generated from an initial structure of A-d(Ap)9A•d(Tp)9T, converting the H2' groups of the purine strand into hydroxyl groups, and replacing the phosphate linkages of the pyrimidine strand with guanidinium linkages. Explicit Cl$^-$ and Na$^+$ counterions were included. The adopted-basis Newton Raphson algorithm was used to energy minimize using CHARMm v. 21.3 until the root mean square derivative reached < 0.5 kcal/mol•Å.

150

at the azido termini were the sugars are again C4'-endo. This is suggestive of a B-type polynucleotide structure as is the relatively long (3.2 Å/nucleotide) axial rise.[7] The structure of d(Tg)$_9$T-azido•r(Ap)$_9$A•d(Tg)$_9$T-azido is compact relative to both d(pT)$_{10}$•d(pA)$_{10}$•d(gT)$_{10}$[8] and d(pT)$_{10}$•d(pA)$_{10}$•d(pT)$_{10}$.[5] The major and minor grooves of d(Tg)$_9$T-azido•r(Ap)$_9$A•d(Tg)$_9$T-azido are 2.1 and 0.3 Å narrower than those of d(pT)$_{10}$•d(pA)$_{10}$•d(pT)$_{10}$ and 0.3 and 1.4 Å narrower than those of d(pT)$_{10}$•d(pA)$_{10}$•d(gT)$_{10}$.

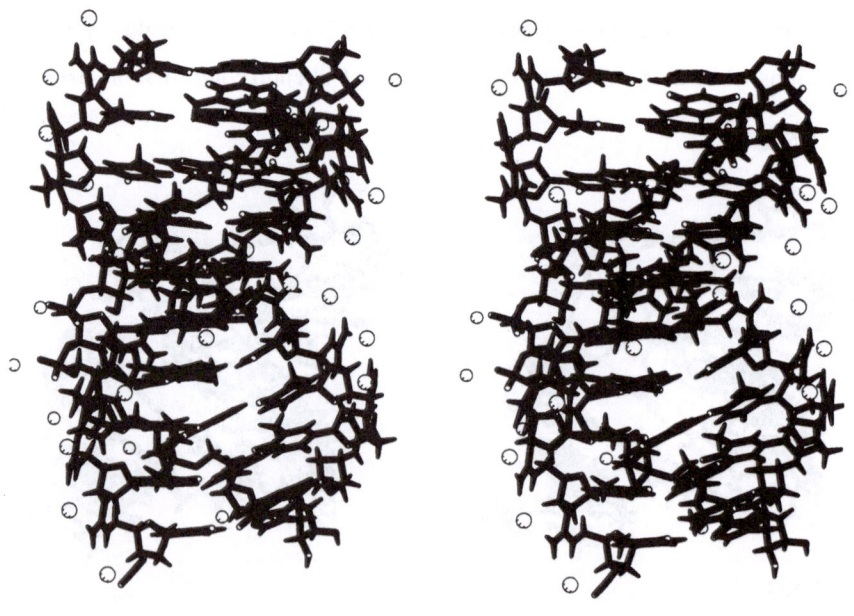

Figure 6. Stereoview of the triple-helical structure of d(Tg)$_9$T-azido•r(Ap)$_9$A•d(Tg)$_9$T-azido.

3. Synthesis Of Thymidyl DNG.

3.1 GENERAL SYNTHETIC PROCEDURE.

The general synthetic strategy for the formation of thymidyl DNG oligomers with a 5-OH terminal begins with the condensation reaction between 3'-amino-3'-deoxythymidine and the thymidyl 5'-isothiocyanate derivative **7** (Scheme 2). Chain extention from the thiourea dimer **8** is then possible by a two-step process involving reduction of the 3'-azido group with H_2S gas in aqueous pyridine, followed by condensation of the resulting amino dinucleoside with another equivalent of **7**. The chain-extending intermediate **7** was synthesized by a seven-step pathway (Scheme 3) starting with 5'-amino-5'-deoxythymidine, which was conveniently prepared by the method of Horwitz et al.[9]

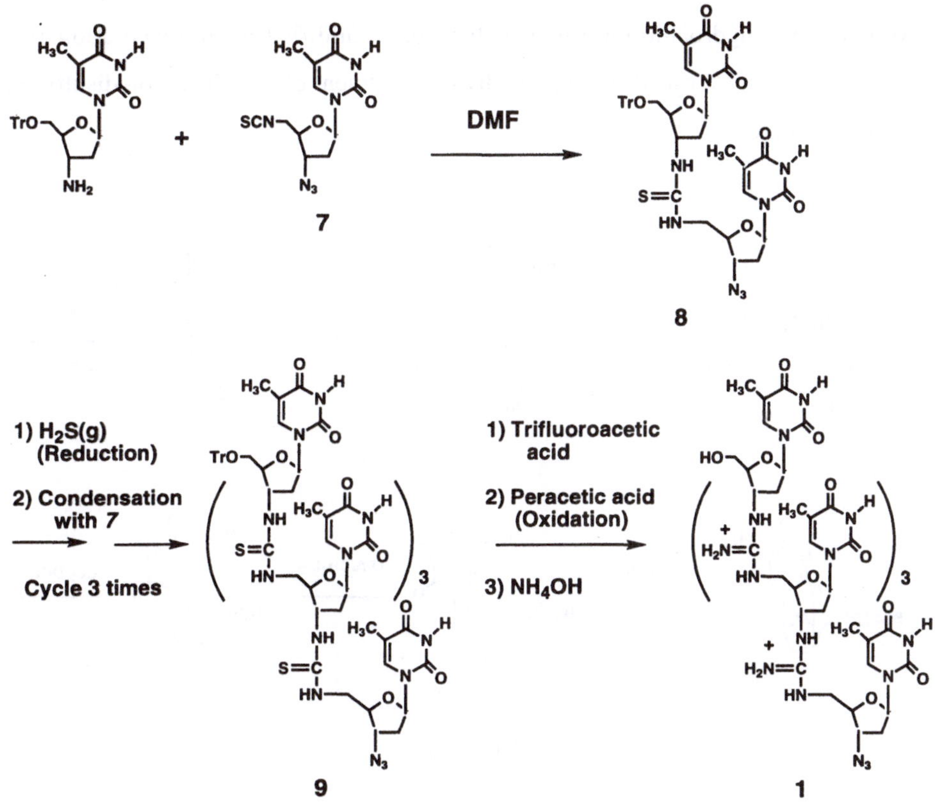

Scheme 2.

Functionalization of the 3'-position with the azido group first required protection of the 5'-amino group. Thus, treatment of 5'-amino-5'-deoxythymidine with di-*tert*-butyl dicarbonate in dry DMF afforded the Boc-protected amino derivative **2** which crystallized from water after evaporation of the DMF. The reaction of **2** with methanesulfonyl chloride in methylene chloride produced the 3'-O -mesyl compound **3** which was then converted to the 2,3'-anhydronucleoside **4** upon treatment with potassium phthalimide in DMF.[10] Ring opening of **4** with sodium azide in DMF-water[10] afforded the desired 3'-azido compound **5**. The IR spectrum of **5** presents an absorbance at 2110 cm^{-1} confirming the presence of the azido functionality. Compound **5** was converted to the isothiocyanate derivative by a two-step process beginning with the acid-catalyzed deprotection of **5** with trifluoroacetic acid in methylene chloride. The product from this reaction, the ammonium trifluoroacetate salt of **6**, proved difficult to solidify and was converted to the hydrochloride salt which readily solidified upon evaporation from ethanol. Compound **7** was then prepared by the treatment of **6** with carbon disulfide in the presence of DCC.

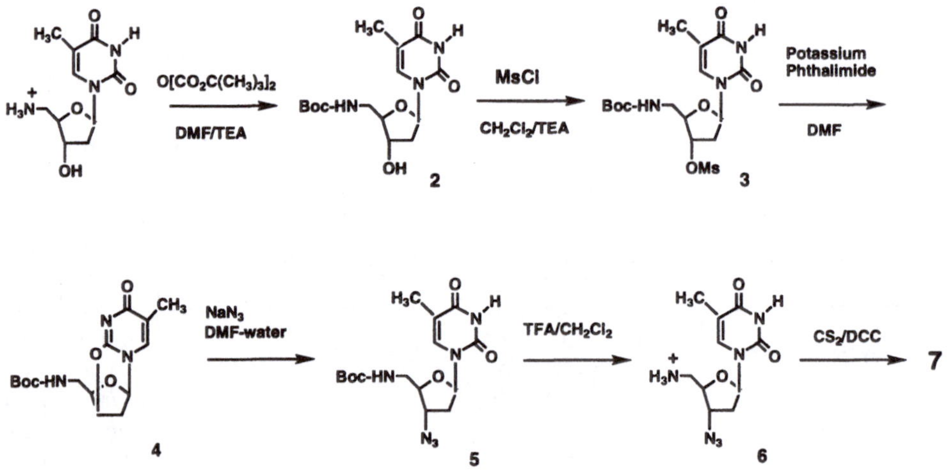

Scheme 3.

The thiourea linkages in **9** were then converted to guanidinium linkages (Scheme 2) by a two-step process involving oxidation of the thiourea linkages with peracetic acid (forming the aminoiminosulfonic acid derivative) followed by amidation to form the DNG oligomer d(Tg)₄T-azido (**1**).

Purification of **1** was carried out by preparative HPLC using an *Alltech WCX* cation exchange column employing 0.80 M ammonium acetate buffer, pH 5.0, as the mobile phase. Ammonium acetate was removed from the isolated DNG sample by repetitive evaporations from water. Evidence for the successful conversion of the thiourea linkages to the corresponding guanidinium linkages was established by ^{13}C NMR (the guanidinium carbons are observed around 155 ppm) and mass spectral analysis {m/e 1328.60 (M + H)$^+$, calc. 1328.56}.

3.2. ELECTROPHORETIC MOBILITY

The electrophoretic behavior of DNG migrating through a polyacrylamide gel was compared to the migration of DNA (Fig. 7). Both dimeric and pentameric DNG molecules migrate towards the cathode while the analogous DNA oligonucleotides travel towards the anode, the longer oligonucleotides migrating at a faster rate. This is as expected for molecules that are essentially identical except for net charge (e.g. +4 vs -4 for 1 and d(Tp)4T, respectively). The relative magnitudes of migration are similar, although not the same, for DNG and DNA with the same number of bases. This slight

Figure 7. The electrophoretic mobilities of the DNG oligonucleotides 1 and d(TgT)-azido (lanes 1 and 2) and the DNA oligonucleotides d(TpT) and d(Tp)4T (lanes 3 and 4). Between 1 and 2 x 10^{-7} moles (in bases) of each oligonucleotide were loaded into adjacent wells in the center (Origin) of the gel.

discrepancy could be due to the increased rigidity of the guanidinium vs the phosphodiester backbone, or to the phosphate moiety's greater mass (38 Daltons per linkage). Since the oligonucleotides are migrating mainly by charge, the difference in mass between DNG and DNA should have a negligible influence on their relative electrophoretic migrations. Thus, the difference in oligonucleotide backbone flexibility is likely the main factor in the differing migration of DNG vs DNA.

4. Thermal Denaturation Analysis of DNG Complexes with DNA/RNA.

4.1. DENATURATION OF DNA•DNG COMPLEXES.[11]

In the thermal denaturation analysis of 1 bound to poly(dA), two distinct hyperchromic shifts were observed in the UV spectra at high ionic strength (μ = 0.22, 0.62, and 1.2; Fig. 8a). In analogy to DNA and RNA thermal denaturation analysis,[12] these transitions correspond to the denaturation of a triple-helical hybrid at 68, 41 and 36 °C followed by denaturation of the duplex at 79, 70 and 71 °C. At μ = 0.12, only one hyperchromic shift was seen centered at 85 °C; the thermal stability is apparently so great that the duplex structure does not denature at near boiling temperatures. This result contrasts with the denaturation profiles of d(Tp)$_{15}$T bound to poly(dA); all have sharp denaturation transitions (39, 41, 52, and 56 °C at μ = 0.12, 0.22, 0.62, and 1.2, respectively; plots not shown). There was no hyperchromic shift in the UV spectra of a solution between ca. 5 - 95 °C which contained 1 and either p(dG)$_{11-18}$, poly(dI), poly(dC) or poly(dT) (examined at both μ = 0.12 and 1.2, pH 7.0, Fig. 8b) that could be attributed to denaturation of a complex with 1 {although there was a hyperchromic shift centered ca. 40 °C that was due to the denaturation of poly(dC):poly(dC); this phenomenon has been reported.[13] This is evidence against DNG binding to DNA in a nonspecific manner. Therefore, we conclude that DNG recognizes its complementary base pairs while dramatically increasing its affinity for DNA.

(a)

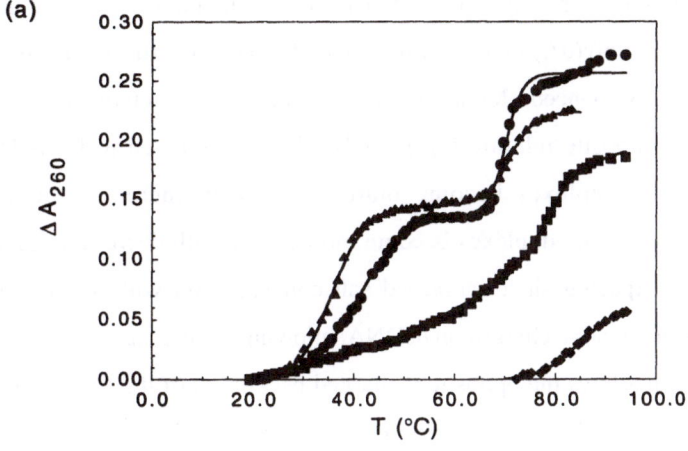

(b)

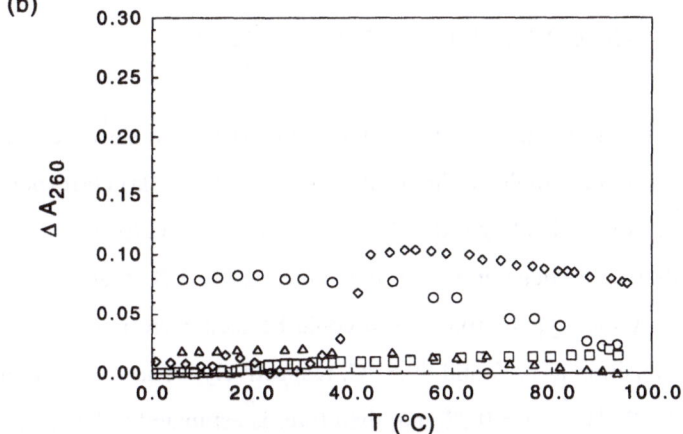

Figure 8. (a) Plots of the change of A_{260} *vs* T (°C) for **1** in the presence of poly(dA) at pH 7.0 (0.01 M K_2HPO_4). The ionic strength (μ) was held constant at 0.12 ($\blacklozenge$), 0.22 ($\blacksquare$), 0.62 ($\bullet$), and 1.20 ($\blacktriangle$) with KCl. (b) Plots of the change of A_{260} *vs* T (°C) for **1** in the presence of $p(dG)_{12\text{-}18}$ (O), poly(dI) ($\square$), poly(dC) ($\lozenge$), or poly(dT) ($\triangle$) are presented at $\mu = 1.2$.

4.1.1. *Ionic Strength Dependency Studies of DNA •DNG Complexes.*

In order to be able to compare our thermal denaturation results for DNG with DNA and modified oligonucleotides, the results were reduced to the unit terms of T_m/base pair and T_m/linkage. The plot of T_m/base pair vs ionic strength (Fig. 9) clearly exemplifies the differences between the binding of DNG to DNA compared to the analogous DNA:DNA complex. Not only does thymidyl DNG have a tremendously higher affinity than thymidyl DNA for poly(dA) under a wide range of ionic strengths, but the effect of ionic strength is more pronounced. Most noteworthy is the observation that the effect of ionic strength has an opposite relationship for DNG:DNA compared to DNA:DNA. Thus, while DNA:DNA duplexes become more stable with increasing ionic strength, DNG:DNA triplexes and duplexes become more stable with decreasing ionic strength. This should be expected since increased salt concentration will effectively mask the opposing rows of negative charges on dsDNA, allowing a more stable duplex. Decreased salt concentration allows the oppositely charged backbones of the DNG:DNA hybrid to be intimately salt paired, thus stabilizing the complex.

4.2. DENATURATION OF RNA•DNG COMPLEXES.[14]

The thermal denaturation analysis of **1** bound to poly(rA) also shows two distinct UV hyperchromic shifts for samples at high ionic strength (Fig. 10a) corresponding to the denaturation of a triple-helical hybrid at 44.4, 50.2, and 53.2 °C ($\mu = 0.62$, 0.82 and 1.2, respectively) followed by denaturation of the duplex at 90.7, 90.4, and 82.4 °C. At $\mu = 0.22$ and 0.42, only one hyperchromic shift could be seen corresponding to a triple- to double-helical transition at 62.5 and 49.5 °C, respectively. The second transitions did not commence by 93 °C at $\mu = 0.22$ and, therefore, is estimated to be ≥ 100 °C; at $\mu = 0.42$, a second transition was not complete by 93 °C and was tentatively assigned to be ca. 100 °C. Similar to the DNA•DNG hybrid, at low ionic strength ($\mu = 0.12$), **1** does not dissociate from poly(rA) at temperatures as high as 93 °C. Evidence against DNG binding to RNA in a nonspecific manner was provided by the observation that no hyperchromic shift was seen in the UV spectra of a solution between ca. 1 - 93 °C which contained **1** and either poly(rG), poly(rC), poly(rU), or poly(rI) (examined at both $\mu = 0.12$ and 1.2, pH 7.0) (Fig. 10b).

4.2.1. *Ionic Strength Dependency Studies of RNA •DNG Complexes.*
The plot of T_m/base pair *vs* ionic strength (Fig. 9) demonstrates the differences between the thermal stability of DNG•RNA and the analogous DNA•RNA complexes. Thymidyl

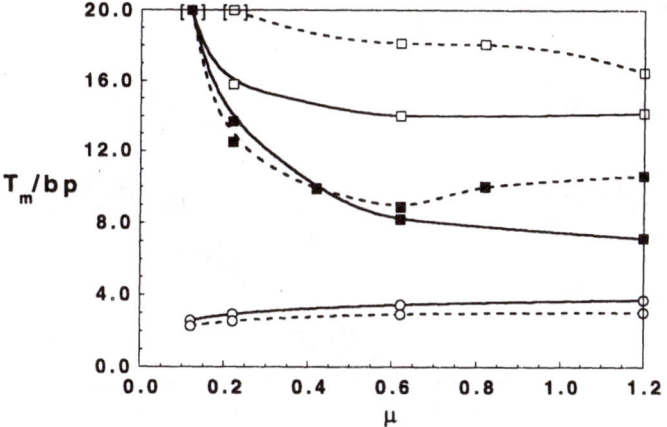

Figure 9. A plot of T_m/base pair vs ionic strength (μ varied with different [KCl]) at pH 7.0 (0.01 M KHPO$_4$ buffer) for denaturation of **1** (■ and ❑ for the first and second transitions, respectively) and d(Tp)$_{15}$T (○) complexed to poly(dA) (——) and poly(rA) (- - -). The data point at the lowest ionic strength for the nucleic complexes with DNG was taken as 100 °C because no hyperchromic transition was apparent up to 93 °C.

DNG binds much tighter than thymidyl DNA to poly(rA) over a wide range of ionic strengths. Again, the effect of ionic strength on the DNG•RNA complexes is opposite and much more pronounced compared to DNA•RNA complexes with the DNG•RNA triplexes and duplexes becoming more stable with decreasing ionic strength.

4.3. EQILIBRIUM COMPLEXES OF THYMIDYL DNG WITH POLY(ADENYL) NUCLEIC ACIDS.

To complement the stoichiometry of **1** binding to poly(rA) and poly(dA) as determined from the thermal denaturation studies, the method of continuous variation[15] was used to generate mixing curves of the absorbance *vs* mol percent of d(Tg) (single guanidinium-linked 2'-deoxyribothymidyl unit). This method is based on the assumption that the decrease in absorbance is proportional to the number of base pairs hydrogen bonded generate mixing curves of the absorbance *vs* mol percent of d(Tg) (single guanidinium-

158

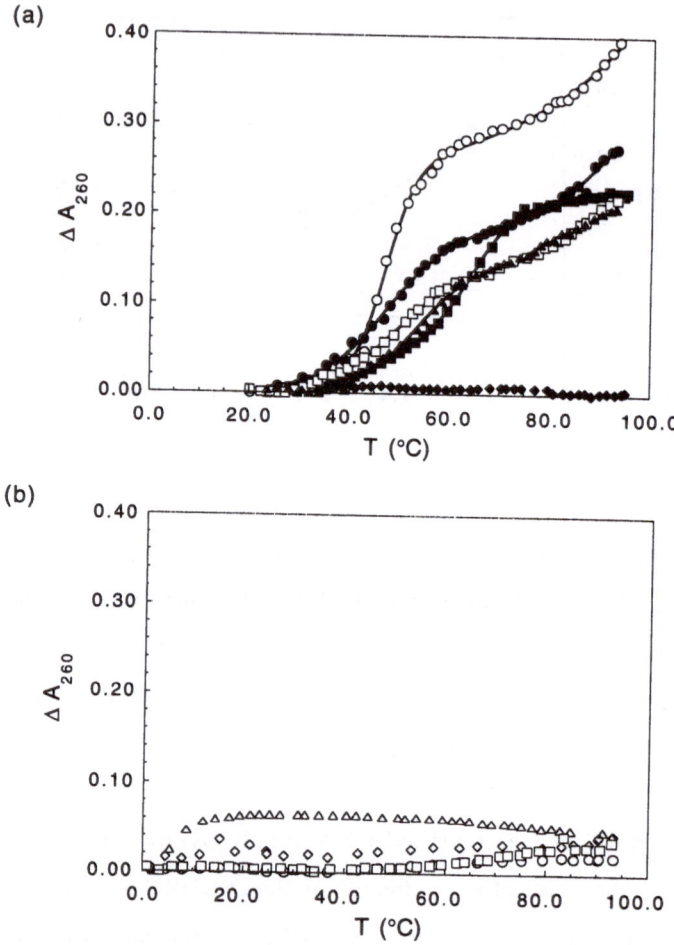

Figure 10. (a) Plots of the change of ΔA_{260} vs T (°C) for 1 in the presence of poly(rA) at pH 7.0 (0.01 M KHPO$_4$). The ionic strength (μ) was held constant at 0.12 ($\blacklozenge$), 0.22 ($\blacksquare$) 0.42 (O), 0.62 ($\bullet$), 0.82 ($\square$), and 1.20 ($\blacktriangle$) with KCl. (b) Plots of ΔA_{260} vs T (°C) for 1 in the presence of poly(rG) (O), poly(rI) ($\square$), poly(rC) ($\lozenge$), or poly(rU) (Δ) at pH 7.0 (0.01 M KHPO$_4$ buffer) and $\mu = 1.2$. The concentration of each oligonucleotide was 4.17×10^{-5} M in bases.

linked 2'-deoxyribothymidyl unit). This method is based on the assumption that the decrease in absorbance is proportional to the number of base pairs hydrogen bonded between the interacting species. Mixtures of 1 with poly(rA) at $\mu = 0.12$ and 30 °C

(Fig. 11) reach a minimum absorbance at a mol fraction of ca. 0.67 d(Tg) to 0.33 of r(Ap) (single phosphate-linked riboadenyl unit). These results indicate that triple stranded complexes were formed which contained two d(Tg) per r(Ap). The fact that the minima occur at the intersection of two straight lines through the data points is indicative of a reversibly formed structure consisting of no unoccupied sites.[16] At 60 and 90 °C, the data points for the mixing curves of d(Tg) with r(Ap) again form two

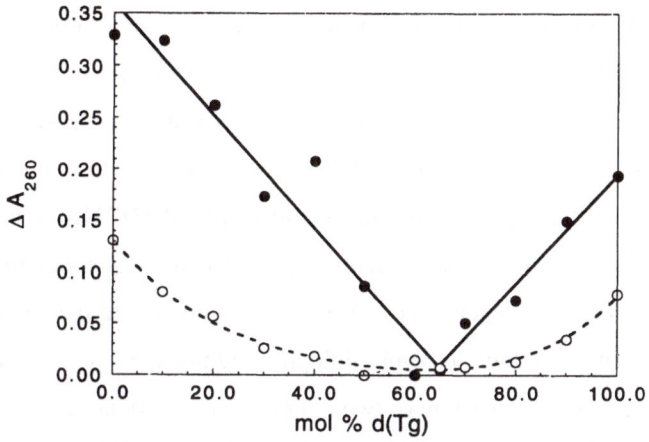

Figure 11. Plots of ΔA_{260} vs mol % of d(Tg) (mixing curves) for **1** with poly(rA) (●) and poly(dA) (○) at 30 °C, pH 7.0, $\mu = 0.12$. The inflection points indicate the stoichiometry of the complexes.

lines that intersect sharply at ca. 67 % (data not shown). This confirms the stable triple helical nature of the DNG•RNA complexes at $\mu = 0.12$ suggested by the lack of hyperchromic shifts in the thermal denaturation studies from 30 to 90 °C.

The mixing curve for d(Tg) with d(Ap) (single phosphate-linked 2'-deoxyriboadenyl unit) at 30 °C and $\mu = 0.12$ is also centered near a mol fraction of ca. 0.67 d(Tg) to 0.33 d(Ap) (Fig. 11). Unlike the experiments with r(Ap), this mixing curve has a rounded minimum indicating that the dissociation reaction is too slow to allow overlapping sequences to adjust such that the maximum number of sites could be occupied[14]. The smaller change in absorbance seen when d(Tg) interacts with d(Ap) compared to r(Ap) (Fig. 10) is also indicative of some binding sites remaining unoccupied. Equilibrium complexes are the same from 30 to 90 °C since the mixing curves at 60 and 90 °C (data not shown) for d(Tg) mixing with d(Ap) are similar to results at 30 °C.

5.   Conclusion.

From the above discussion, the following conclusions can be drawn about DNG: (i) thymidyl DNG is specific for its complementary tracts of adenine bases and does not interact with guanylic, cytidylic, or uridylic tracts; (ii) due to electrostatic attractions in the place of electrostatic repulsions, DNG binds to RNA with a much greater affinity than DNA to RNA; (iii) the thermal stability of DNG•RNA hybrid structures is attenuated by increasing salt concentrations; (iv) DNG appears suited for use as either an antigene or an antisense agent since, from results obtained to date, it invariably forms triple helical structures under near-physiologic conditions (37 °C, pH 7.0, and $\mu = 0.22$); (v) while a triple helix consisting of two d(Tg) to one r(Ap) has a similar stability to the corresponding DNG$_2$•DNA complex, a DNG duplex with RNA is more stable than the corresponding DNG•DNA complex; and (vi) molecular modeling suggests that the DNG strands take on the general conformation of the nucleic acid backbone to which they bind and that the overall structure of the hybrid complexes is more compact than that found in the homopolymeric RNA complexes.

Recently we have completed the synthesis of the positively charged RNA analog polyadenosyl RNG.   Currently work is focused toward comparing the properties of RNG and DNG.

6.   References

1.   (a) Uhlmann, E. and Peyman, A. (1990) Antisense oligonucleotides: A new therapeutic principle, *Chemical Reviews*, **90**, 543-584. (b) Crooke, S. T. (1992) Therapeutic applications of oligonucleotides, *Annu. Rev. Pharmacol. Toxicol.* **32**, 329-

376. (c) Crooke, S. T. (1992) Progress toward oligonucleotide therapeutics: pharmacodynamic properties *FASEB J.* **7**, 533-539. (d) Cook, P. D. (1993) *Antisense Research and Applications*; CRC, Boca Raton, pp 149-187. (e) Sanghvi, Y. S.; Cook, P. D. (1993) *Nucleosides and Nucleotides as Antitumor and Antiviral Agents*; Plenum, New York, ; pp 311-324.

2. (a) Nielsen, P. E., Egholm, M., Berg, R. H.; Buchardt O. (1991) Sequence-selective recognition of DNA by strand displacement with a thymine-substituted polyamide, *Science* , **254**, 1497-1500. (b) Almarsson, Ö.; Bruice, T. C.; Kerr, J.; Zuckermann, R. (1993) Molecular mechanics calculations of the structures of polyamide nucleic acid DNA duplexes and triple helical hybrids, *Proc. Natl. Acad. Sci. USA* , **90**, 7518-7522.

3. Brooks, B. R., Bruccoleri, R. E., Olafson, B. D., States, D. J., Swaminathan , S. and Karplus, M. (1983) CHARMm: A program for macromolecular energy, minimization and dynamics calculations, *J. Comput. Chem.* **4**, 187-217.

4. Camerman, N., Fawcett, J. K. and Cameraman, A. (1976) Molecular structure of a deoxyribose-dinucleotide, sodium thymidylyl-(5'-3')-thymidylate-(5') hydrate (pTpT), and possible structural model for polythymidylate, *J. Mol. Biol.* **107**, 601-621.

5. Raghunathan, G., Miles, H. T. and Sasisekharan, V. (1993) Symmetry and molecular structure of a DNA triiple helix: $d(T)_n \cdot d(A)_n \cdot d(T)_n$, *Biochemistry* **32**, 455-462.

6. Almarsson, O. and Bruice, T. C. (1993) Peptide nucleic acid (PNA) conformation and polymorphism in PNA-DNA and PNA-RNA hybrids, *Proc. Natl. Acad. Sci. USA* **90**, 9542-9546.

7. Saenger, W. (1984) *Principles of Nucleic Acid Structure*, Springer-Verlag, pp. 228-241.

8. Dempcy, R. O., Almarsson, O. and Bruice, T. C. (1994) Design and synthesis of deoxynucleic guanidine: a polycation analogue of DNA, *Proc. Natl. Acad. Sci. USA* **91**, 7864-7868.

9. Horwitz, J. P., Tomson, A. J., Urbanski, J. A. and Chua, J. (1962) Nucleosides. I. 5'-Amino-5'deoxyuridine and 5'-deoxythmidine, *J. Org. Chem.* **27**, 3045-3048.

10. Glinski, R. P., Khan. M. S., Kalamas, R. L., Sporn, M. B. (1973) Nucleotide synthesis. IV. Phosphorylated 3'-amino-3'-deoxythymidine and 5'-amino-5'-deoxythymidine and derivatives. *J. Org. Chem.* **38**, 4299-4305. The 3'-azido analog in this reference was reduced with Pd/C, $H_2(g)$ in ethanol.

11. (a) Dempcy, R. O., Browne, K. A. and Bruice, T. C. (1995) Synthesis of the polycation thymidyl DNG, its fidelity in binding polyanionic DNA/RNA, and the stability and nature of the hybrid complexes, *J. Amer. Chem. Soc.* **117**, 6140-6141. (b) Dempcy, R. O. Browne, K. A. and Bruice, T. C. (1995) Synthesis of a thymidyl pentamer of deoxyribonucleic guanidine and binding studies with DNA homopolynucleotides, *Proc. Natl. Acad. Sci. USA* **92**, 6097-6101.

12. Riley, M., Maling, B. and Chamberlin, M. J. (1966) Physical and chemical characterization of two- and three-stranded adenine-thymine and adenine-uracil homopolymer complexes, *J. Mol. Biol.* **20**, 359-389.

13. Akinrimisi, E. O., Sander, C. and Ts'o, P. O. P (1963) Properties of helical polycytidylic acid, *Biochemistry* **2**, 340-344.

14. Browne, K. A., Dempcy, R. O. and Bruice, T. C. (1995) Binding studies of cationic thymidyl deoxyribonucleic guanidine to RNA homopolynucleotides, *Proc. Natl. Acad. Sci. USA* **92**, 7051-7055.

15. Job, P. (1928) Formation and stability of inorganic complexes in solution, *Ann. Chim.* **9**, 113-203.

16. Felsenfeld, G. (1958) Theoretical studies on the interaction of synthetic polyribonucleotides, Biochim. Biophys. **29**, 133-144.

DESIGN OF PHOTOACTIVATABLE DNA-CLEAVING AMINO ACIDS WITH HIGH SEQUENCE SELECTIVITY

Guanine-Guanine Stacking Rule for Electron-Transfer from DNA

I. SAITO, M. TAKAYAMA, H. SUGIYAMA,
AND T. NAKAMURA

*Department of Synthetic Chemistry and Biological Chemistry, Faculty of
Engineering, Kyoto University
Kyoto 606, Japan*

1. Introduction

A number of attempts have so far been made for the design of artificial DNA-cleaving
molecules that are chemically stable and activatable by photostimulation, since such
photoactivable DNA-cleaving molecules can be used to probe nucleic acid structure, as
designed "photonucleases", as photo-footprinting agents, and as a potent photodrug for
cancer phototherapy.[1,2] While the photochemical DNA cleavage has been observed
with various types of DNA binding molecules, the detailed chemistry associated with
photoinduced DNA cleavage is not well understood.[1] Our laboratory has demonstrated
several different types of photochemical DNA-cleaving molecules[2], including so called
"photo-Fenton reagent" [3,4] and a light-inducible DNA alkylating agent[5] as typically
exemplified below.

B. Meunier (ed.), DNA and RNA Cleavers and Chemotherapy of Cancer and Viral Diseases, 163–176.
© 1996 Kluwer Academic Publishers. Printed in the Netherlands.

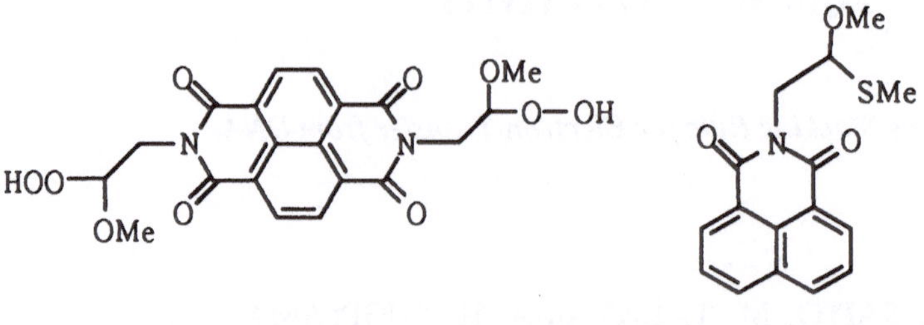

photo-Fenton reagent   light-inducible DNA alkylating agent

We are particularly interested in the design of photoactivable DNA-cleaving amino acids that can be used as a DNA-cleaving moiety for the design of DNA-cleaving polypeptides and hybrid molecules. We describe herein novel water-soluble L-lysine derivatives possessing a naphthalimide chromophore that can induce efficient and highly sequence selective cleavage of double-stranded DNA upon photoirradiation at 320-380 nm.[6] By incorporating a nitro group into the naphthalimide ring, the selectivity of DNA cleavage was dramatically changed from a specific cleavage at the 5' side of guanine-guanine (-GG-) sequences to a thymine (T) specific cleavage.

2. Photoactivatable DNA-Cleaving Amino Acids (PCA)[6]

In our efforts to design practically useful DNA-cleaving amino acids, we incorporated a 1,8-naphthalimide chromophore into the ε-amino group of L-lysine. Treatment of Boc-L-lysine methyl ester with commercially available 1,8-naphthalic anhydride followed by deprotection provided water-soluble L-lysine derivative 1 having an intense UV absorption (log ε at 366 nm). Nitro-substituted lysine derivatives 2 (log ε at 366 nm) and 3 (log ε at 366 nm) were also synthesized in a similar fashion, since certain aromatic nitro compounds have been known to cleave DNA by photoirradiation.[7]

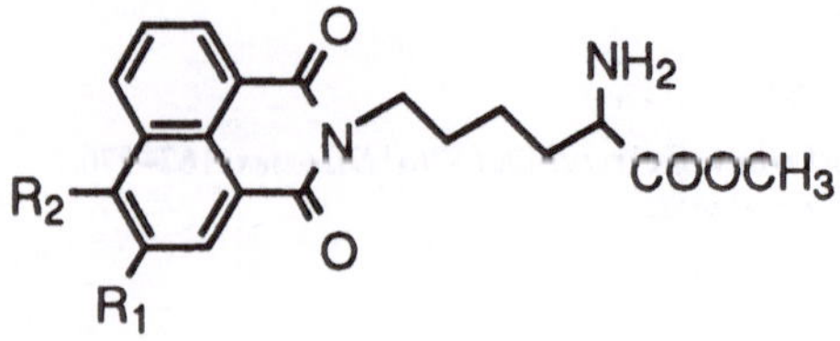

1   $R_1 = R_2 = H$

2   $R_1 = NO_2$, $R_2 = H$

3   $R_1 = H$, $R_2 = NO_2$

The DNA-cleaving properties of **1**, **2**, and **3** were examined using supercoiled circular pBR322 (form I) DNA under photoirradiation with a transilluminator (366 nm) at 0 °C for 1 h. In each case, efficient single-strand breaks and a small amount of double-strand breaks were observed. The base and sequence specificity of DNA cleavage was analyzed using ^{32}P-end-labeled double-stranded DNA fragments. Lysine derivative **1** induced highly specific DNA cleavage at the 5' side of 5'-GG-3' steps with a very weak cleavage at the 5' side of 5'-GA-3' steps after piperidine treatment of the photoirradiated mixture: no cleavage was observed at other sites, including single G residues. Control experiments established that photoirradiation, piperidine treatment and double-helical structure of DNA are all indispensable for the specific 5'-GG-3' cleavage. In marked contrast, under the same conditions 3-nitro derivative **2** photonicked the double-stranded DNA preferentially at T residues with a relatively weak cleavage at 5'-GT-3' steps after piperidine treatment: neither 5'-GG-3' nor G cleavage has been observed in this case. In the photoiradiation of 4-nitro derivative **3** both 5'-GG-3' and T cleavage occurred almost equally after piperidine treatment. Thus, the sequence selectivity of the double strand cleavage is highly dependent upon substituents and the substitution pattern on the aromatic ring. In none of these cases occurred the DNA cleavage without piperidine treatment.

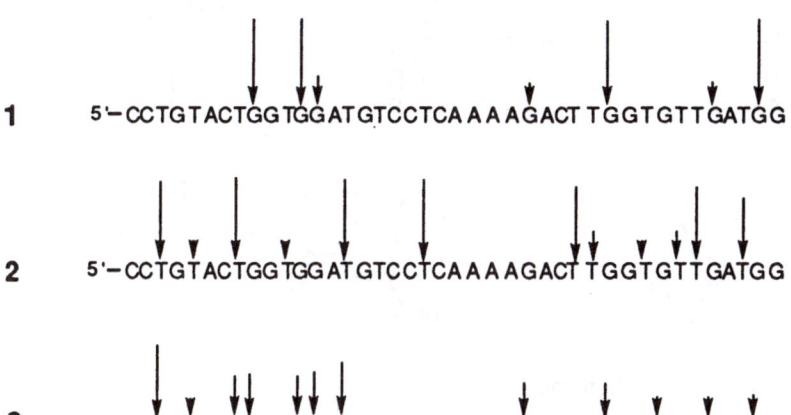

Figure 1 Histogram representation of the photo-induced cleavage of ^{32}P-5'-end-labeled DNA fragments in the presence of **1**, **2** and **3**. Arrow represents the extent of cleavage by photoirradiation (366 nm, 20 min at 0 °C) followed by piperidine treatment.

The high sequence-selectivity observed for **1** and **2** was completely lost in the photoreaction of heat-denatured single-stranded DNA. Thus, almost equal cleavage occurred at all G residues in the photoreaction of **1** with single-stranded DNA, whereas the cleavage at both T and G residues occurred in the presence of **2**. The equal G cleavage of single-stranded DNA by **1** may be due to the oxidation of guanine base with singlet oxygen, since the single-strand breaks induced by **1** were enhanced more than 2-fold in D$_2$O and inhibited by the addition of singlet oxygen quencher, NaN$_3$, whereas the 5'-GG-3' specific cleavage of double-stranded DNA was not affected in D$_2$O.

To get insight into the mechanism of the T-specific cleavage by **2**, we examined the photoreaction of **2** with T monomer and T-containing deoxyoligonucleotides. Irradiation of **2** in the presence of 3',5'-O-dibenzoylthymidine (**4**) (5 equiv) in acetonitrile gave 5-formyldeoxyuridine (d^{f}U) derivative **5** (14%) and deoxyuridine **6** (17%) together with reduction product **7** (36%) (Scheme 1). Deoxyuridine **6** was proven to be derived from further photoirrdiation of **5** via photodecarbonylation.

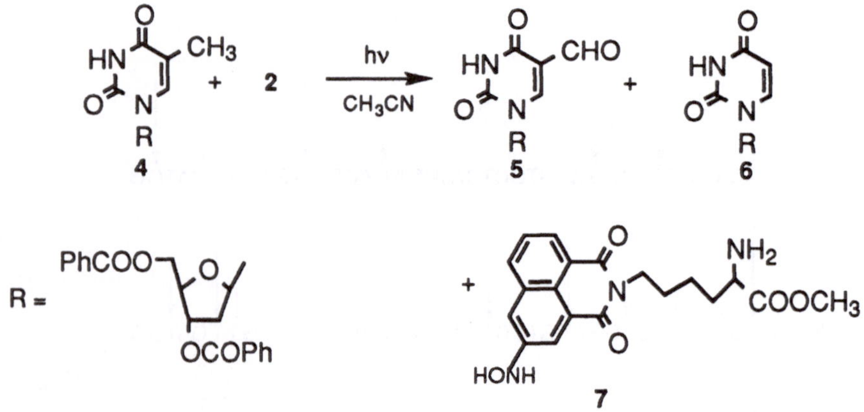

Scheme 1

Photoirradiation of **2** in the presence of deoxytrinucleotide d(ATA) (2 equiv) in sodium cacodylate buffer (pH 7.0) gave d(A^{f}UA) in 65% yield. The structure of d(A^{f}UA) was confirmed by converting it to d^{f}U and dA (2 equiv) by enzymatic digestion with snake venom phosphodiesterase (s.v. PDE) and alkaline phosphatase (AP). Treatment of isolated d(A^{f}UA) with 1 M piperidine at 90 °C for 20 min gave 5-formyluracil (^{f}U) (52%) and Ap plus pA. Likewise, photoirradiation of **2** (100 mM) and poly[dA]/poly[dT] (200 mM) in sodium cacodylate buffer (pH 7.0) at 0 °C for 1 h followed by enzymatic digestion with s.v. PDE and AP produced d^{f}U (55%) and deoxyuridine (11%). These results clearly indicate that the first step of the T-specific cleavage of double-stranded DNA by **2** is an oxidative transformation of T methyl group into formyl group initiated by the hydrogen abstraction from T methyl group by photoexcited nitro group. Heating ^{f}U-containing sites in DNA (90 °C, 20 min) eventually resulted in a DNA strand scission via the mechanism shown in Scheme 2 as evidenced by the thermal degradation of d(A^{f}UA).

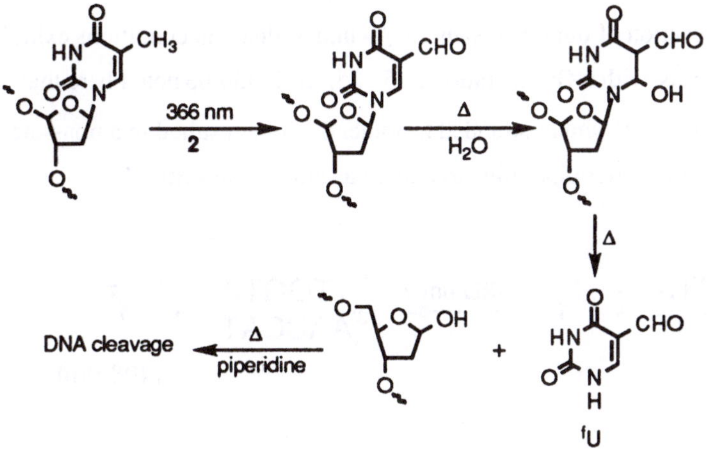

Scheme 2

3. Photoinduced DNA Cleavage via Electron Transfer[8]

The $5'$-GG-$3'$ specific cleavage induced by **1** was assumed to proceed via electron transfer from the electron-rich -GG- steps to photoexcited **1**, since naphthalimide derivatives such as **1** are well known as an electron-acepting photosensitizer . In order to obtain more quantitative data on the -GG- specific photocleavage, we examined the photoreaction of **1** with oligodeoxynucleotides containing a -GG-step in the middle.[8]

Irradiation of **1** and duplex hexamer $5'T_1T_2G_3G_4T_5A_6$ (**8**) / $5'$TACCAA (**9**) in sodium cacodylate buffer (pH 7.0) with a transilluminator (366 nm) at 0 °C followed by treatment with piperidine (90 °C, 20 min) and subsequent dephosphorylation with alkaline phosphatase (AP), produced TT and GTA as the major products together with small amounts of TTG and TA, while the complementary strand **8** was unchanged (Scheme 3). The formation of TT and GTA from **8** implies that hexamer **8** was degraded by reacting with photoexcited **1** to give an alkali-labile site at G_3, whereas TTG and TA were derived from the photoreaction at G_4 as a minor pathway. The G_3/G_4 ratio (84 : 16) determined by HPLC was not significantly changed whether the photoirradiation was conducted under aerobic conditions or under a nitrogen atmosphere. Riboflavin-sensitized photoreaction of duplex hexamer **8 / 9** under identical conditions exhibited a similar G_3 selectivity with a G_3/G_4 ratio of 75 : 25. It should be noted here that photoirradiation of **1** with single-stranded hexamer **8** alone resulted in a non-selective cleavage at G_3 and G_4 after piperidine treatment as already described.[6]

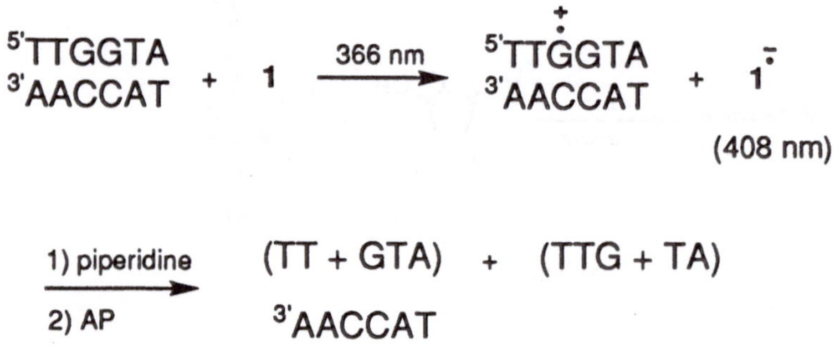

$$5'\text{TTGGTA} \atop 3'\text{AACCAT} \quad + \quad \mathbf{1} \quad \xrightarrow{366\ nm} \quad {}^{5'}\overset{+}{\underset{\cdot\cdot}{\text{TTGGTA}}} \atop {}^{3'}\text{AACCAT} \quad + \quad \mathbf{1}^{\bar{\cdot}}$$

$$(408\ nm)$$

$$\xrightarrow[\text{2) AP}]{\text{1) piperidine}} \quad \begin{array}{c} (TT + GTA) \ + \ (TTG + TA) \\ {}^{3'}\text{AACCAT} \end{array}$$

Scheme 3

The limiting quantum yield ($\phi = 3.0 \times 10^{-4}$) for the disappearance of **8** in duplex **8 / 9** in the presence of **1** was obtained under anaerobic conditions using monochrometer (350 nm) by changing oligomer concentrations. Photoirradiation of **1** with other deoxyhexanucleotides containing a -GT- or a -GC- step in the middle, such as self-complementary duplex (TAC<u>GT</u>A)$_2$, (AT<u>GC</u>AT)$_2$ and (TA<u>GC</u>TA)$_2$, under similar conditions resulted in no appreciable consumption of the hexamers, whereas (ATC<u>GA</u>T)$_2$ was reacted with a quantum yield of 1.5×10^{-4} to give ATC and AT almost quantitatively after piperidine and AP treatment. Photoreaction of **1** with duplex heptamer TT<u>GGG</u>TA (**10**)/ TACCCAA (**11**) under the conditions proceeded more efficiently ($\phi = 5.2 \times 10^{-4}$) to give G$_3$- and G$_4$-cleavage products from **10**. Thus, the reactivity of G-containing duplex oligomers toward photoexcited **1** increased in the order, -GGG- > -GG- > -GA- >> -GC-, -GT-.

The laser flash photolysis studies of transient intermediates in the photoreaction of **1** with DNA oligomers were next examined. Fortunately, the luminescence and transient properties of 1,8-naphthalimide derivatives have already been studied, and the triplet-triplet absorption (λ_{max} 475 nm) and the radical anion (λ_{max} 415 nm, $\varepsilon = 27000 \pm 2500$ dm^3 mol^{-1} cm^{-1} in dry acetonitrile) of N-phenyl-1,8-naphthalimide have already been assigned by time-resolved spectroscopy.[9] The nanosecond laser flash photolysis was then conducted with argon-purged solutions containing **1** and various donors. The laser flash photolysis of **1** (0.1 mM) and DABCO (1.0 mM) in acetonitrile-water (1 : 1) resulted in the formation of the radical anion **1**$^{-\cdot}$ (λ_{max} 408 nm) as a transient species which decayed on a time scale longer than that of the triplet (λ_{max} 475 nm) as previously reported.[9] A very similar transient-time profile has been obtained in the laser flash photolysis of **1** in the presence of duplex hexamer **8 / 9** in acetonitrile-water at pH 7.0 (Figure 2). The growth of the absorption of **1**$^{-\cdot}$ at 408 nm occurred exactly in the same time interval as did the triplet decay, implying that the triplet state is the precursor of the radical anion. The quenching rate constant of the triplet state of **1** by **2** giving **1**$^{-\cdot}$ via electron transfer was estimated to be 5.3×10^7 M^{-1} s^{-1}. It is noteworthy that (i) the triplet state of **1** and radical anion **1**$^{-\cdot}$ are the only transients detected on the time scale, and that (ii) **1**⁻· decayed according to second-order kinetics ($k_r = 5.6 \times 10^9$ M⁻¹ s⁻¹). A similar result has also been obtained in the laser flash photolysis of **1** in the presence of calf thymus DNA.

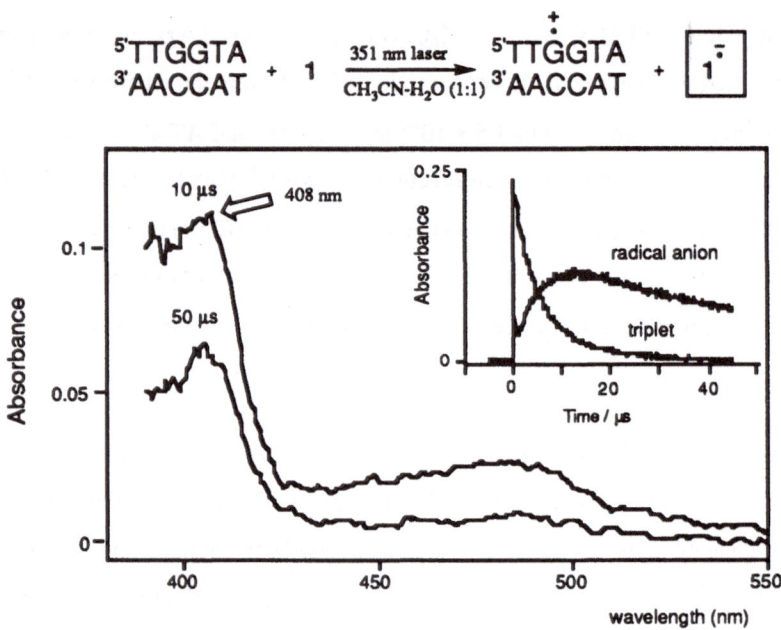

Figure 2. Transient absorbance spectra of the intermediate radical anion (**1**⁻·) recorded at 10 and 50 μs delays after a 351 nm excimer laser pulse. [**1**] = 0.1 mM, [**2**] and [**3**] = 4 mM (strand concentration) in acetonitrile-water (1:1). Inset : Transient-time profiles following laser flash excitation. Triplet decay at 475 nm and the radical anion built-up at 408 nm and decay.

Electron transfer from a -GG- step in duplex DNA to triplet excited **1** has now been demonstrated for the first time by direct observation of the electron transfer intermediate.[8] While the electron loss center created in DNA has long been known to ultimately end up at guanine due to its lowest ionization potential among the nucleobases,[10] the present studies indicate that the most electron-donating sites in duplex DNA are the guanine residues located 5' to guanine.

4. Guanine-Guanine Stacking Rule for Electron-Donating Sites in B DNA

There has long been great interest in the one-electron oxidations of DNA in connection with DNA damage caused by ionizing radiation, oxidizing agents and photooxidation with endogeneous photosensitizers.[11] As is well-known, guanine is the most readily oxidizable base among the nucleobases as evident from the experimentally observed ionization potentials (IP)[12] (adiabatic, G = 7.77 eV, A = 8.26 eV, C = 8.68 eV, T = 8.87 eV) and the aqueous-phase oxidation potentials (E^0_{ox} vs SCE, GMP = 1.30 V, AMP = 1.40 V, TMP = 1.50 V, CMP = 1.65 V). It has thus been suggested that the electron loss center created in DNA ultimately ends up at guanine residues via hole migration due to its lowest IP.[10] While certain types of photoinduced DNA cleavage being occurred at G residues are suggested to proceed via one-electron transfer mechanism,[13] there has so far been no direct evidence for the electron-transfer process.

In our recent paper we reported direct evidence for electron transfer from a guanine base in duplex DNA to an electron-accepting photosensitizer by means of excimer laser falsh photolysis and demonstrated that the most readily oxidizable sites in duplex DNA are the G residues located 5' to G, due to the π-stacking intereaction of the two bases.[8] We have now examined the photo-induced DNA cleavage reaction of ^{32}P-end labeled DNA fragments using four different electron-accepting photosensitizers, riboflavin[14, 16], water-soluble benzophenone derivatives 12 and 13.

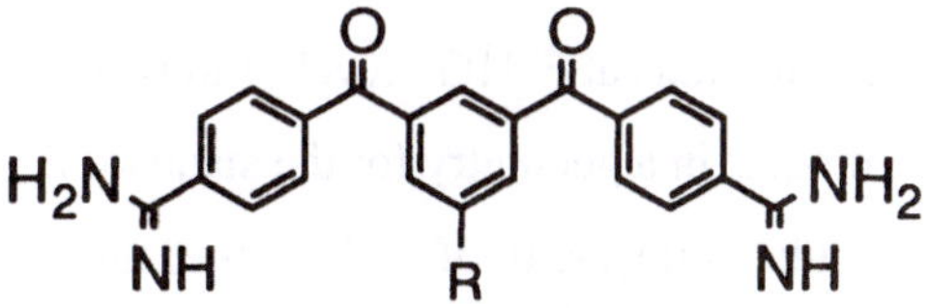

12 R = H

13 R = tert-butyl

Careful examination of the DNA cleavage data obtained from photoirradiation (366 nm) and subsequent piperidine treatment (90 °C, 2 min) reavealed a general trend for the susceptibility of G-containing sites to photoinduced one-electron oxidation, regardless of the structures of the photosensitizers used (Figure 3). i) The DNA

cleavage occurred most easily at the 5' side of $5'$-GG-$3'$ steps with a very weak
cleavage at -GA- steps; no cleavage occurred at other sites including single G residues
(2). ii) Among the -GG- containing sites, DNA clevage occurred more intensively at
the following sequences, -GGGG- > -GGG- > -CGG- ~ -AGG- > -TGG-.

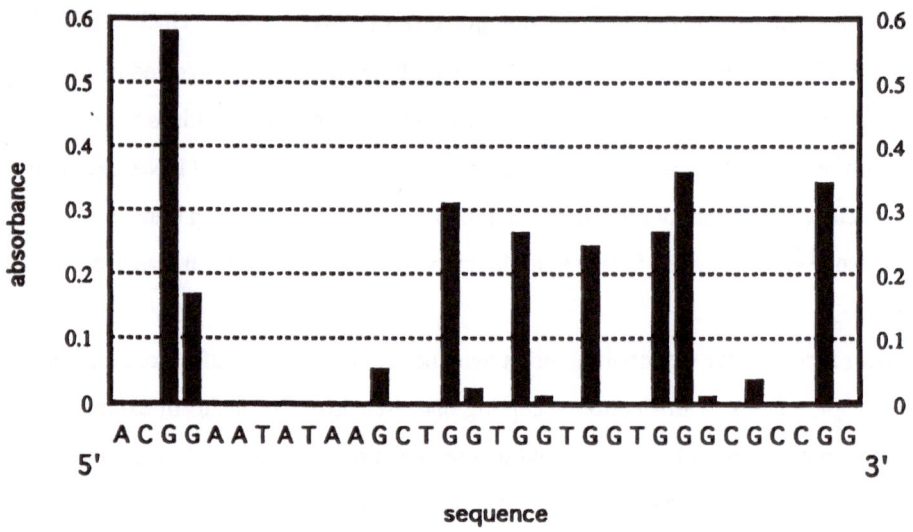

Figure 3. Photo-induced clevage of ^{32}P-5'-end-labeled DNA fragment in the presence of **1**.
After photoirradiation (366 nm, 20 min at 0 °C) and subsequent piperidine treatment (90 °C, 20
min), the DNA fragment was analyzed by gel electrophoresis. The bands were assayed by laser
densitometry.

We have then carried out *ab initio* MO calculations at 6-31G* level of lowest
ionization potentials of stacked nucleobase models in a geometry for the standard B
DNA strucure at the 6-31G* level.[8] The calculated lowest IP of such stacked nucleobase
models are in the following order: $5'$-GGG (7.07 eV) < $5'$-GG (7.28 eV) < $5'$-GA (7.51
eV) < $5'$-GC (7.68 eV) ~ $5'$-GC (7.68 eV) < G (7.75 eV). This order is in good
agreement with the experimentally observed reactivity toward one-elctron oxidation. Of
special importance is that the HOMO of the stacked $5'$-GG- or $5'$-GGG- site is always
localized ovewhelmingly on the G at the 5' side, which is also compatible with the
observed 5' side selectivity for DNA cleavage. The susceptibility of each G of
-$5'$G$_1$G$_2$G$_3$- sequence to the photocleavage increased in the order, G$_2$ > G$_1$ >> G$_3$

with a $G_2 : G_1$ ratio of 58 : 42 This is inconsistent with the calculated HOMO. However, the *ab initio* calculations indicated that the most stable cation radical is -G_1-$G_2^{+\cdot}$-G_3- due to the stacking stabilization by the two G bases (figure 4). By considering the stability of G cation radical and the HOMO, the most susceptible sites of $(G)_n$ sequences to one-electron oxidations are the G being stacked with two G from both sides.

(a)   (b)

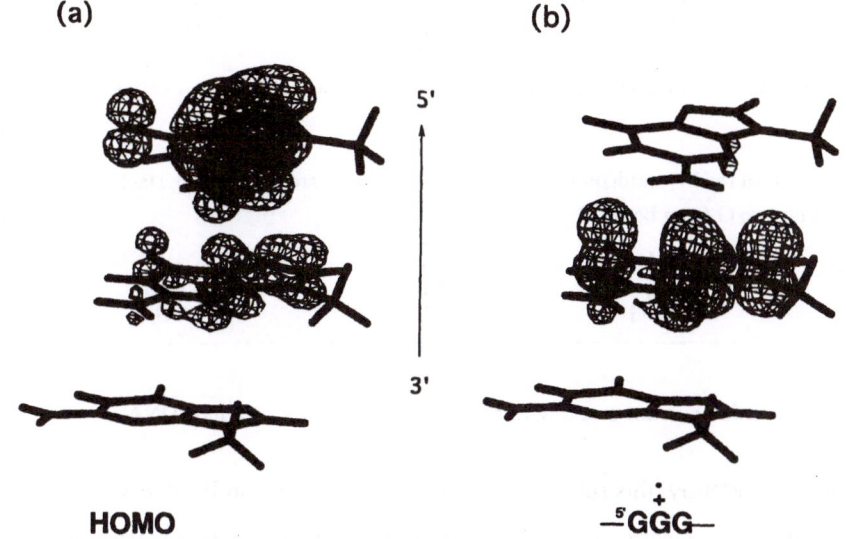

HOMO $-^5GGG-$

Figure 4. (a) The HOMO and (b) the most stable cation radical of $^{5'}$-GGG-$^{3'}$ obtained by ab initio (6-31G*) calculation.

On the basis of these experimental results and *ab initio* calculations, we propose a very important and general rule for predicting most readily oxidizable sites in B DNA toward one-electron oxidation. The rule, referred to as "Guanine-Guanine (G-G) Stacking Rule" (Figure 5), shows the DNA sites which are most susceptible to one-electron oxidations, such as those in ionizing radiation and various types of chemical, biological and photochemical oxidations.

1. Electron-donating ability of G-containing sequences increases in the following order.
5'-GGGG- > 5'-GGG- > 5'-GG- >> 5'-AG- ~ 5'-GA- > -G-
2. The largest HOMO of (G)n sequences is always localized on the G at 5' side.
5'-GGGG-
5'-GGG-
5'-GG-
3. Electron-transfer from the G at 3' side of (G)n sequences is least likely to occur.
4. The most readily oxidizable sites of (G)n sequences are the G being stacked with two G from both sides
5'-GGGG-
5'-GGG-
5'-GG-

Figure 5. Guanine-guanine stacking rule for electron-donating sites in B DNA.

Most importantly, this rule shows the electron-rich sites in B DNA which are most susceptible to one-electron oxidations. The G-G stacking rule is also very important in understanding HOMO-LUMO or donor-acceptor type intereactions of DNA, *e.g.*, charge-transfer (CT) intereaction of electron-accepting molecules with DNA. In such a HOMO-LUMO or a CT interaction, the 5' side of -GG- or -GGG- sequence is the most electron-releasing and therefore most strongly interacting site.

References

1. For a review, Morrison, H. Ed. (1990) *Bioorganic Photochemistry, vol 1, Photochemistry and the Nucleic Acid,* John Wiley and Sons, New York.
2. Saito, I. (1992) Photochemistry of highly organized biomolecules: Sequence-selective photoreaction of DNA, *Pure & Appl. Chem.* **64,** 1305-1310.
3. Matsugo, S., Kawanishi, S., Yamamoto, K., Sugiyama, H., Matsuura, T. and Saito, I. (1991) Bis-(hydroperoxy)naphthaldiimide as a "Photo-Fenton Reagent": Sequence-Specific Photocleavage of DNA, *Angew. Chem. Int. Ed. Engl.* **30,** 1351.

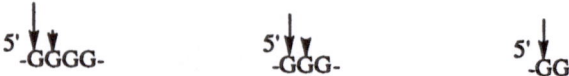

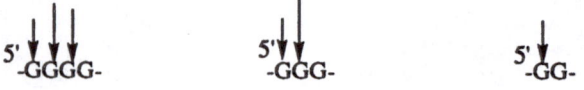

4. Matsugo, S. Kodaira, K. and Saito, I. (1993) Transfecting activity of photoirradiated φX 174 DNA in the presence of hydroperoxynaphthalimides, *Biooragnic & Medicinal Chem. Lett.* **3**, 1671-1674, and references therein.

5. Saito, I., Takayama, K. and Sakurai, T. (1994) Photogeneration of carbocation via intramolecular electron transfer: photoinduced DNA alkylation, *J. Am. Chem. Soc.* **116**, 2653-2654.

6. Saito, I., Takayama, K., Kawanishi, S. (1995) Photoactivatable DNA-cleaving amino acids: Highly sequence-selective DNA photocleavage by novel L-lysine derivatives, *J. Am. Chem. Soc.* **117**, 5590-5591.

7. For an example, Nielsen, P. E., Jeppesen, C., Egholm, M., and Buchardt, O. (1988) Photochemical cleavage of DNA by nitrobenzamides linked to 9-aminoacridine, *Biochemistry*, **27**, 6338-6343, and references therein.

8. Saito, I., Takayama, M., Sugiyama, H., Nakatani, K., Tsuchida, A. and Yamamoto, M. (1995) Photoinduced DNA cleavage via electron transfer: Demonstration that guanine residues located 5' to guanine are the most electron-donating sites, *J. Am. Chem. Soc.* **117**, 6406-6407.

9. Demeter, A., Biczok, L., Berces, T., Wintgens, V., Valat, P. and Kossanyi, J. (1993) laser photolysis studies of transient processes in the photoreduction of naphthalimides by aliphatic amines, *J. Phys. Chem.* **97**, 3217-3224.

10. Steenken, S., Pedro, L. (1993) Electron transfer in di-(deoxy)nucleoside phosphates in aqueous solution: Rapid migration of oxidative damage to guanine, *J. Am. Chem. Soc.* **115**, 2437-2440.

11. Steenken, S. (1989) Purine bases, nucleosides, and nucleotides: Aqueous solution redox chemistry and transformation reactions of their radical cations and e⁻ and OH adducts, *Chem. Rev.* **89**, 503-520.

12. Orlov, V. M., Smirnov, A. N., and Varshavsky, Y. M. (1976) Ionization potentials and electron-donor ability of nucleic acid bases and their analogues, *Tetrahedron Lett.* 4377-4378.

13. For an example. Cadet, J., Berger, M., Buchko, G. W., Joshi, P. C., Raoul, S., and Ravanat, J-L. (1994) A novel and predominant radical oxidation product of 3',5'-di-O-acetyl-2'-deoxyguanosine, J. Am. Chem. Soc. 116, 7403-7404.

14. Ito, K., Inoue, S., Yamamoto, K. and Kawanishi, S. (1993) 8-Hydroxydeoxyguanosine formation at the 5' site of 5'-GG-3' sequences in double-stranded DNA by UV radiation with riboflavin, *J. Biol. Chem.* **268,** 13221-13227.

RATIONAL DESIGN OF A NEW CLASS OF DNA CLEAVER BASED ON PREDICTION OF MINOR GROOVE BINDING INTERACTIONS AND NMR STRUCTURAL STUDIES

Amanda Wilton[‡], Seyed Sadat-Ebrahimi[‡], John A. Parkinson[§], Julie Andrews[‡], Kenneth T. Douglas[‡1]

‡ Department of Pharmacy, University of Manchester, Manchester, M13 9PL, U.K.

§ Department of Chemistry, University of Edinburgh, West Mains Road, Edinburgh EH9 3JJ, Scotland

Abstract

Variants of Hoechst 33258 (1) have been designed and synthesised to contain catecholic sites, their structures being based on high-field NMR spectroscopic data for the solution structures of d(CGCGAATTCGCG)$_2$ with Hoechst 33258 (1) and its meta-hydroxy analogue, meta-Hoechst (2) . In the present study, the predicted solution structure of the 3,4-dihydroxy Hoechst (cat-Hoechst, 3)

---

[1] To whom communications should be addressed

B. Meunier (ed.), DNA and RNA Cleavers and Chemotherapy of Cancer and Viral Diseases, 177–193.

was confirmed by high resolution NMR spectroscopic techniques including DQFCOSY, TOCSY and NOESY. Under these conditions **3** bound to duplex d(CGCGAATTCGCG)2 in a similar manner to **2,** with the 3-hydroxy group directed down into the minor groove in an orientation consistent with the formation of the predicted hydrogen bonds with the C=O group of C9 and NH$_2$ group of G4′, as found for **2**. Molecular modelling indicated that in this orientation it would be difficult for a Cu^{2+} ion to enter the chelating dihydroxy site of the catechol of **3** whilst **3** remained bound to DNA and that, if free **3** pre-formed a complex with Cu(II), the resulting copper:catechol site would clash with the DNA minor groove. To surmount this problem, an additional hydroxy-group was introduced to give **4** (3,4,5-trihydroxy-Hoechst), in which the 3-hydroxy group was planned to make the hydrogen bond(s) to DNA described above, with the 4,5-dihydroxy (catechol) site pointing out of the minor groove, allowing Cu(II) binding in the DNA:ligand complex. In agreement with this, **4** was found to cleave plasmid DNA on activation by Cu^{2+}, whereas **1-3** did not cleave under identical conditions.

1. Introduction

The design of sequence-specific minor groove directed ligands presents several chemical challenges. Whatever its free solution conformation, the ligand must be able to adapt its shape to the helical contours of the minor groove and its charge distribution must remain appropriate. Strongly binding minor groove ligands fit intimately to the groove. The natural minor groove antibiotics, netropsin and distamycin, have been used as prototypes and have spawned synthetic-sequence readers, most notably the isolexins and lexitropsins [1,2].

Relative to enzymology, the field of rational ligand design for DNA is in its infancy. One feature of ligand design which has become clear from enzyme-based work, once the details of binding have been elucidated, is that attempting to structurally modify a lead inhibitor with too many new potential binding sites usually results in a compromise for their utilisation. In extreme cases even a small change in ligand structure can cause major changes in the bound conformation at an enzyme site. This is lucidly illustrated for dihydrofolate reductase by the 180° twist found in the conformation at the pteridine ring for bound dihydrofolate compared to methotrexate [3,4].

A major question which our group is addressing is the **predictability** of the design process for DNA-directed ligands. Can specific interactions be designed into a ligand, in an isolated and testable fashion, and be studied in structural and energetic detail? Added to this, we asked whether we could incorporate a novel chemical cleavage system, intimately juxtaposed to the minor groove.

1.1 THE FIRST STAGE : THE DESIGN OF BINDING INTERACTIONS

To introduce sites capable of specific and predicted interactions with DNA, and also suitable for detailed structural testing, required a molecule whose binding to DNA was already well-defined and could be studied in depth. We chose the Hoechst 33258 framework (**1**) as its binding to DNA has been studied by many physical techniques, such as X-ray diffraction [5], fluorescence spectroscopy

[6]. Our own laboratory had developed a very detailed 3D model of the solution structure of the Hoechst 33258 complex with d(CGCGAATTCGCG)$_2$ by use of NMR spectroscopy [7].

Hoechst 33258 (**1**) was known to bind DNA tightly (K$_{ass}$ ~ 6x10^8 bp^{-1}) [8] and consideration of the model we had built from the NOESY-derived distance restraints combined with molecular modelling showed that the phenolic group was located in the minor groove [7]. It was predicted [9] from this model that using a meta-OH phenolic group (i.e. using meta-Hoechst, **2**) would give additional specific hydrogen-bonds to C9 and G4' in the duplex d(CGCGAATTCGCG)$_2$. This is shown in Figure 1.

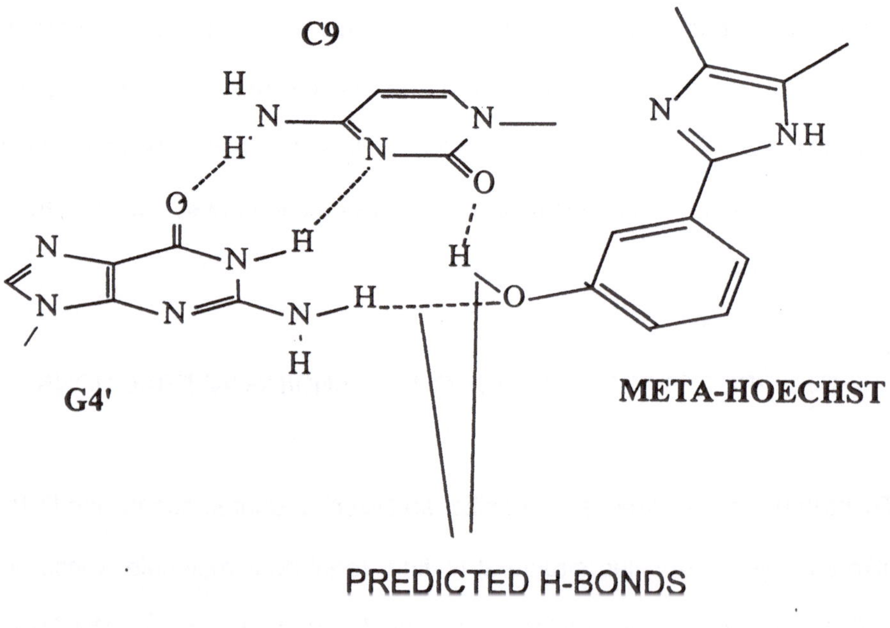

PREDICTED H-BONDS

Figure 1. Diagram of the phenolic region of meta-Hoechst (2) bound in the 1:1 minor groove complex with d(CGCGAATTCGCG)$_2$

By use of 2D NMR spectroscopy (NOESY, DQFCOSY, TOCSY, ROESY), the solution binding details of meta-Hoechst and d(CGCGAATTCGCG)$_2$ were determined [10]. From these studies it was clear that the meta-hydroxy phenolic group was frozen in the minor groove with the -OH group directed towards the floor of the groove (see Figure 2). For the Hoechst 33258 complex with DNA the phenolic group was free to rotate in the groove [11], but this was found not to be the case for meta-Hoechst [10].

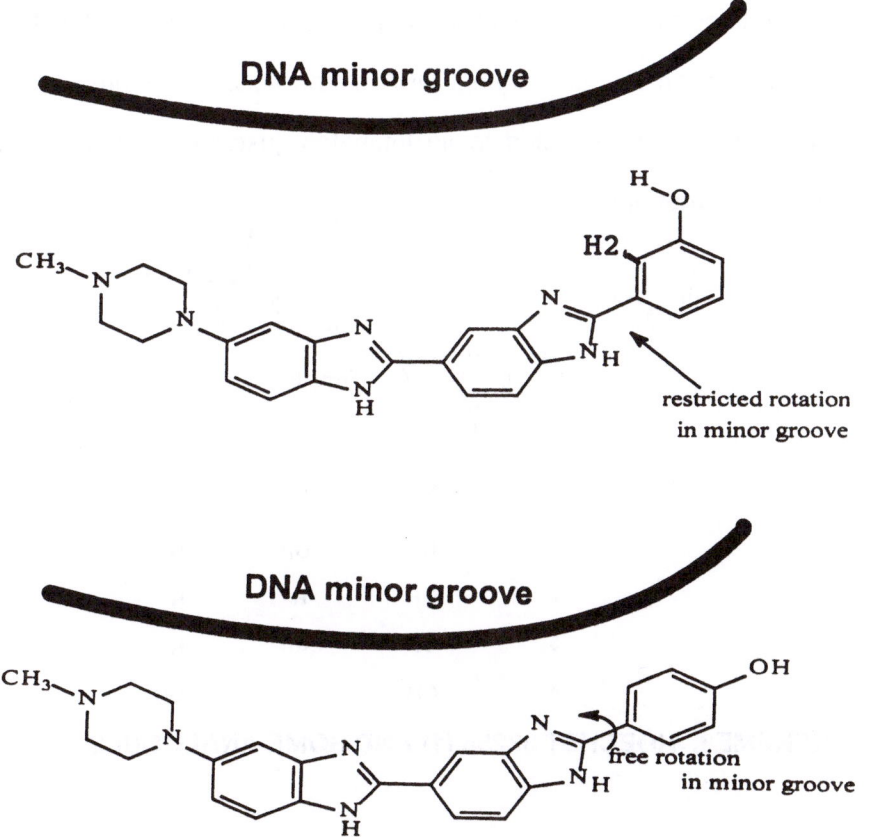

Figure 2. Upper. Orientation of meta-Hoechst (3) in the minor groove of d(CGCGAATTCGCG)$_2$. Lower. In contrast to 3, Hoechst 33258 (1) has free rotation of the bond between the phenol and benzimidazole rings.

1.2 SECOND STAGE : INTRODUCTION OF A NEW DNA CLEAVAGE METHOD

The above analysis of the <u>meta</u>-Hoechst: d(CGCGAATTCGCG)$_2$ model [10] showed that the phenolic OH group of the ligand may act both as H-bond donor (to C^9) and acceptor (from G$^{4'}$). A 3,4-dihydroxy analogue (see Scheme I; 3, X=Y=OH, Z=H) would be expected to behave similarly, but the presence of a catechol (or related function) introduces the potential for Cu(II)-ion stimulated free radical production and consequent DNA strand scission. Moreover, the radical-generating site is predicted to lie intimately juxtaposed to the minor groove cleavage target of DNA.

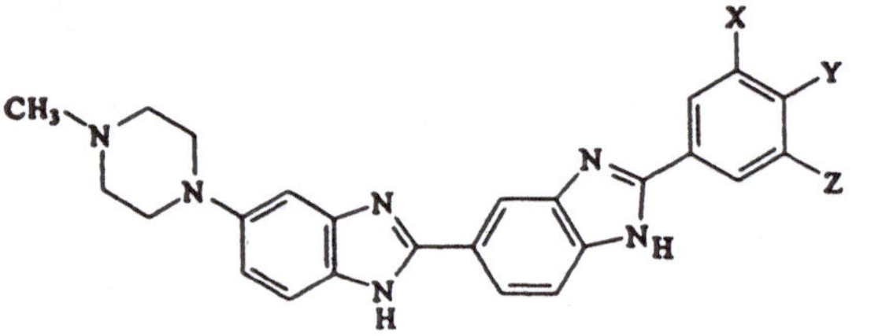

	X	Y	Z
1	H	OH	H
2	OH	H	H
3	OH	OH	H
4	OH	OH	OH

SCHEME I. HOECHST 33258 (1) AND SOME ANALOGUES

To develop these aspects we synthesised the catechol equivalent of **1** and **2**, which we shall refer to as cat-Hoechst (**3**) and the 3,4,5-trihydroxy analogue (**4**, trihydroxy-Hoechst). We here report high field NMR spectroscopic characterisation of the solution structure of the cat-Hoechst (3): d(CGCGAATTCGCG)$_2$ complex and preliminary analysis of the cleavage potential of 3 and 4 towards duplex plasmid DNA.

2. Experimental

2.1 MATERIALS AND METHODS

Cat-Hoechst (3) was synthesised as described for 2 [12] replacing ethyl m-hydroxybenzimidate by ethyl 3,4-dihydroxybenzimidate in the last step to yield product (3) as a dark-brown solid, which was purified by chromatography over silica h (1:1 MeOH:EtOAc solvent linearly changing to 100% MeOH) with isolation of fractions corresponding to the slow-moving yellow spot (tlc, uv visualisation) running before the slowest of the red spots. The yellow powder obtained on solvent removal was further purified by preparative tlc (silica gel plates, MeOH:EtOAc, 1:1) to give the product as a yellow powder, m.p. >250°C; δ_H (270 MHz, D$_2$O + DCl), 2.74 (s, 3H, N-CH$_3$), 2.83 [m, 4H, 2(N-CH$_2$)], 3.38 [m, 4H, 2(N-CH$_2$)], 6.25 (d, 1H, Ar-H, J = 8.9 Hz), 6.42 (s, 1H, Ar-H), 6.46 (s, 1H, Ar-H), 6.60 (d, 1H, ar-H, J = 8.2 Hz), 6.70 (d, 1H, Ar-H, J = 9.2 Hz). 6.98 (s, 1H, Ar-H), 7.01 (s. 1H, Ar-H), 7.17 (brs, 2H, Ar-H). FAB mass spectrometry (m/z, percent relative intensity); 433 [(M + H)$^+$ + 2, 4%]calculated 4.27% 442 [(M + H$^+$) + 1, 30%], calculated 29.97%, 441 [(M + H)$^+$, 100%]. Accurate mass measurement; calculated for $C_{25}H_{24}N_6O_2$, 441.20374; measured 441.2031 (1.4 ppm error).

For the synthesis of 2-[2-(3,4,5-trihydroxyphenyl)-6-benzimidazolyl]-6-(4-methyl-1-piperazinyl) benzimidazole (3,4,5-trihydroxy Hoechst) (4) a suspension of 2-(3,4-diaminophenyl)-6-(4-methyl-1-piperazinyl)benzimidazole [12] (640 mg, 2 mmol) and 3,4,5-trihydroxybenzaldehyde (314.5 mg, 2 mmol) in nitrobenzene (15 ml) was gradually heated to 140°C on an oil bath, the colour changing to brown and then darkening, and then stirring the mixture at 140°C for 24 hours. The cooled dark solution was triturated with ether and the precipitate filtered off and washed with ether to give dark brown solid, a hot methanolic solution of which was decolourized with charcoal, filtered and evaporated to give a yellow-red precipitate, which was purified by the method used for cat-Hoechst (3) R_f = 0.15 in methanol. δ_H [270 MHz, $(CD_3)_2$ SO], 2.25 (s, 3H, N-CH$_3$), 3.13 (brs, 4H, N-CH$_2$), 3.43 (brs, 4H, N-CH$_2$), 6.93 (brs, 2H, Ar-H), 7.18 (s, 1H, Ar-H), 7.47 (d, 1H, Ar-H, J = 8 Hz), 7.56 (m, 1H, Ar-H), 7.97 (m, 1H, Ar-H), 8.22 (d, 1H, Ar-H, J = 8 Hz), 8.32 (s, 1H, Ar-H), 9.00 (d, 1H, Ar-H, J = 8 Hz), 12.55 (brs), 12.75 (brs). FAB mass spectrum (m/z, percent relative intensity); 457 [(M + H)$^+$, 100%]. Accurate mass measurement: calculated for $C_{25}H_{25}N_6O_3$, 457.1988; measured 457.1987 (0.2 ppm error).

2.2 NMR SPECTROSCOPY

The synthesis and purification of the oligonucleotide, (d(CGCGAATTCGCG)$_2$), and sample preparation for NMR experiments were as described previously [10]. Creation of the 1:1 Cat-Hoechst/DNA complex was monitored by ^{1}H NMR spectroscopy by assessing the thymine methyl resonances of the oligonucleotide. A 1:1 ratio of ligand to oligonucleotide was judged to exist when the two thymine methyl resonances of the free DNA (δ1.59, 2xT^{8}CH$_3$; δ1.28, 2xT^{7}CH$_3$) had been completely replaced by four separate thymine methyl resonances (δ1.432, T^{8}CH$_3$; δ1.409, T$^8{'}$CH$_3$; δ1.345, T^{7}CH$_3$; δ1.26, T$^7{'}$CH$_3$). Standard pulse sequences were used to acquire sets of two-dimensional NMR data for resonance assignment and for structural analysis including DQFCOSY, TOCSY and NOESY data according to the details described for the meta-Hoechst:DNA complex [10]. Under these solution conditions, the cat-Hoechst:DNA complex behaved in a closely similar manner to the meta-Hoechst:DNA complex. All NMR data were accumulated on a Varian VXR600S (Unity) 600MHz NMR spectrometer on a dual (inverse) ^{1}H (X) probehead. Data were processed using the VNMR 4.3 and Tripos Sybyl 6.1 software package incorporating the Triad NMR module.

2.3 DNA CLEAVAGE CONDITIONS

The plasmid DNA used was pMA802. This plasmid contains the 649bp EcoRI - HindIII insert of pMT702 cloned into expression vector pMAC5-14 [13]. Cleavage conditions were based on conditions developed for Cu(II):RSH cleavage of DNA [14,15] and Chelex-treated water was used to prepare solutions. In a typical cleavage experiment, plasmid DNA (1μg) was incubated for 30 min in 10mM sodium phosphate buffer, pH 8.0, with appropriate concentrations of **1-4** (typically up to 250μM) with the addition or absence of Cu(II) (up to 150μM) added as an aliquot (1μl) from a stock solution in water (to maintain solubility it is preferable to add the Cu(II) solution to the DNA solution). For electrophoresis, at appropriate times reaction was stopped by loading the incubation mixtures onto agarose (1%) gels for electrophoresis (60 mV, 1hour) in TAE buffer. Bands were visualised by staining with ethidium bromide and u.v. transillumination. The electrophoretic positions of open circular and linear DNA were confirmed by partial and complete EcoRI restriction digestion respectively.

3. Results and Discussion

3.1 NMR STRUCTURAL STUDIES

One-dimensional titration of a solution of d(CGCGAATTCGCG)$_2$ with increasing amounts of **3,** monitored by following the resonances of the thymine methyl groups, showed a tightly bound complex. The two-dimensional NOESY data for all non-exchangeable protons, when fully assigned for the complex, revealed that the region correlating spatial relationships between the ligand and the DNA duplex was almost identical to that for data acquired on the meta-Hoechst:DNA complex [10].

Analysis of the intermolecular nOes between the catecholic ring and the DNA duplex showed that H2 of **3** was orientated toward the floor of the minor groove with nOes to A^6'H2 and A^5'H2 (see Figure 2). NOes from protons on the convex edge of **3** were only seen to oligonucleotides 4' and 5' protons (e.g. H6 to C^{9}5"H), indicating that they are directed towards the bulk solvent (as was the case for **2** with this oligonucleotide [10]).

3.2 DNA CLEAVAGE

When **1** or **2** were incubated with pMA802 plasmid DNA in the absence or presence of Cu(II) ions (up to 150μM) we could find no evidence of DNA scission. The molecular model (see Figure 3) indicated that if **3** were already bound to DNA, Cu(II) ions could not enter the chelating arms of the catechol unit. If the Cu(II) complex of **3** were pre-formed, the Cu(II):catecholic site would compromise the closely contoured fit of the Hoechst framework to the minor groove in this region.

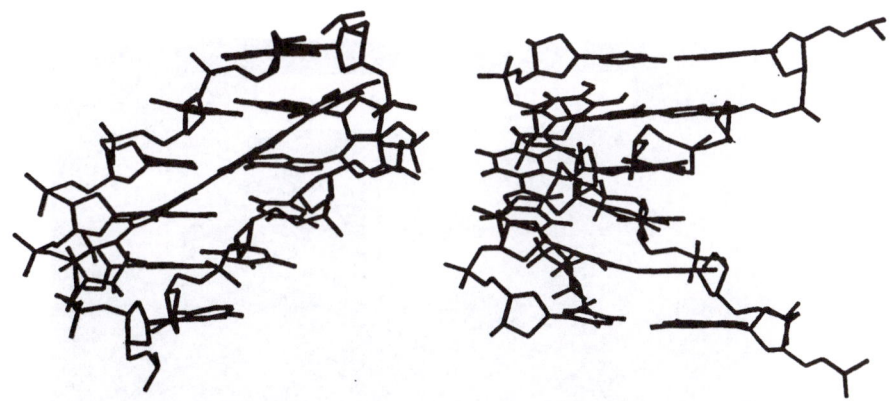

Figure 3. Orthogonal views of cat-Hoechst (3) bound in the minor groove of d(CGCGAATTCGCG)₂. The left-hand view looks down into the minor groove showing the ligands with its piperazinyl group at the lower left and the catecholic site at the upper right, with its 3-hydroxy group clearly directed into the groove bottom. The right-hand view shows the ligand wending its way up the minor groove, on the left edge of the duplex, with its catecholic site at the top between the top two base-pairs. The structures were calculated using NMR-derived distance-restraints and the TRIAD software of Tripos SYBYL 6.1

The model also indicated that if the Hoechst molecule were to be oriented with its catecholic unit directed **out** of the minor groove towards bulk solvent a Cu(II) test this we added an additional OH group to the phenolic ring to give **4** in which the 3-OH group would hold the phenolic ring tightly into the minor groove through interactions with C9 and G4' (as found here for **3** and previously [10] for **2**). This would leave an exo-directed catecholic site, the 4,5-dihydroxyphenyl, to effect Cu(II)-cleavage. Figure 4 shows the dependence on the concentration of trihydroxyHoechst (**4**) of the cleavage of supercoiled pMA802 plasmid DNA by 150μM $CuSO_4$ on incubation in 10 mM sodium phosphate buffer, pH 8.0.

Figure 4. Agarose gel electrophoresis of the cleavage of pMA802 DNA by 3,4,5-trihydroxy Hoechst (4). Incubation mixtures contained plasmid DNA and Cu(II) in pH 8.0 phosphate buffer with the following additions of 4 for the lanes indicated: lane 1, none; lane 2, 8μM; lane 3, 4μM; lane 4, 2μM; lane 5, 1μM.

There was no detectable scission in the absence of added Cu(II) for **4** (250µM) incubated with plasmid DNA. However, in the presence of 150µM Cu(II), progressive concentration-dependent cleavage occurs. Even at 1µM Hoechst derivative, after 30 minutes there was already extensive cleavage of supercoiled DNA, with formation of open circular DNA. Cleavage of supercoiled DNA by **4** under the influence of Cu(II) was shown to be time-dependent.

Cu(II)-induced cleavage of DNA has been reported for several systems. A benzene metabolite (1,2,4-benzenetriol) has been reported to damage human DNA, possibly by auto-oxidation in the presence of copper ions, but the species responsible for the damage was not clear [16,17].   Hydroquinone shows copper-ion-dependent DNA strand cleavage, with other metal ions (Fe(III), Mn(II), Cd(II), Zn(II)) being considerably less active [18].   Partial protection was afforded by singlet oxygen scavengers, but not hydroxyl scavengers. Other polyhydroxy-compounds which lead to Cu(II)-promoted DNA damage include flavanols and procyanidins, again by a mechanism largely specific to Cu(II) [19].   These classes of copper-dependent damage to DNA mostly involve oxygen radical or active oxygen species effecting radical abstraction reactions at the deoxyribose backbone, but Cu(II) in the presence of hydrogen peroxide causes damage to the bases themselves also [20].   Both types of Cu(II)-induced effects can occur in the same system, as we have found for the Cu(II)-thiol DNA cleavage system [15,21], and it will be of interest to define the detailed mechanism of DNA strand scission for the trihydroxyHoechst:Cu(II) system, given the very intimate nature of the ligand binding to the minor groove for this cleaver, based on the high resolution

solution structure that we have determined for the 3,4-dihydoxy Hoechst:DNA complex. The experimentally demonstrated close contact of the phenolic units of **1-3** with the minor groove of DNA [7,10,22] means that the radicals once generated need diffuse only short distances to act and the details of the cleavage mechanism should be readily probed.

4. Acknowledgements.

We are grateful to the EPSRC for time on the 600 MHz instrument at Edinburgh (KTD), to the Wellcome Trust for computing facilities and to the Iranian Government (SSE).

5. References

1. Kopka, M. L., C. Yoon, D. Goodsell, P. Pjura, and R. E. Dickerson. 1985 The molecular origin of DNA-drug specificity in netropsin and distamycin. *Proc. Natl. Acad. Sci. USA* **82**, 1376-1380.

2. Goodsell, D. and R. E. Dickerson. 1986 . *J. Med. Chem.* **29**, 727-733.

3. Matthews, D. A. 1978 Dihydrofolate reductase from *Lactobacillus casei*. X-ray structure of the enzyme methotrexate.NADPH complex. *J. Cell. Biochem.* **253**, 6946.

4. Fontecilla-Camps, J. C. 1979 Absolute configuration of biological tetrahydrofolates. A crystallographic determination. *J. Am. Chem. Soc.* **101**, 6114.

5. Teng, M-K., N. Usman, C. A. Frederick, and A. H-J. Wang. 1988 The molecular structure of the complex of Hoechst 33258 and the DNA dodecamer d(CGCGAATTCGCG). *Nucl. Acids Res.* **16**, 2671-2690.

6. Jin, R. and K. J. Breslauer. 1988 Characterization of the minor groove environment in a drug-DNA complex: Bisbenzimide bound to the poly[d(AT).poly[d(AT)duplex. *Proc. Natl. Acad. Sci. USA* **85**, 8939-8942.

7. Parkinson, J. A., J. Barber, K. T. Douglas, J. Rosamund, and D. Sharples. 1990 Minor-Groove Recognition of the Self-Complementary Duplex (CGCGAATTCGCG)$_2$ by Hoechst 33258: A High-Field NMR Study. *Biochemistry* **29**, 10181-10190.

8. Loontiens, F. G., P. Regenfuss, A. Zechal, L. Dumortier, and R. M. Clegg. 1990 Binding characteristics of Hoechst 33258 with calf thymus DNA, Poly[d(A-T)], and d(CCGGAATTCCGG): Multiple stoichiometries and determination of tight binding with a wide spectrum of site affinities. *Biochemistry* **29**, 9029-9039.

9. Ebrahimi, S. E. S., J. A. Parkinson, K. T. Fox, J. H. McKie, J. Barber, and K. T. Douglas. 1992 Studies of the interaction of a meta-hydroxy analogue of Hoechst 33258 with DNA by melting temperature, footprinting and high-resolution ^{1}H NMR spectroscopy. *J. Chem. Soc. Chem. Commun.* 1398-1400.

10. Parkinson, J. A., S. E. Ebrahimi, J. H. McKie, and K. T. Douglas. 1994 Molecular design of DNA-directed ligands with specific interactions: Solution NMR studies of the interaction of a meta-hydroxy analogue of Hoechst 33258 with d(CGCGAATTCGCG)$_2$. *Biochemistry* **33**, 8442-8452.

11. Parkinson, J. A., J. Barber, B. A. Buckingham, K. T. Douglas, and G. A. Morris. 1992 Hoechst 33258 and its complex with the oligonucleotide d(CGCGAATTCGCG)$_2$: ^{1}H NMR assignments and dynamics. *Mag. Res. Chem.* **30**, 1064-1069.

12. Ebrahimi, S. E. S., M. C. Bibby, K. R. Fox, and K. T. Douglas. 1995 Synthesis, DNA-binding, footprinting and *in vitro* antitumour studies of a meta-hydroxy analogue of Hoechst 33258. *Anti-Cancer Drug Design* (In Press)

13. Andrews, J., S. J. Minter, and R. W. Davies. 1991 Production of mutant dihydrofolate reductases of *Lactobacillus casei* for nuclear magnetic resonance spectroscopy. *Gene* **100**, 219-224.

14. Reed, C. J. and K. T. Douglas. 1989 Single-strand Cleavage of DNA by Cu(II) and Thiols: a powerful Chemical DNA-cleaving System. *Biochem. Biophys. Res. Commun.* **162**, 1111-1117.

15. Reed, C. J. and K. T. Douglas. 1991 Chemical cleavage of plasmid DNA by glutathione in the presence of Cu(II) ions. *Biochem. J.* **275**, 601-608.

16. Kawanishi, S., S. Inoue, and M. Kawanishi. 1989 Human DNA damage induced by 1,2,4-benzenetriol, a benzene metabolite. *Cancer Res.* **49**, 164-168.

17. Rao, G. S. and K. P. Pandya. 1989 Release of 2-thiobarbituric acid reactive products from glutamate or deoxyribonucleic acid by 1,2,4-benzenetriol or hydroquinone in the presence of copper ions. *Toxicology* **59**, 59-65.

18. Li, Y. and M. A. Trush. 1993 DNA damage resulting from the oxidation of hydroquinone by copper. Role for a Cu(II)/Cu(I) redox cycle and reactive oxygen generation. *Carcinogenesis* **7**, 1303-1311.

19. Shirahata, S., H. Murakami, K. Nishiyama, K. Yamada, G-I. Nonaka, I. Nishioka, and H. Omura. 1989 DNA breakage by flavan-3-ols and procyanidins in the presence of cupric ion. *J. Agric. Food Chem.* **37**, 299-303.

20. Aruoma, O. I., B. Halliwell, E. Gajewski, and M. Dizdaroglu. 1991 Copper-ion-dependent damage to the bases in DNA in the presence of hydrogen peroxide. *Biochem. J.* **273**, 601-604.

21. John, D. C. A. and K. T. Douglas. 1993 Sequence-dependent reactivity of linear DNA to chemical cleavage by Cu(II):thiol combinations including cysteine or glutathione. *Biochem. J.* **289**, 463-468.

22. Leupin, W., D. Bur, A. Dorn, Y-H. Ji, A. Labhardt, A. Fede, M. Billeter, and K. Wüthrich. 1994 Bis-benzimidazole derivatives as DNA ligands: design based on the solution structure of a Hoechst 33258-DNA complex with subsequent molecular modelling. *Actual. Chim. Thér.* **21**, 153-170.

BINARY SYSTEMS OF OLIGONUCLEOTIDE CONJUGATES FOR SEQUENCE SPECIFIC ENERGY-TRANSFER SENSITIZED PHOTOMODIFICATION OF NUCLEIC ACIDS

V.V. VlASSOV, M.I. DOBRIKOV, S.A. GAIDAMAKOV, E.K. GAIDAMAKOVA, T.I. GAINUTDINOV and A.A. KOSHKIN
Institute of Bioorganic Chemistry, Siberian Division of Russian Academy of Sciences
8, Lavrentiev Ave., Novosibirsk 630090 Russia

Improvement of specificity is an important consideration in the design of reactive derivatives of oligonucleotides for biological and therapeutic applications. We are developing a methodology based on using binary systems of oligonucleotide conjugates which form reactive species when assembling on target nucleotide sequences. In this chapter we describe for the first time the design of a binary system which becomes photoactivatable in controlled conditions upon assembling on its target. The binary system consists of two oligonucleotides, one of which contains a photosensitizing group and the second oligonucleotide is equipped with a photoreactive group. Binding of the oligonucleotides to adjacent sequences in the target nucleic acid brings the groups to a close proximity which allows efficient photosensitized activation of the reagent molecules and modification of the target. The binary system demonstrates high efficiency and specificity in reaction with specific single stranded DNA target. The results evidence that using binary systems of oligonucleotide conjugates can be a perspective approach for development of specific efficient and low toxic oligonucleotide-based therapeutics.

1. Introduction

Reactive derivatives of oligonucleotides have become useful tools for investigating structure and functions of nucleic acids. Oligonucleotides coupled with reactive groups may serve as potent inactivating ligands of specific nucleic acids in vivo and thus may potentially be used in therapeutic applications [1]. The traditional oligonucleotide derivatives contain chemical groups capable of reacting with nucleic acids under physiological conditions or groups generating highly reactive diffusing species which damage nucleic acids. The reactions of these groups, however, are generally uncontrolled and oligonucleotide derivatives with traditional reactive groups can affect nontarget nucleic acids and biopolymers other than nucleic acids, which may lead to undesirable side effects. Another problem is related to the sequence specificity of oligonucleotide-target interaction. *In vivo*, hybridization has to occur in a narrow window of physiological conditions. Under these conditions long oligonucleotides needed for recognition of unique targets in eucaryotic cells can form numerous nonperfect complexes with nontarget nucleic acids. Therefore the properties of long oligonucleotides conjugated to traditional reactive groups hinder their use to any purposeful application in vivo.

B. Meunier (ed.), DNA and RNA Cleavers and Chemotherapy of Cancer and Viral Diseases, 195–207.
© 1996 Kluwer Academic Publishers. Printed in the Netherlands.

We propose here a new approach for the design of oligonucleotide based conjugates for sequence specific chemical modification of nucleic acids. Instead of using long oligonucleotides with highly reactive groups, we propose to design binary systems of oligonucleotides conjugated to relatively inactive precursor groups that are capable of assembling into an active complex when the two components are juxtaposed to each other due to simultaneous binding to a target nucleic acid. The advantage of the system is a higher modification specificity because it is determined by recognition of two oligonucleotide components which bind to the target independently. Each of the oligonucleotide components is short enough to avoid formation of nonspecific complexes in physiological conditions. Since the components bear relatively low reactive groups which are activated only upon complex formation, they can be expected to produce less nonspecific effects due to interactions with nontarget biopolymers. At the first step of our studies we have developed a relatively simple system of oligonucleotide conjugates where a photoreactive group linked to one oligonucleotide is activated by a juxtaposed photosensitizer group which is attached to another oligonucleotide.

2. Materials and methods

Absorption spectra were taken on Specord M 40 (Karl Zeiss, Jena, Germany). Fluorescence spectra were recorded on spectrofluorimeter MPF-4 (Hitachi, Japan) in the wavelength range from 300 to 500 nm for both excitation and emission spectra.

2.1. PREPARATION OF OLIGONUCLEOTIDE DERIVATIVES

Oligonucleotides were synthesized by Dr. T. Abramova (this Institute) by the triester method. After deprotection, oligonucleotides were purified by HPLC on Nucleosil RP-18 (5-20 μ) columns.

Synthesis of the photoreagent N-(β -alanyl) p-azidotetrafluoro benzalhydrazone hydrochloride R (Fig. 1) has been described previously [2]. Photoreagent R was attached to the 3'-terminal phosphate of oligonucleotide 2 according to described method [3]. Cetyltrimethylammonium salt of oligonucleotide 2 (0.15 μmol) was dissolved in 40 μl of DMSO. Mixture of 2.0 mg (7.6 μmol) of triphenylphosphine and 1.7 mg (7.6 μmol) of 2, 2'-dipyridyl disulfide was added and, after 10 min, 1 mg (4.5 μmol) of 4-N,N-dimethylaminopyridine was added to the reaction mixture. After 15 min incubation at 20°C the nucleotide material was precipitated by the addition of 1.5 ml of acetone containing 2% LiClO$_4$. Precipitate was washed with dry acetone and dissolved in 60 μl of 0.01M potassium phosphate buffer pH 10.3 containing 1.8 μmol of the reagent R. The solution was incubated at 20°C for 1 h, then the reaction product was precipitated by addition of 1.5 ml of acetone containing 2% of LiClO$_4$. The oligonucleotide derivative 4 was isolated by reverse-phase HPLC. Yield of the product 4 was 80%.

Synthesis of the benzanthranyl oligonucleotide derivative 1 was carried out in two steps. First, ethylenediamine linker group was attached to the terminal phosphate of the oligonucleotide, as described in [3]. Then cetyltrimethylammonium salt of the oligonucleotide 3 with the aminolinker (70 nmol) was dissolved in 50 μl of DMSO solution of N-hydroxysuccinimide ester of 1,2-benzo-3-fluoro-9-methylanthra-10-nyl acetic acid (1 μmol). After 4 h incubation at 20°C, the oligonucleotide was precipitated by addition of 15 volumes of acetone containing 2% lithium perchlorate. The oligonucleotide derivative was finally purified by HPLC on Nucleosil RP-18 columns.

The conjugate was eluted from the column at higher concentration of acetonitrile, as compared to the parent oligonucleotide, because of the hydrophobic fluorescent group. The yield of the product **1** was 65%. UV spectra - λ, (nm), (ε, $l \cdot M^{-1} \cdot cm^{-1}$): 258 (160000), 307 (55000), 367 (15000). Small (2 nm) red shift of absorption spectrum maximum of the benzanthranyl group and a considerable hypochromicity (24%) was observed upon attachment of the sensitizer to the oligonucleotide due to stacking interactions with the nucleobases.

2.2. PREPARATION OF SINGLE-STRANDED DNA TARGET *T2*

Phage M13BMR4, containing fragment -215/+83 of bacterial cytochrom P-450 gene CYP-102, was constructed by insertion of the RsaI-HindIII fragment of plasmid pBM3 [4] into SalI site of phage M13mp18. Single-stranded DNA of the bacteriophage M13BMR4 was isolated by phenol extraction of the phage particles as described in [5]. The DNA was digested with restriction endonuclease HspAI for 24 h at 37°C. The reaction mixture contained 50 µg of the DNA and 10 activity units of HspAI enzyme (Sibenzyme Co, Novosibirsk) in 200 µl of buffer recommended by the enzyme manufacturer. The digests were characterized by electrophoresis on 4% polyacrylamide gels in native conditions, in 100 mM Tris-borate buffer pH 7.5 containing 1 mM EDTA. The band, containing the target sequence was identified by hybridization with oligonucleotide AGAAGTTCATCAGTTAAAGA overlapping the sequence recognized by the components of the binary system.

2.3. PHOTOCHEMICAL MODIFICATION OF DNA TARGETS

Reaction solutions contained 5'-end labeled *T1* target (1 nM-100 µM) and 50 µM oligonucleotide derivatives **1 - 4**, in 2 mM sodium phosphate pH 7.1 containing 200 mM NaCl. Samples (5 µl each, in immunological 60-wells plates Medpolymer Co, Sankt Petersburg) were irradiated at 4°C - 37°C, with UV light of a 120 W high pressure mercury lamp through glass filters providing bands (λ, nm, W, mW/cm^2): 300-315, 0.1; 330-365, 0.25; 365-390, 0.2; 390-420, 0.15; 425-455, 0.2 or through filters allowing irradiation with visible light, λ =400-480; W, 0.5 mW/cm^2. At the end of the irradiation, the samples were added to 10 µl of the loading buffer (95% deionized formamide containing 0.1% bromophenol blue) and electrophoresed on a denaturing polyacrylamide gel (13% acrylamide, 0.5 % bis-acrylamide, 7 M urea). The gels were autoradiographed and quantitation of reaction products was carried out using scanning densitometer Ultroskane-XL (LKB, Sweden).

Reactions with the target *T2* were carried out similarly. Concentration of the target *T2* was 0.1 µM.

3. Results and discussion

The organization of the binary oligonucleotide system tested in this study is summarized in Fig. 1. The system is composed of two oligonucleotides complementary to adjacent nucleotide sequences in the single stranded DNA target. The length of the oligonucleotides was chosen taking into account that they form duplexes with melting point near 37°C. Under physiological conditions, oligonucleotides of this length are not expected to form nonperfect complexes with partially complementary nucleotide sequences.

OLIGONUCLEOTIDES:

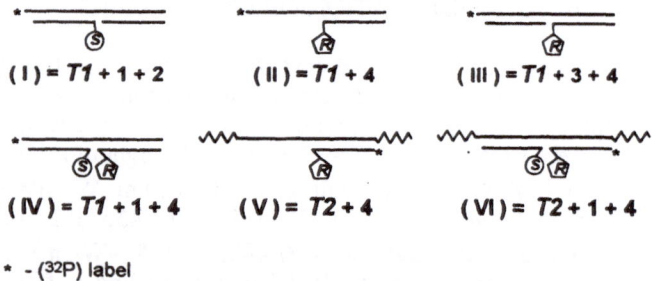

INVESTIGATED DUPLEXES:

(I) = *T1* + 1 + 2   (II) = *T1* + 4   (III) = *T1* + 3 + 4

(IV) = *T1* + 1 + 4   (V) = *T2* + 4   (VI) = *T2* + 1 + 4

* - (^{32}P) label

Figure 1. Target single stranded DNAs *T1, T2* and complementary oligonucleotides alone and conjugated to components of the binary photoactivatable pair, photosensitizing group **S** and photoreactive group **R**. Compositions of the investigated complexes are shown in the bottom of the figure.

One of the components, **1**, bears a fluorescent sensitizer β-(1,2-benzo-3-fluoro-9-methylanthranyl-10-methylcarbamido)-ethylamino- group **S** at its 5'-terminal phosphate. The second component, **4**, bears a photoreactive p-azidotetrafluorobenzalhydrazone-β-alanyl residue **R** at its 3'-terminal phosphate. The photosensitizer was chosen taking into account spectral properties of the photoreactive group and high quantum yield of the benzoanthranyl derivatives fluorescence [6] The used perfluoroaromatic azido group is known to react with nucleic acids upon UV irradiation [7], presumably via formation of singlet nitrene [8,9]. Maximum UV absorbancy of the fluorescent sensitizer is 367 nm which is considerably a longer wavelength as compared to the optimal absorbancy for activation of the azido group (303 nm) (Fig. 2).

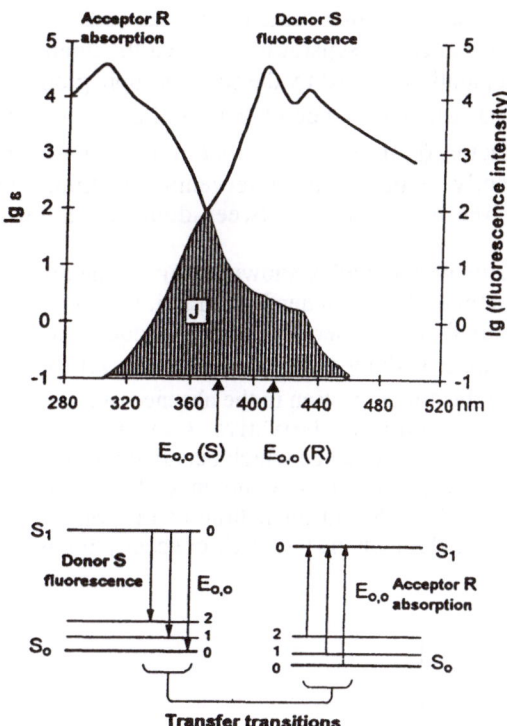

Figure 2. Absorption spectrum of the azide **R**, fluorescence spectra of the benzanthracene **S** and energy levels of the donor and acceptor chromophores, showing the origin of **J**, the overlap integral (striped area).

The energy absorbed by the sensitizer group can be transferred to a closely juxtaposed azido group by the known mechanism of singlet-singlet Forster excitation energy transfer [10,11] or electron transfer mechanism [12]. Energy transfer from the sensitizer **S** to azide **R** leads to formation of the excited singlet azide, which than undergoes chemical conversions typical of arylnitrenes [13].

The electronic spectrum of the azide **R** has a weak long-wavelength absorption tailing at least to 450 nm. This feature indicates the possible existence of an unbound excited singlet state [14]. The relatively small overlapping of the absorption spectrum of **R** and the emission spectrum of the sensitizer allows efficient energy transfer only in the case when distance between the chromophores is very small [15]. The process is extremely sensitive to the distance between the groups also because the efficiency of the energy transfer depends on the inverse sixth power of the distance between donor and acceptor moieties [14]. The rate constant for the Forster energy transfer (k_t) can be estimated from equation

$$k_t = 1/\tau_0 \cdot (R_0/R)^6$$

where τ_0 is the inverse of the first order decay rate constant of the excited donor in the absence of acceptor and R, the separation between the donor and acceptor groups. R_0 is the characteristic separation defined by the spectroscopic properties of the system: $R_0^6 = C \cdot k^2 \cdot \phi_0 \cdot n^{-4} \cdot J$, where C is a combination of physical and mathematical constants, k^2 an orientation factor, ϕ_0 the quantum yield of the donor in absence of the acceptor, n, the effective refractive index of the intervening medium and J, an overlap integral expressing the degree of resonance between donor emission and acceptor absorption spectra.

In the complementary complex shown in Fig. 1, the distance R between the groups cannot exceed the sum of their Van der Vaals radii, which is of the order of magnitude of R_0, by factor of more than 10. Therefore, from the equation it follows, that photolysis of the azide in the complex with the target and the sensitizer can be accelerated by factor of 10^6, as compared to the direct reaction in the absence of sensitizers.

Binding of benzanthranyl derivative **1** to the target **TI** was investigated by measuring quenching of fluorescence which occurs upon formation of the duplexes **I** and **IV**. Results of these experiments are shown in Fig. 3. It is seen that the duplexes formation results in at least 5-fold quenching of the sensitizer **S** fluorescence and that dependence of the quenching on the target **TI** concentration is linear in range 0.1-3μM .

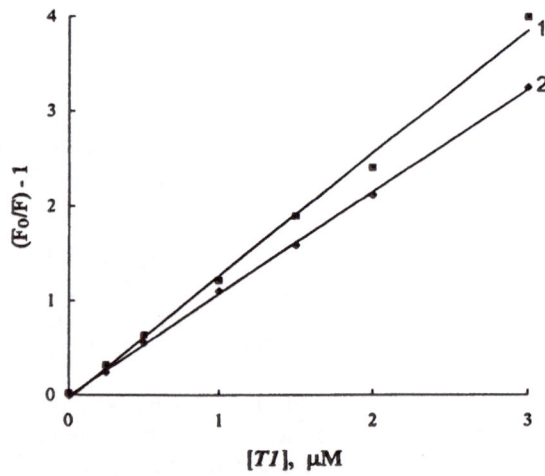

Figure 3. Stern-Folmer quenching of the benzanthracene S fluorescence $(F_0/F-1)$ upon formation of duplexes **I** and **IV**, lines 1 and 2, respectively. (λ_{ex} = 365-369 nm, λ_{em} = 405-409 nm).

The association constants for complexes **I** and **IV** were found to be $1.0 \cdot 10^6$ M^{-1} and $1.2 \cdot 10^6$ M^{-1} , respectively. Similarity of the estimated association constants indicates that binding of oligonucleotides to the target occurs independently. Apparently, conjugated chromophores interfere with the end- to end interaction of the bound oligonucleotides, which is observed usually when two oligonucleotides bind to adjacent nucleotide sequences.

It was found, that irradiation of the target DNA in the presence of the photoreactive oligonucleotide derivative leads to covalent crosslinking of the oligonucleotide to the target and that piperidine treatment cleaves the DNA at positions of the modified nucleotides. Results of experiments on modification of the target *T1* with the photoreactive oligonucleotide derivative **4** in duplexes **III** and with the binary system of oligonucleotide derivatives **1+4** in duplex **IV** are shown in Fig. 4.

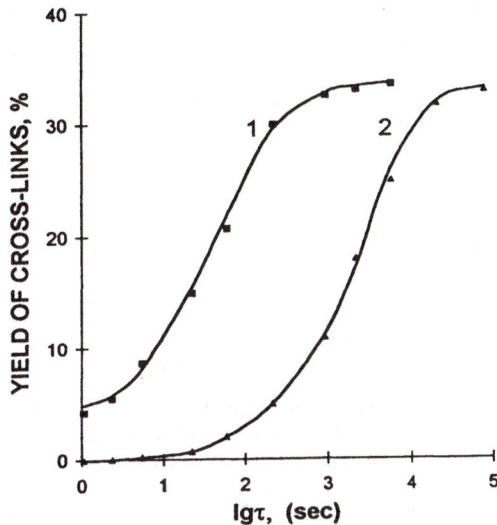

Figure 4. Kinetics of photomodification of the target DNA *T1* . The solution was irradiated with UV light at 4°C. Curve 1, reaction in duplex **IV** (sensitized photomodification); curve 2, reaction in duplex **III** (direct photomodification). Concentrations of the 5'-end labeled target and the oligonucleotide derivatives were 50 μM each. τ, irradiation time.

Under identical conditions the binary system modifies DNA 140 times faster than the derivative **4** alone. Apparently, this is the consequence of a highly efficient energy transfer from **S** to **R**. The final reaction yields in these conditions were similar in both duplexes and reached 33%. From these results it follows that at sufficiently short irradiation times, photolysis of the azide occurs only within the duplex structure **IV**.

Fig. 5 displays results of experiments with duplexes **III** and **IV** in conditions, where excess reactive oligonucleotide conjugate was present in solution. In the case of duplex **III**, the reaction yield first increases and then plateaus at 33% with the increase of the ratio of the photoreagent and the target concentrations [4]/[*T1*]. In the case of the binary system (duplex **IV**), the modification yield increases with increase of the ratio [4]/[*T1*] and reaches finally 65%. This dependence of the crosslinking yield on the ratio [4]/[*T1*] can be explained by the expected selective activation of the photoreagent complexed to its target in conditions where the reagent **4** in solution remains intact. In conditions of excess of the reagent **4**, exchange between the intact reagent **4** molecules present in solution and the reagent molecules unproductively photolysed within the complex results in repeated activation of the photoreagent molecules bound to the same nucleotide sequence and increases the modification yield. These results bear evidence that using the binary system based on energy transfer allows a considerable increase of the modification efficiency.

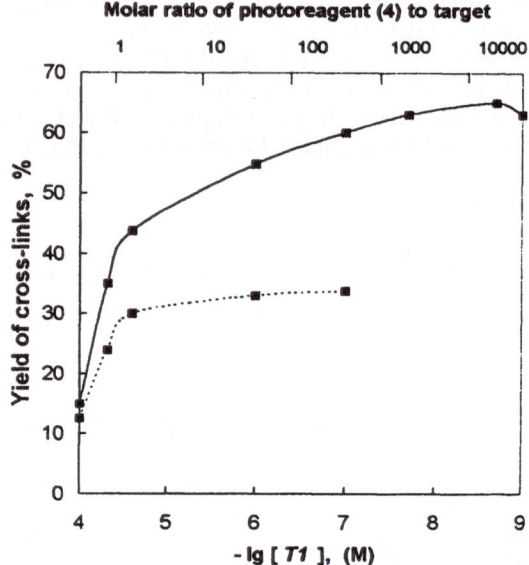

Figure 5. Effect of the ratio of concentrations [reagent]/[target] on the photoinduced crosslinking in duplexes **III** and **IV**. Concentration of the target *T1* was varied. Concentrations of **1** and **4** were 50 μM. The solution was irradiated with UV light at 4°C.

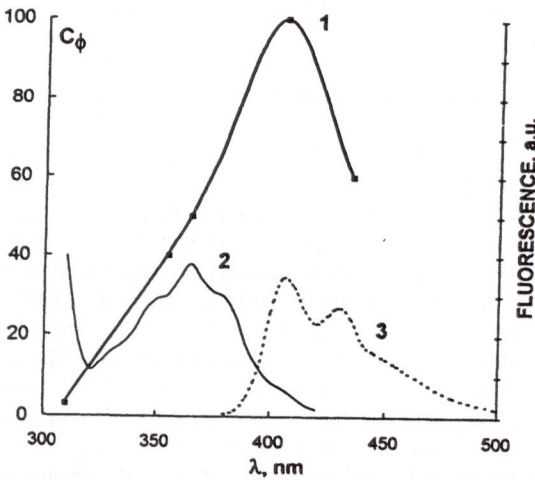

Figure 6. Spectral dependence of the transfer efficiency coefficient (C_ϕ) and the excitation and emission spectra of oligonucleotide derivative **1**. Curve 1- transfer efficiency coefficient $C_\phi = V_{sens}/V_{direct}$. Curves 2 and 3, excitation ($\lambda_{em} = 405$-409 nm) and emission ($\lambda_{ex} = 365$-369 nm) spectra of the benzanthranyl oligonucleotide derivative **1**.

Results of investigation of the spectral photosensitivity of the binary system are shown in Fig. 6. The fluorescence excitation is maximised by the broad band centred at 367 nm, whereas the fluorescence emission is observed with two maxims at 407 and 430 nm. The transfer efficiency coefficient $C_\phi = V_{sens}/V_{dir}$ (where V_{sens} and V_{dir} are the rates of the sensitized and the direct reactions, respectively) is maximal in the visible light region (405 nm) and red shifted as compared to the excitation (absorption) spectrum of S. Coincidence of the C_ϕ maximum with the fluorescence 0-0 band is an indication in favour of the singlet-singlet energy transfer mechanism. The spectral properties of the system suggest that it can be activated by irradiation with visible light (400-480 nm).

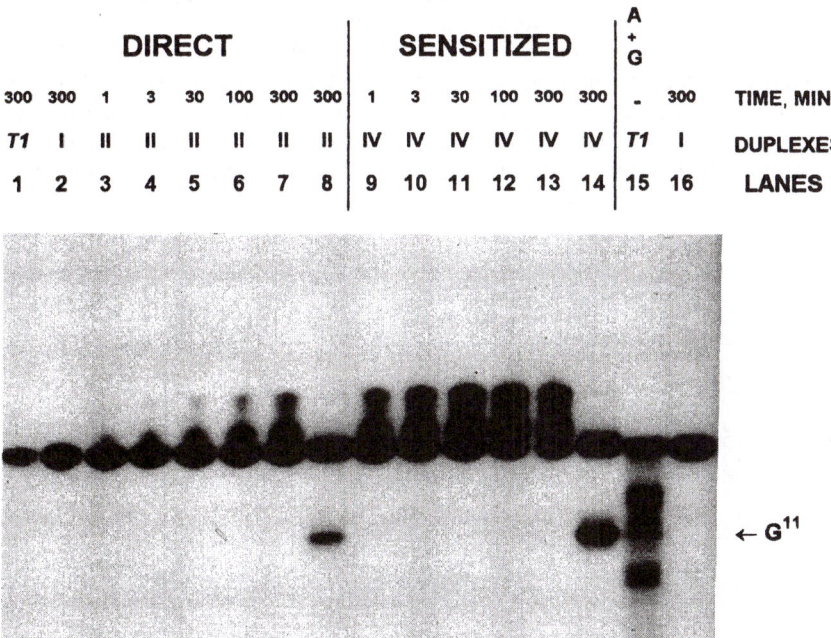

Figure 7. Direct and the sensitized photomodification of the target DNA *T1* in duplexes **III** and **VI** at 4°C. Autoradiograph of a 13% denaturing polyacrylamide gel. The samples were irradiated with visible light (λ= 400-480 nm). Concentration of the target DNA was 1 μM and concentrations of the oligonucleotide derivatives were 50 μM each: Lane 1, control, *T1* target irradiated in the absence of reagents for 300 min; Lanes 2, 16 controls, duplex I , irradiated for 300 min; Lanes 3 -8, duplex III; Lanes 9-14, duplex **IV**. Irradiation times were 1 min, 3 min, 30 min, 100 min and 300 min for the lanes 3,9; 4,10;, 5,11; 6,12 and 7,13, respectively. Lanes 8 and 14 contained the samples similar to those analyzed in lines 7, 13, respectively, subjected to piperidine treatment after irradiation. Lane 15, Maxam-Gilbert A+G reaction.

In theory, the arylazido group can be activated by using fluorophores absorbing at wavelengths as long as 750 nm, where mammalian cells are relatively transparent. Therefore synthesis of corresponding sensitizers and arylazido derivatives will provide a possibility for development of binary oligonucleotide conjugates for affecting cellular nucleic acids.

Results of a typical experiment on *T1* DNA modification using visible light are shown in Fig. 7. In the duplex **IV** the sensitized reaction proceeds 300 times faster, than the direct photomodification in the duplex **III**, and provides higher crosslinking yield (more than 70%). Electrophoretic analysis of the target *T1* after direct and sensitized photomodifications followed by piperidine treatment (Fig. 7, lanes 8 and 14) indicates that in both cases the reaction occurs at guanosine G11.

In [16, 17] it has been suggested that quenching of fluorescence of polynuclear aromatic dyes bound to DNA involves electron transfer mechanism with the photoexcited dye acting as an electron acceptor. In our case, fluorescence of S is quenched efficiently (see Fig. 3) but no reaction with the *T1* target occurs in the absence of **R** (Lanes 1 and 15). Therefore, it seems that in the case of the investigated system, electron transfer with formation of reactive cation radicals does not occur.

Earlier we have shown by the pyridine-ylide method [2], that direct photolysis of azide **R** yields singlet nitrene as the primary product. Similarity of the final modification yields in the absence of excess of the photoreagent (Fig. 4) and identical position specificity of the direct and the sensitized reactions (Fig. 7, lanes 8 and 14) are in accordance with the energy transfer mechanism leading to production of singlet nitrenes.

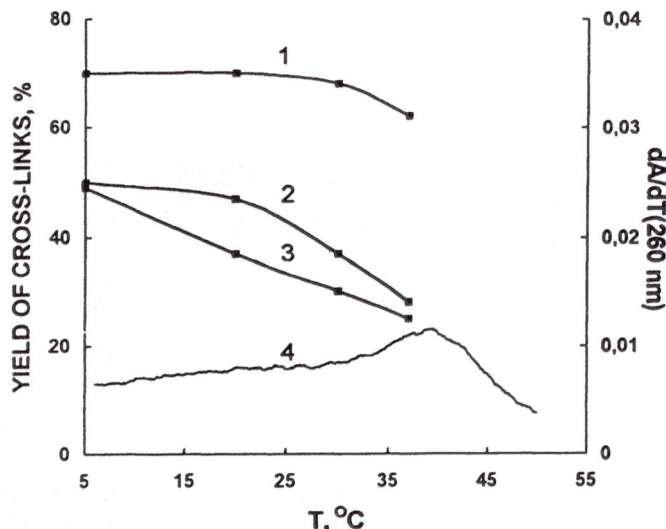

Figure 8. Temperature dependence of the direct and sensitized photomodifications of the *T1* target. Curve 1, sensitized photomodification in duplex **IV**; curves 2 and 3, direct photomodifications in duplexes **III** and **II**, respectively. Concentrations: *T1*, 0.1 µM; 1, 50 µM; **4**, 50 µM. Curve 4, differential melting curve of the duplex **II**, concentrations of oligonucleotides, 50 µM each.

Results of studies on the temperature dependence of the *T1* target photomodification using visible light are shown in Fig. 8. At 37°C, near melting temperatures of the duplexes **II-IV**, the sensitized photomodification is only slightly less efficient than at 4°C, while the efficiency of the direct photomodification is decreased considerably. Apparently, in the first case, at higher temperatures, the accelerated exchange between the reagent molecules in solution and the molecules in the complex compensated to some extent the effect of partial dissociation of the complex.

High specificity and efficiency of the binary system is illustrated by the data of experiments on modification of the DNA target *T2* with [^{32}P]- labeled oligonucleotide derivative **1**. *T2* target is a complex mixture of single stranded DNA fragments, one of which contains nucleotide sequence identical to that of the target *T1*. In Fig. 9, it is seen, that under irradiation, the binary system reacts only with the target fragment, in conditions where no detectable direct nonsensitized reaction occurs.

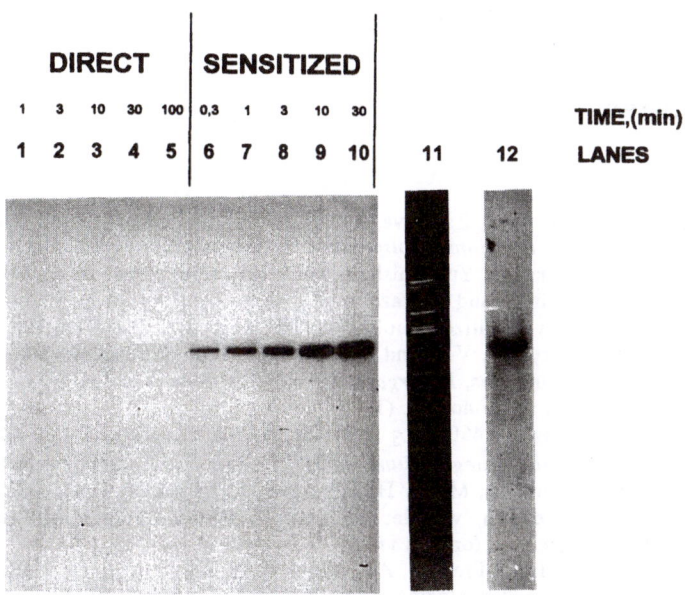

Figure 9. Direct and sensitized photomodifications of the target T2 using [^{32}P]- labeled oligonucleotide derivative 1. The samples were irradiated with visible light at 37°C. Autoradiograph of a 13% denaturing polyacrylamide gel. Lanes 1-5 - direct photomodification in duplex **V**, lanes 6-10 - sensitized photomodification in duplex **VI**. Irradiation time: 0.3 min (lane 6); 1 min (lanes 1, 7); 3 min (lanes 2, 8); 10 min (lanes 3, 9); 30 min (lanes 4, 10); 100 min (lane 5). 11, the same gel, staining with ethidium bromide. 12, autoradiograph of the 4% native polyacrylamide gel, in which fragments of the T2 target were resolved and hybridized to the radiolabeled 20-mer AGAAGTTCATCAGTTAAAGA.

Results of the present study demonstrate that the strategy of design of the binary systems indeed works. It can be used for development of various specific and efficient modular systems activatable upon assembling at the target nucleic acids. Thus, two chelated metal ions can be delivered to certain phosphodiester linkage by two oligonucleotide conjugates which will allow a cooperative action of the metal ions, needed for efficient cleavage of the linkage (see chapter by M. Komiyama). Two oligonucleotides can be equipped with peptides or organic molecules mimicking fragments of catalytic centres of nucleolytic enzymes. Upon binding of the conjugates, the catalytic centre will be formed in vicinity of the phosphodiester bond to be cleaved. Juxtaposition of oligonucleotides in binary systems can be used for delivery of an activating group to a latent reactive group, for *in situ* generation of active species upon the complex formation. Thus, delivery of a strongly nucleophilic group by one oligonucleotide can be used for triggering reaction of a synthetic enediyne molecule conjugated to another oligonucleotide. Apparently, such smart oligonucleotide- based binary reagents can find applications as tools for superspecific chemical modification of nucleic acids and in design of efficient nontoxic therapeutics.

Acknowledgements

The present work was supported by grants from the International Science Foundation, from US Department of Energy (Human Genome program), INSERM (grant 94EO 08)) and Ministry of Science of Russia.

References

1. Knorre, D.G., Vlassov, V.V., Zarytova, V.F., Lebedev, A.V. and Fedorova, O.S. (1994) *Design and targeted reactions of oligonucleotide derivatives* , CRC Press, Boca Raton.
2. Dobrikov, M.I., Doodko, R. Yu., Shishkin, G.V. (in press) Reagents for specific modification of biopolymers. VI. Oximes and hydrazone of p-azidotetrafluorobenzaldehyde: synthesis, spectral properties, direct and sensitized photolysis, *Bioorgan.Khim.*. (In Russian).
3. Godovikova, T.S., Zarytova, V.F. and Khalimskaya, L.M. (1986) Reactive phosphamidates of mono- and dinucleotides, *Bioorgan. Khim.* 12, 475-481.
4. Ruetinger, R., Wen, L. and Fulco, A. (1989) Coding nucleotide, 5'-regulatory, and deduced amino acid sequences of P450$_{BM-3}$, a single peptide cytochrom P 450:NADFH-p450 reductase from *Bacillus megaterium.*, *J. Biol. Chem.*. 264, p.10987-10995.
5. Messing, J., Gronenborn, B., Muller-Hill, B., and Hofschneider, P. H. (1977) Filamentous coliphage M13 as a cloning vehicle: insertion of a HindII fragment of the lac regulatory region in M13 replicative form *in vitro* , *Proc. Natl. Acad. Sci. USA* 74, 3642-3646.
6. Shahbar, M., Harvey, R.G., Prakash, A.S., Boal, T.R., Zegar, I.S. and LeBreton, P.R. (1983) Fluorescence and photoelectron studies of the intercalative binding of benz[a]antracene metabolite models to DNA, *Biochem. Biophys. Res. Commun.* 112, 1-7
7. Levina, A.S., Berezovskii, M.V., Venjaminova, A.G., Dobrikov, M.I., Repkova, M.N. and Zarytova, V.F. (1993) Photomodification of RNA and DNA fragments by oligonucleotide reagents bearing arylazido group, *Biochimie* 75, 25-27.
8. Schnapp, K.A., Poe, R., Leyva, E., Saundararajan, N. and Platz, M. S. (1993) Exploratory photochemistry of fluorinated aryl azides. Implications for the design of photoaffinity labeling reagents, *Bioconj. Chem.*. 4, 172-177.
9. Schnapp, K.A. and Platz, M.S. (1993) A laser flash photolysis study of di-, tri- and tetrafluorinated phenylnitrenes; implications for photoaffinity labeling, *Bioconj. Chem.* 4, 178-183 .
10. Forster, Th. (1965) Delocalised excitation and excitation transfer, in O. Sinanoglu (ed.), *Modern Quantum Chemistry, pt III.*, Academic Press, N.-Y, pp. 93-137.
11. Lakowicz, J.R. (1986) Energy transfer, in M.G. Kuzmin (ed.), *Principles of Fluorescence Spectroscopy* , Mir Press, Moskow, pp. 262-344.
12. Leyshon, L.J. and Reiser, A. (1972). Sensitized photodecomposition of phenyl azide and α-naphthyl azide, *J. Chem. Soc. Faraday Trans. II*, 11, 1918-1927.
13. Shields, C.J., Farley, D.E., Schuster, G.B., Buchardt, O. and Nielsen, P.E. (1988) Competitive singlet-singlet energy transfer and electron transfer activation of aryl azides: Application to photo-cross-linking experiments, *J.Org.Chem.* 53, 3501-3507.

14. Dale, R.E., Novros, J., Roth, S., Edidin, M. and Brand, L. (1981) Application of Forster long-range exitation energy-transfer, in G.S. Beddard and M.A. West (eds.), *Fluorescent probes.*, Academic Press, L.- N.Y- Toronto, pp. 159-182.

15. Teale, F.W.J. and Constable, D. (1981) Intermolecularly-quenched thiol-specific fluorescence probes. In G.S. Beddard and M.A. West (eds.), *Fluorescent Probe.s*, Academic Press, L. N.-Y. Toronto, pp. 1-20.

16. Sharifian, H.A., Puin, C.-H., Jiang, F.-B. and Park, S.-M. (1985) Interactions of several polycyclic aromatic hydrocarbons with DNA based molecules studied by absorption spectroscopy and fluorescence quenching, *J. Photochem.* **30**, 229-244.

17. Geacintov, N.E., Zhao, R., Kuzmin, V.A., Kim, S.K. and Pecora, L.J. (1993) Mechanisms of quenching of the fluorescence of the benzo[a]pyrene tetraol metabolite model compound by 2'-deoxynucleotides, *Photochem. and Photobiol.* **58**, 185-194.

Mechanism of Oxidative DNA Cleavage

TARGETED OXIDATIVE DNA CLEAVAGE MEDIATED BY OLIGONUCLEOTIDE - CATIONIC METALLOPORPHYRIN CONJUGATES.

J. BERNADOU, G. PRATVIEL and B. MEUNIER
Laboratoire de Chimie de Coordination du CNRS,
205 route de Narbonne, 31077 Toulouse cedex, France

1. Introduction

Oxidative DNA cleavage mediated by transition metal complexes can be ascribed either (i) to an indirect role of the transition metal complex, *i.e.* catalytic generation of superoxide anion from oxygen via an electron transfer or catalytic formation of hydroxyl radicals from hydrogen peroxide, or (ii) to direct DNA breaks produced by high-valent transition metal-oxo complexes.[1,2] Examples of DNA damage due to high-valent iron, copper, nickel and ruthenium species are now available. The present paper is focused on the high efficient and specific DNA cleavage mediated by manganese cationic porphyrin complexes when associated to oxygen atom donors like potassium monopersulfate or magnesium monoperphthalate. Within fifteen years, manganese and other metallated porphyrins have been extensively used to mimic heme-enzymes in oxygenation reactions, olefin epoxidation and alkane hydroxylations.[3,4] More recently cationic derivatives in this series showed nuclease activity[5-13] with applications in molecular biology and potential interest in antitumor or antiviral fields. After a recalling on methods of oxidative activation of metalloporphyrin and mechanisms of DNA cleavage, we present our recent attempts in order to tailor the nuclease activity of these metalloporphyrin cleavers on specific DNA sequences, in both antisense and antigene strategies.

2. Mechanism of activation of the parent compound Mn-TMPyP

Manganese porphyrins with peripheral positive charges, *e.g.* Mn-*meso*-tetra(4-*N*-methylpyridiniumyl)porphyrin (Mn-TMPyP, **1**, Scheme 3), exhibit a strong interaction

B. Meunier (ed.), DNA and RNA Cleavers and Chemotherapy of Cancer and Viral Diseases, 211–223.
© *1996 Kluwer Academic Publishers. Printed in the Netherlands.*

with DNA that "bring" in the vicinity of the target a powerful oxidizing species after oxidative activation of the metal center. The active oxidative species in the case of DNA cleavage is probably the same as that involved in catalytic oxygenation and oxidation reactions described for metalloporphyrins in general, namely a high-valent metal-oxo porphyrin complex able to hydroxylate a C-H bond or to epoxidize an olefin.[3] The active species formed from cationic metalloporphyrins in order to perform oxidative cleavage of DNA is generated in the presence of oxygen atom donor compounds like iodosylbenzene,[6] hydrogen peroxide,[9] magnesium monoperphthalate[10] or, the most efficient, potassium monopersulfate[8,9,13,14].

All collected data on mechanism of DNA cleavage by the "Mn-TMPyP/KHSO$_5$" system (see below) strongly support the hypothesis of a non-diffusible metal-oxo complex as active species in the abstraction of H-atoms from DNA sugar units.

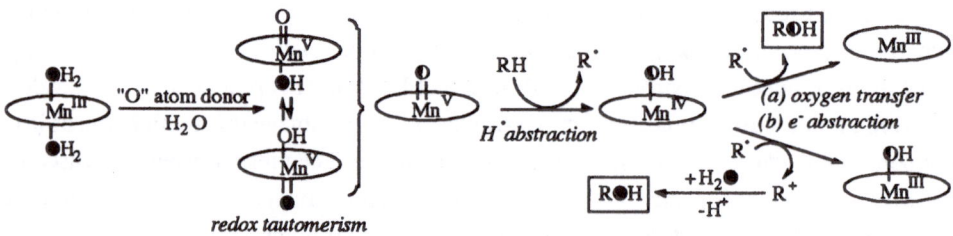

Scheme 1. Two possible oxidative pathways for porphyrin ligands: (a) P-450 route; (b) diverted P-450 route. R-H stands as example for the C1'-H bond of a deoxyribose unit of DNA.

The activation reaction is described in Scheme 1. The question was on a direct oxygenation of the DNA substrate by the rebound of the "HO equivalent" carried by the activated intermediate TMPyP-M^{IV}-OH and the carbon centered radical on deoxyribose (Scheme 1, route a, P-450 route) or an electron transfer (Scheme 1, route b, "diverted P-450" route). Recent studies on the origin of the oxygen atom incorporated within the oxidized sugar residue at the site of the lesion helped us to answer this question.[15] Half of the oxygen atom incorporated within 5-methylene-2-furanone (5MF) after oxidation at C1' (see below for the formation of this DNA sugar residue) came from the solvent and half from KHSO$_5$, the primary oxidant. These data are in agreement with a redox tautomerism mechanism[16] involving the metal-oxo species and the trans axial hydroxo ligand (Scheme 1, on the left). The present observations indicate that, at least in the case of the oxidation of the C-H bond at the 1' position of DNA sugars, only high-valent Mn-oxo-TMPyP was involved, via a chemistry mimicking cytochrome P-450. The Mn^V=O species abstracts the H1'-atom generating a radical which quickly reacts with the oxygen atom of Mn^{IV}-OH to form the carbon-oxygen bond present at C1' of the

5MF precursor. The alternative e⁻ abstraction route (Scheme 1, route b) which should give 100 % oxygen incorporation from the solvent can be discarded.

3. Positioning of Mn-TMPyP with respect to the target

Mn-TMPyP belongs to the class of cationic metalloporphyrins that do not intercalate between DNA base-pairs [6,13,17-23] but that bind in the minor groove of AT rich regions of DNA. The Mn-porphyrin framework is devoid of any H-bonding donor/acceptor capacity and intercalation is precluded by the presence of axial ligand(s) on manganese. So the ability of Mn-TMPyP to select AT rich regions of DNA is apparently electrostatic/steric in origin. It has been proposed that this cationic metalloporphyrin is attracted by the high negative potential at the surface of the minor groove[24-26] of A·T rich sequences. Considering the size of Mn-TMPyP, it could span over 5 to 6 base-pairs in the minor groove of B-form DNA, but the preferred cleaving site is a three consecutive A·T base-pairs sequence creating a suitable "box" for highly selective DNA cleavage.[6,13,27] Beside $(A \cdot T)_3$ cleaving sites, some secondary reacting sequences can also be noted. They consist of one base-pair change in the $(A \cdot T)_3$ site (one G·C base-pair over three, no matter the position of the G·C bp). This secondary reactivity is especially observed on both sides of a main cleaving sequence when drastic cleaving reactions are performed.[12,13] The reactivity at these secondary sites is one order of magnitude less than for $(A \cdot T)_3$ sites.

In order to address the sequence selectivity, we prepared hybrid molecules with the metalloporphyrin as DNA cleaving moiety covalently attached to oligodeoxyribo-nucleotides (ODNs) as specific DNA sequence recognizing agent. In particular, this strategy should lead to a family of synthetic restriction enzymes with tailored specificity which might be useful in the analysis of complex genomes (molecular and cellular biology) or could enhance the ability of an antisense or antigene ODN to block the expression of targeted genes (potential medical applications). Examples are presented in Section 5.

4. Mechanisms of DNA cleavage by activated Mn-TMPyP

Despite the usual reference to chemical nuclease for cationic manganese porphyrin derivatives, a hydrolytic mechanism is not involved. In fact these systems are mainly able to achieve phosphodiester backbone scission on DNA by sugar oxidation,[23] accompanied, as main consequence, by the loss of the base and a part or the totality of

214

the damaged sugar. Two main mechanisms were proposed to explain the breaks observed in the presence of Mn-TMPyP/KHSO$_5$ system: C1' and C5' hydroxylations on deoxyribose units of DNA.

More recently, our cleavage experiments with ODN-metalloporphyrin conjugates showed another type of oxidative lesion, probably involving oxidation of guanine bases.

4.1. HYDROGEN ABSTRACTION at C1'.

H1' atom abstraction is supposed to be the first step of the DNA oxidative attack by Mn-TMPyP (as for enediyne compounds[28] and copper orthophenanthroline complexes[29]) producing a C1' centered radical which quickly reacts with Mn^{IV}-OH to give the 1'-hydroxylated sugar (Scheme 1 and 2). The next key steps of the reaction cleavage are (i) the release of free base associated to the formation of an oxidized abasic site, (ii) the first β-elimination inducing the strand break that leaves a metastable α,β-unsaturated lactone at the 3'-end and a 5'-phosphate end, (iii) the second and fast β-elimination allowing then to recover the 3' phosphorylated ending DNA fragments and 5-methylene-2-furanone (5-MF) as main sugar residue.[11,30,31]

In the course of various DNA and oligonucleotide cleavage reactions with Mn-TMPyP/KHSO$_5$, the following stable reaction products were identified: 5'- and 3'-phosphorylated termini, free bases and 5-MF. More often, another stable product can be observed (furfural) that is indicative of a C5' chemistry which will be described in the next Section. With Mn-TMPyP/KHSO$_5$ as cleaving reagent, C1' represents the main oxidation target in the cleavage of single-stranded DNA and G-C polymers or G-C rich DNA.[30]

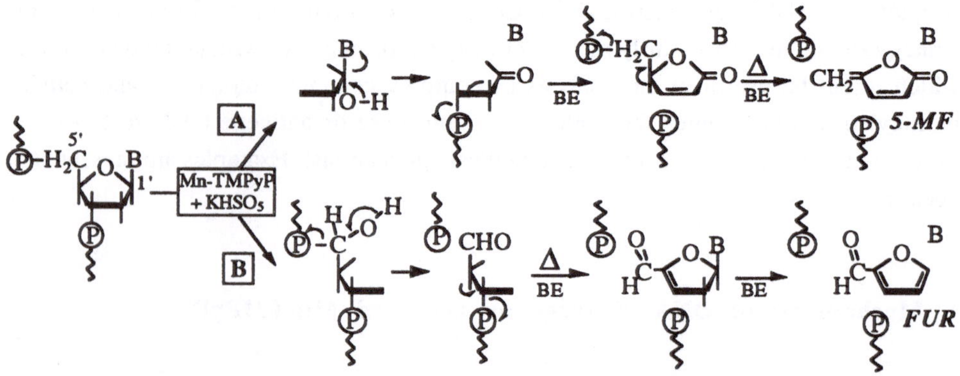

Scheme 2. Proposed mechanism for the cleavage of DNA after hydroxylation of C-H bonds on the 1' and 5' positions of deoxyribose by Mn-TMPyP/KHSO$_5$ (Δ = thermal step; B = base; BE = β-elimination).

4.2. HYDROGEN ABSTRACTION AT C5'

Up to now oxidative chemistry at C5' of deoxyribose has only been demonstrated with enediyne compounds[28] and with Mn-TMPyP[13,30,32]. Either enediynes or the cationic metalloporphyrin Mn-TMPyP initiates the oxidation of the deoxyribose by abstraction of H5' giving a C5' centered radical. Direct hydroxylation by Mn^{IV}-OH follows giving the 5'-OH derivative (Schemes 1 and 2). After a spontaneous cleavage, 3'-phosphate end and 5'-aldehyde ending derivative (direct break of the DNA backbone) are formed. Further evolution (first β-elimination) produces a second break on the DNA backbone with release of a 5'-phosphate end and a α,β-unsaturated aldehyde. At last a second β-elimination gives rise to free base and furfural (FUR) as sugar degradation product.[13,33] HPLC analysis of products released in solution after a heating step allowed the observation and characterization of free bases, oxidized nucleosides and FUR. Besides C1'-H target, C5'-H oxidative activation constitutes the main mechanism for cleavage of A·T rich sequences of double stranded DNA, with selective oxidations on both nucleosides at 3'-sides of the " (A·T)$_3$ box". Determination of the ratio 5-MF/FUR constitutes an index of the respective attacks at C1' and C5' on deoxyribose units.[30]

4.3. NUCLEOBASE DAMAGE

In the course of the study of the nuclease activity of a hybrid "manganese-tris-methylpyridiniumylporphyrin-oligonucleotide" molecule as reactive antisense oligonucleotide,[34] we could observe an alternative mode of oxidative cleavage event. Heating treatment at 90 °C in the presence of 1 M piperidine revealed cleaved DNA fragments clearly corresponding to DNA breaks at the only guanine residues located in the targeted zone of interaction of the vectorized manganese porphyrin with DNA. The exact nature of the damage still remains to be elucidated but it is clear that guanine lesions revealed by the piperidine treatment strongly suggest that the active manganese-oxo species is not only acting as a deoxyribose cleaver as observed when the cationic manganese porphyrin is not tethered to an oligonucleotide.

5. Oligonucleotide-metalloporphyrin conjugates: a strategy to target both single- and double-stranded DNA

The cationic metalloporphyrin entity was covalently linked to the 5'-end of ODNs[35] in order to modulate the sequence selectivity of cleavage. This type of conjugates, ODN-M-POR (POR stands for porphyrin), should allow to target the cleavage reactivity of the metalloporphyrin onto a short defined sequence (complementary to the oligonucleotide part of the molecule) of single-stranded DNA (ssDNA). Alternatively, in order to cleave double-stranded DNA (dsDNA), the cationic metalloporphyrin was covalently linked to triplex-forming oligonucleotides.

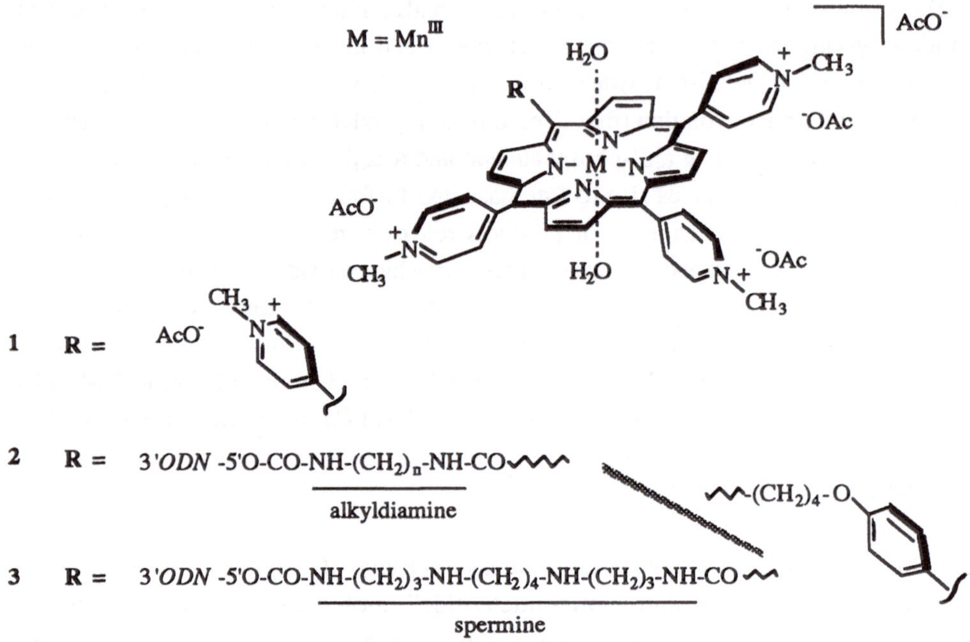

1 R =

2 R = 3'*ODN* -5'O-CO-NH-(CH$_2$)$_n$-NH-CO$\sim\sim\sim$

   alkyldiamine $\sim\sim$-(CH$_2$)$_4$-O

3 R = 3'*ODN* -5'O-CO-NH-(CH$_2$)$_3$-NH-(CH$_2$)$_4$-NH-(CH$_2$)$_3$-NH-CO$\sim\sim$

   spermine

Scheme 3. General structures of cationic metalloporphyrin derivatives. **1**: Mn-*meso*-tetra(4-*N*-methylpyridiniumyl)porphyrin (Mn-TMPyP); **2** and **3**: oligonucleotide-metalloporphyrin conjugates with an alkyldiamine linker and a spermine linker, respectively.

5.1. SELECTIVE SS DNA CLEAVAGE (ANTISENSE STRATEGY)

A hybrid "manganese-tris-methylpyridiniumylporphyrin-oligonucleotide" molecule (**2**, Scheme 3) was shown to recognize and cut selectively its complementary ssDNA strand even in the presence of a large excess of random DNA.[34,36,37] The cleavage of the target ssDNA occured in the predicted zone, in the vicinity of the metalloporphyrin site and could be observed at 10 nM concentration of conjugate and 10 nM of target (see legend of Figure 1 for some experimental details). As it can be noted on Figure 1, lane 1, the decrease of the intensity of the band of the full length material was accompagnied by the appearance of a smear of cleavage products (compare lane 1 and the control lane 5). After piperidine treatment (lanes 2 and 4) the smear and a part of the full length material were transformed to DNA fragments all resulting from alkali-labile lesions at guanine residues. The cleavage products observed on lane 2 probably result from

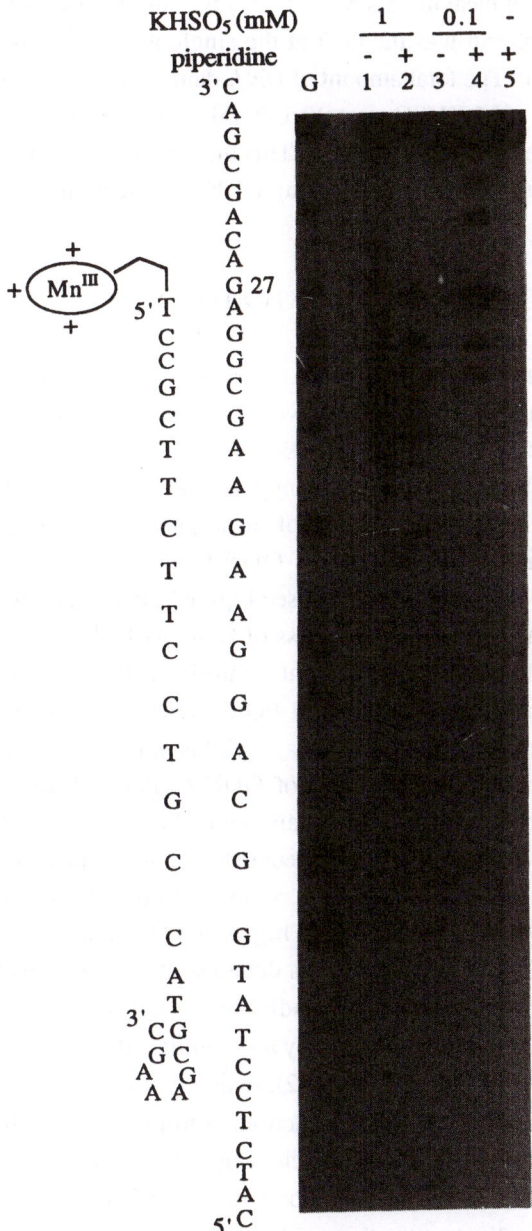

Figure 1. Selective cleavage of a 35-mer ssDNA target by an ODN-M-POR conjugate. 10 nM 5'[^{32}P] 35-mer target and 10 nM ODN-M-POR were annealed in the presence of an excess of herring testes DNA (0.4 mM b) in 100 mM NaCl, 50 mM phosphate buffer pH 7. The cleavage reaction, initiated by addition of KHSO$_5$ (1 or 0.1 mM), lasted 1 h at 4 °C. KHSO$_5$ was destroyed by addition of 10 mM Hepes buffer. Piperidine treatment, when performed, consisted in 1 h further incubation at 90 °C with 1 M piperidine. [34]

consecutive cleavage events on the same strand (the target ssDNA was cleaved more than one time due to catalytic potential of the cleaving agent[3,15,16]) because the first site of oxidative attack by the metalloporphyrin was the G27 at the single and the double strand junction (compare lanes 2 and 4). The total amount of DNA damage mediated by the ODN-M-POR in the presence of 1 mM KHSO$_5$ reached, in this example, 90 % of target ssDNA and 50 % in the presence of 0.1 mM KHSO$_5$. This high yield of selective cleavage is remarkable compared to similar systems involving ODN-chemical nuclease conjugates.[38-48]

5.2. SELECTIVE DS DNA CLEAVAGE (TRIPLE HELIX STRATEGY)

When the manganese-tris(methylpyridiniumyl)porphyrin was attached to an ODN able to bind to dsDNA by triple-helix formation, the conjugate was then able to cleave its dsDNA target. The nature of the linker between the metalloporphyrin and the ODN proved of primary importance for the cleavage efficiency of the ODN-M-POR molecule.[49,50] When the linker is a spermine instead of an aliphatic diamine (3 instead of 2, Scheme 3), the cleavage efficiency of the ODN-M-POR compound increased significantly: degradation of the target was 80 % (see Figure 2, lane 2 and the legend for experimental details) with a 100-fold molar excess of ODN-M-POR (1 µM) with respect to target (10 nM). In the same experimental conditions, the yield of cleavage was only 40 % with the corresponding ODN-M-POR having an aliphatic linker. This 80 % value represents one of the best cleavage yields reported in the literature.[51-55] The difference between the two types of ODN-M-POR (2 or 3, Scheme 3) may be due to hydrophobic repulsion of the aliphatic linker with DNA compared to spermine which is kown to contact favorable electrostatic interactions with DNA and to stabilize triple-helix structures. Furthermore we could verify that the Tm of triple-helix formed with spermine-linked ODN-M-POR was higher (42 °C) than for the aliphatic diamine-linked ODN-M-POR (30 °C). Our results demonstrate that efficient cleavage of dsDNA can be achieved in physiological conditions by the means of a triplex-forming ODN linked to a cationic metalloporphyrin by a spermine tether.

DNA fragments detected in these conditions (Figure 2), mainly at the T_{15}, C_{16}, T_{20} and A_{21} positions i.e. near the triplex-to-duplex junction, comigrated with the corresponding Maxam-Gilbert fragments; the pattern of cleavage observed on both strands (not shown) revealed an asymmetric strand distribution with a shift to the 5'-end. These preliminary results probably indicate that the oxidative reaction occurs by hydroxylation of sugar C-H bonds available within the major groove (H_S2', H3' and H_R5').[1]

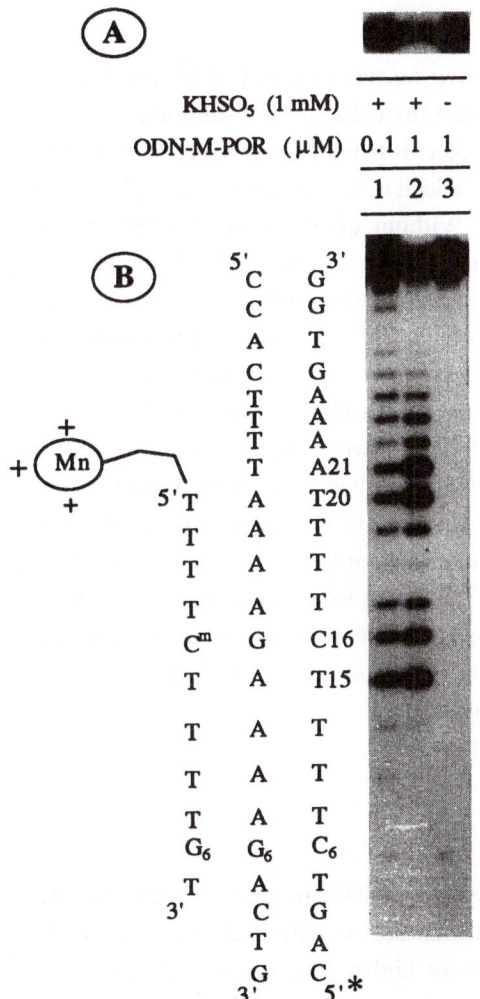

Figure 2. Selective cleavage of a dsDNA target by an ODN-M-POR conjugate with a spermine linker. 10 nM duplex target (5'[^{32}P] on the pyrimidine strand) and 0.1 or 1 µM of ODN-M-POR were annealed in 25 mM Tris/HCl buffer pH 7, 100 mM NaCl, 10 mM MgCl$_2$. The cleavage reaction, initiated by addition of 1 mM KHSO$_5$, lasted 1 h at room temperature. A: underexposed autoradiogram of the full length material of the B part of this figure.[50]

5.3. STRUCTURAL MODIFICATIONS OF ODN-M-POR IN ORDER TO IMPROVE THEIR POTENCY

In view of cell culture or *in vivo* assays, an easy way to increase the metabolic stability of an antisense oligonucleotide, is to add a stable 3'-mini-hairpin at the 3'-end of its ODN sequence in order to protect it from 3'-exonucleases degradation.[56-59] The 5'-GCGAAAGC 8-mer was thus added to the sequence of the ODN-M-POR molecules (see one example on Figure 1). Without question we checked that the 3'-mini-hairpin structure was very efficient in protecting the ODN-M-POR conjugates toward the degradation by 3'-exonuclease.[34] We addressed the question of the interference of the 3'-loop on the DNA cleaving efficiency of these modified type of ODN-M-POR. We found that the presence of the 3'-loop slightly decreased the binding affinity of the conjugated ODN with its complementary ssDNA (for a 19-mer ODN sequence, the Tm of the conjugate bearing a 3'mini-loop is lowered by 6 °C compared to the Tm of the non-protected analogue). The cleaving efficiency was slightly reduced by the presence of the 3'-mini-hairpin.[34] But this decrease of reactivity may be compensated by the higher metabolic stability of these modified antisenses for experiments in cell culture medium or *in vivo*. Addition of this 3'-mini-hairpin with three consecutive A bases should be restricted to antisense oligonucleotides that do not contain three consecutive T bases near the 3'-end, in order to avoid the formation of a secondary larger loop as described in ref [34] that preclude any antisense effect.

6. Conclusion

Development of artificial cleavage of DNA recently appears as a challenging area in the rational design of future antitumoral or antiviral agents as well as in the field of molecular biology. The need for highly efficient (in terms of yield of cleavage) and highly specific (sequence specificity on large DNA fragments) cleaving reagents is obvious and considerable achievements have already been done. Using oligonucleotide-cationic metalloporphyrin conjugates we showed that a remarkably efficient selective cleavage *in vitro* can be obtained both in the antisense and the antigene strategy. This is encouraging in vue of future applications of ODN-M-POR conjugates as tools in molecular biology or as reactive antisense or antigene oligonucleotides able to degrade , respectively, the messenger RNAs or the genomic sequence of a given gene *in vivo*.

The authors are deeply indebted to the work of many co-workers and collaborators whose names are listed in the reference list of this review article. We are also grateful to the CNRS, the 'Association pour la Recherche contre le Cancer' (ARC, Villejuif), the

'Agence Nationale de Recherches sur le Sida' (ANRS, Paris), the 'Région Midi-Pyrénées' and Genset (Paris) for financial support.

References

1. Pratviel, G., Bernadou, J., and Meunier B. (1995) *Angew. Chem. Int. Ed. Engl.* **3 4**, 746-769.

2. Meunier, B., Pratviel, G., and Bernadou J. (1994) *Bull. Soc. Chim. Fr.* **1 3 1**, 933-943.

3. Meunier, B. (1992) *Chem. Rev.* **9 2**, 1411-1456.

4. McMurry, T.J. and Groves, J.T. (1985) *Cytochrome P-450: Structure, Mechanism and Biochemistry* (Ed.: P. Ortiz de Montellano), Plenum Press, New-York, pp. 1-28.

5. Fiel, R.J., Beerman, T.A, Mark, E.H., and Datta-Gupta, N. (1982) *Biochem. Biophys. Res. Commun.* **1 0 7**, 1067-1074.

6. Ward, B., Skorobogaty, A., and Dabrowiak, J. C. (1986) *Biochemistry* **2 5**, 6875-6883.

7. Fouquet, E., Pratviel, G., Bernadou, J., and Meunier, B. (1987) *J. Chem. Soc., Chem. Commun.* 1169-1171.

8. Dabrowiak, J.C., Ward, B., and Goodisman, J. (1989) *Biochemistry* **2 8**, 3314-3322.

9. Bernadou, J., Pratviel, G., Bennis, F., Girardet, M., and Meunier, B. (1989) *Biochemistry* **2 8**, 7268-7275.

10. Pratviel, G., Bernadou, J., Ricci, M., and Meunier, B. (1989) *Biochem. Biophys. Res. Commun.* **1 6 0**, 1212-1218.

11. Pratviel, G., Pitié, M., Bernadou, J., and Meunier, B. (1991) *Nucleic Acids Res.* **1 9**, 6283-6288.

12. Pitié, M., Pratviel, G., Bernadou, J., and Meunier B. (1993) *The Activation of Dioxygen and Homogeneous Catalytic Oxidation* (D. H. R. Barton, A. E. Martell, and D. T. Sawyer, eds.), Plenum Press, New-York, pp. 333-346.

13. Pitié, M., Pratviel, G., Bernadou, J., Meunier, B. (1992) *Proc. Natl. Acad. Sci. USA* **8 9**, 3967-3971.

14. Ward, B., Rehfuss, R., and Dabrowiak, J.C., (1987) *J. of Biomol. Struct. & Dynamics* **4**, 685-695.

15. Pitié, M., Bernadou, J., and Meunier, B. (1995) *J. Am. Chem. Soc.* **1 1 7**, 2935-2936.

16. Bernadou, J., Fabiano, A-S., Robert, A., and Meunier, B., (1994) *J. Am. Chem. Soc.* **1 1 6**, 9375-9376.

17. Byrnes, R.W., Fiel, R.J., Datta-Gupta, N. (1988) *Chem. Biol. Interact.* **6 7**, 225-241.

18. Pasternack, R.F., Gibbs, E.J. (1989) *Metal-DNA Chemistry* (Ed.: T. D. Tullius), ACS Symposium Series 402, pp 59-73.

19. Raner, G., Ward, B., and Dabrowiak, J.C. (1988) *J. Coord. Chem.* **1 9**, 17-23.

20. Fiel, R.J. (1989) *J. of Biomol. Struct. & Dynamics* **6**, 1259-1273.

21. Marzilli, L.G. (1990) *New J. Chem.* **1 4**, 409-420.

22. Bromley, S.D., Ward, B., and Dabrowiak, J.C. (1986) *Nucleic Acids Res.* **1 4**, 9133-9148.

23. Ward, B., Skorobogaty, A., and Dabrowiak, J.C. (1986) *Biochemistry* **2 5**, 7827-7833.

222

24. Hui, X., Gresh, N., and Pullman, B. (1990) *Nucleic Acids Res.* **1 8**, 1109-1114.

25. Lavery, R. and Pullman, B. (1985) *J. Biomol. Struct. Dynamics*, **2**, 1021-1032.

26. Weiner, P.K., Langeridge, R., Blaney, J.M., Schaefer, R., and Kollman, P.A. (1982) *Proc. Natl. Acad. Sci. USA*, **7 9**, 3754-3758.

27. Dabrowiak, J.C., Ward, B., and Goodisman, J. (1989) *Biochemistry*, **2 8**, 3314-3322.

28. Goldberg, I.H. (1991) *Acc. Chem. Res.* **2 4**, 191-198.

29. Goyne, T.E. and Sigman, D.S. (1987) *J. Am. Chem. Soc.* **1 0 9**, 2846-2848.

30. Pratviel, G., Pitié, M., Bernadou, J., and Meunier, B. (1991) *Angew. Chem. Int. Ed. Engl.* **3 0**, 702-704.

31. Bernadou, J., Lauretta, B., Pratviel, G., and Meunier, B. (1989) *C. R. Acad. Sci. Paris,* **309 III**, 409-414.

32. Pratviel, G., Duarte, V., Bernadou, J., and Meunier, B. (1993) *J. Am. Chem. Soc.* **1 1 5**, 7939-7943.

33. Pratviel, G., Pitié, M., Périgaud, C., Gosselin, G., Bernadou, J., and Meunier, B. (1993) *J. Chem. Soc. Chem. Commun.*, 149-151.

34. Mestre, B., Pratviel, G., Meunier,B. (1995) *Bioconjugate Chem.* **6**, 466-472.

35. Casas, C., Lacey, C.J., and Meunier, B. (1993) *Bioconjugate Chem.* **4**, 366-371.

36. Pitié, M., Casas, C., Lacey, C.J., Pratviel, G., Bernadou, J., and Meunier, B. (1993) *Angew. Chem. Int. Ed. Engl.* **3 2**, 557-559.

37. Pratviel, G., Bigey, P., Bernadou, J., and Meunier, B. (1995) *Metal and Genetics* (B. Sarkar, ed.), Marcel Dekker, New-York, pp. 153-171.

38. Dreyer, G.B. and Dervan, P.B. (1985) *Proc. Natl. Acad. Sci. USA* **8 2**, 968-972.

39. Chu, B.C.F. and Orgel, L.E. (1985) *Proc. Natl. Acad. Sci. USA* **8 2**, 963-967.

40. Chen, C.H.B. and Sigman, D.S. (1986) *Proc. Natl. Acad. Sci. USA* **8 3**, 7147-7151.

41. Bergstrom, D.E. and Gerry, N.P. (1994) *J. Am. Chem. Soc.* **1 1 6**, 12067-12068.

42. Perrin, D.M., Mazumder, A., Sadeghi F., and Sigman, D.S. (1994) *Biochemistry* **3 3**, 3848-3854.

43. François, J-C., Thuong N.T., and Hélène, C. (1994) *Nucleic Acids Res.* **2 2**, 3943-3950.

44. Le Doan, T., Perrouault, L., Chassignol, M., Thuong, N.T., and Hélène, C. (1987) *Nucleic Acids Res.* **1 5**, 8643-8659.

45. Ortigao, J.F., Ruck, A., Gupta, K.C., Rosch, R., Steiner, R., and Seliger, H. (1993) *Biochimie* **7 5**, 29-34.

46. Fedorova, O.S., Savitskii, A.P., Shoikhet, K.G., and Ponomarev, G.V. (1990) *FEBS Lett.* **2 5 9**, 335-337.

47. Frolova, E.I., Ivanova, E.M., Zarytova, V.F., Abramova, T.V., and Vlassov, V.V. (1990) *FEBS Lett.* **2 5 9**, 101-104.

48. Groves, J.T. and Kady, I.O. (1993) *Inorg Chem.* **3 2**, 3868-3872.

49. Bigey, P., Pratviel, G., and Meunier, B. (1995) *J. Chem. Soc. Chem. Comm.*, 181-182.

50. Bigey, P., Pratviel, G., and Meunier, B. (1995) *Nucleic Acids Res.*, **2 3**, 3894-3900.

51. Dervan, P.B. (1992) *Nature* **3 5 9**, 87-88.

52. Strobel, S.A., Moser, H.E., and Dervan, P.B. (1988) *J. Am. Chem. Soc.* **1 1 0**, 7927-7929.

53   Perrouault, L., Asseline, U., Rivalle, C., Thuong, N.T., Bisagni, E., Giovannangeli, C., Le Doan, TL., and Hélène, C. (1990) *Nature* **3 4 4**, 348-360.

54   Praseuth, D., Perrouault, L., Le Doan, T.L., Chassignol, M., Thuong, N.T., and Hélène, C. (1988) *Proc. Natl. Acad. Soc. USA* **8 5**, 1349-1353.

55.   Shimizu, M., Inoue, H., and Ohtsuka, E. (1994) *Biochemistry* **3 3**, 606-613.

56.   Hirao, I., Nishimura, Y., Tagawa, Y.-i., Watanabe, K., and Miura K.-i. (1992) *Nucleic Acids Res.* **2 0**, 3891-3896.

57.   Hirao, I., Kawai, G., Yoshizawa, S., Nishimura, Y., Ishido, Y., Watanabe, K., and Miura, K.-i. (1994) *Nucleic Acids Res.* **2 2**, 576-582.

59.   Khan, I.M. and Coulson J.M. (1993) *Nucleic Acids Res.* **2 1**, 2957-2958.

PULSE RADIOLYSIS AND LASER PHOTOLYSIS STUDIES ON NUCLEOTIDE RADICALS. THE PROTONATION STATE OF THE RADICALS.

S. Steenken
Max–Planck–Institut für Strahlenchemie
D–45413 Mülheim, Germany

Abstract

Changes in the oxidation state of the DNA bases, induced by oxidation (ionization) or by reduction (electron capture), have drastic effects on the acidity or basicity, respectively, of the molecules. Since in DNA every base is connected to its complementary base in the other strand, any change of the electric charge status of a base in one DNA strand that accompanies its oxidation or reduction may affect also the other strand via proton transfer across the hydrogen bonds in the base pairs. The free energies for electron transfer to or from a base can be drastically altered by the proton transfer processes that accompany the electron transfer reactions. Electron–transfer (ET) induced proton transfer sensitizes the base opposite to the ET–damaged base to redox damage, i.e. damage produced by separation of charge (ionization) has an increased chance of being trapped in a base pair. Of the two types of base pair in DNA, A–T and C–G, the latter is more sensitive to both oxidative and reductive processes than the former.

Proton transfer induced by ET does not only occur between the heteroatoms (O and N) of the base pairs ("intra–pair proton transfer"), but also to and from adjacent water molecules in the hydration shell of DNA ("extra–pair proton transfer"). These proton transfers can involve carbon and as such are likely to be irreversible. It is the A–T pair which appears to be particularly prone to undergo such irreversible reactions.

A. Introduction

In DNA, due to the pairing of the bases, the protonation state of a base radical does not only depend on the "intrinsic" acidity or basicity of the radical but also on that of the complementary base in the opposite strand, which is the "natural" proton acceptor or donor ("intra-pair proton transfer").[1] This is in contrast to the situation in aqueous solution where the proton exchange partner is always the same, i.e. bulk H_2O. In DNA the situation is really even more complex due to the fact that at least some of the O and N atoms between which exist intra-pair hydrogen bonds are involved in hydrogen bonds with water

B. Meunier (ed.), DNA and RNA Cleavers and Chemotherapy of Cancer and Viral Diseases, 225–247.

molecules in the hydration shell of DNA.[2] In the following, these bonds will be called "extra-pair hydrogen bonds". In other words, in DNA in principle both intra- and extra-pair proton transfers will have to be considered in order to be able to predict the protonation state and thereby the reactivity of a particular one-electron oxidized or reduced base. Obviously, this is not at all an easy task, particularly since the free energies of (de)protonation reactions depend strongly on the orientations of proton donor and acceptor.[3] In the following, an attempt at predicting protonation states of base radicals in their base pairs will nevertheless be made, using essentially acidity/basicity and reactivity data obtained from aqueous phase studies[1] and combining these with structural data from solid (single crystal) phase accumulated[4] using ESR/ENDOR techniques. It is hoped that this approach will improve the understanding of the mechanisms of damage to DNA induced not only by ionizing radiation and high energy light but also by chemical oxidizing or reducing agents.

B. Changes of acidity/basicity resulting from removal/addition of one electron from/to a molecule

1. Electron removal (One–electron oxidation)

1.1 Deoxyadenosine (A)

Deoxyadenosine is not easily oxidized in aqueous solution. One–electron removal from this molecule requires[5] the use of the strong oxidant $SO_4^{\bullet-}$ ($E^0 = 2.5 - 3.1$ V/NHE[6])[1,7,8] or photolysis with 193 nm light.[9] The $SO_4^{\bullet-}$ radical, which can easily be produced from $S_2O_8^{2-}$ by reaction with e^-_{aq} or by photolysis (see eq 1),

$$S_2O_8^{2-} \quad \begin{cases} \xrightarrow{+\ e^-_{aq}} & SO_4^{\bullet-} + SO_4^{2-} \\ \xrightarrow{h\ \nu} & 2\ SO_4^{\bullet-} \end{cases} \qquad (1)$$

reacts with A with the rate constant 3.2×10^9 M^{-1} s^{-1}.[1] In this reaction, the N^6–centered radical $A(-H)^\bullet$ ($\equiv A(N^6\text{-H})^\bullet$ is produced[1,7,8,10] by electron removal followed by deprotonation from N^6 of the resulting radical cation, $A^{\bullet+}$. Also in single crystals, $A^{\bullet+}$ deprotonates to give the neutral radical, $A(-H)^\bullet$, even at 10 K,[11-13] which is evidence for the large intrinsic driving force for this reaction.

A(−H)[.] is also produced by reaction in aqueous solution by the OH radical, via addition of [.]OH at C4 of A followed by dehydration of the OH adduct[14,15], analogous to the case of deoxyguanosine (see eq 4). However, this is not the <u>only</u> reaction of OH[.]. An additional ≈ 50% of the OH radicals react by attachment at C8 followed by opening of the imidazole ring.[15]

It was found that the absorption spectrum of the species produced in aqueous solution by reaction with photochemically generated $SO_4^{.-}$, identified as A(−H)[.],[7,8,15] does not change from pH ~ 11 down to pH ~ 1, from which it was concluded[1] that the pK_a of $A^{.+}$ is ≤ 1.[16] Taking pK_a ($A^{.+}$) ≤ 1 and pK_a (A) ≥ 14,[17] ΔpK_a ($A/A^{.+}$) results as ≤ −13. This means that the acidity of the adenine moiety[18] is increased by ≥ 13 orders of magnitude by its one–electron oxidation. This ΔpK_a corresponds to a differential driving force for deprotonation of the radical cation compared to its parent of ≥ 18 kcal/mol.

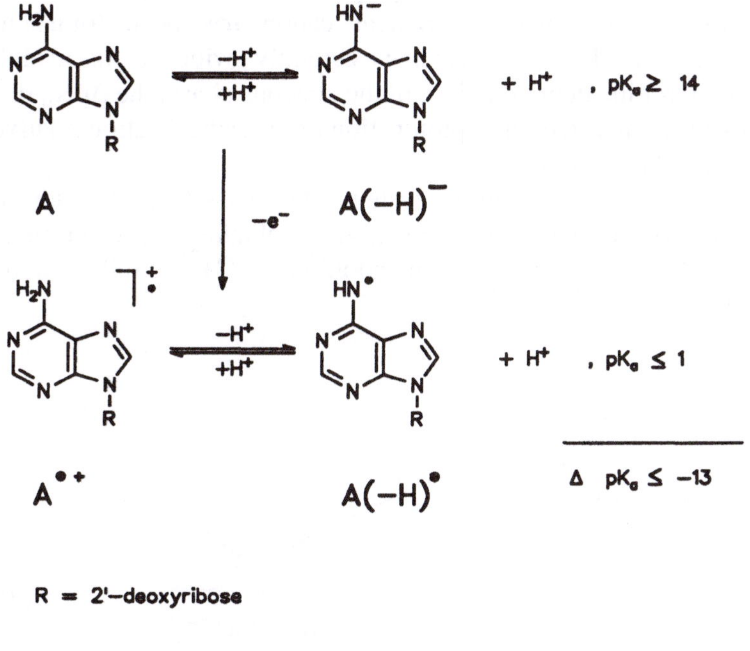

(2)

R = 2'−deoxyribose

1.2 Deoxyguanosine (G).

In aqueous solution, this compound is a weak acid with a pK_a of 9.5. On removal of one electron by use of the oxidants $SO_4^{.-}$,[19] $Br_2^{.-}$,[19] or Tl(II)[5a,19] or by 193 nm photoionization,[9] the radical cation is formed which has a pK_a of 3.9, as determined by time–resolved optical and conductance techniques.[19]

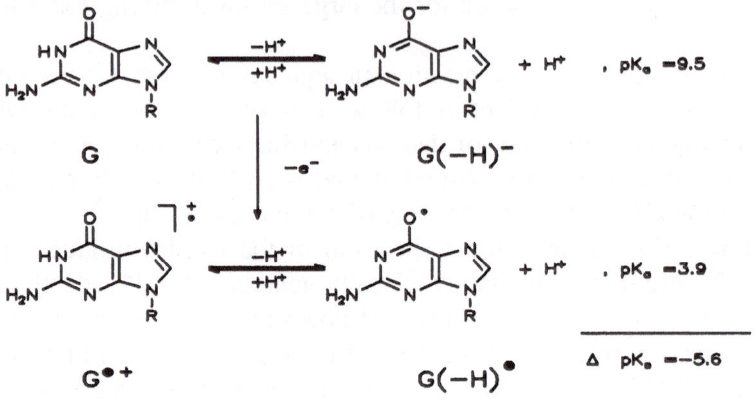

(3)

R = 2'-deoxyribose

The difference in the pK$_a$ values of the parent, G (pK$_a$ = 9.5) and that of its radical cation G$^{\bullet+}$ (pK$_a$ = 3.9) is -5.6. This means that removal of one electron from the molecule increases its acidity by 5.6 orders of magnitude.

Also in the crystal state, the radical cation has been found to deprotonate, even at ~ 10 K,[20,21] which is certainly evidence for a high intrinsic driving force for the deprotonation. In the aqueous phase, the ΔpK$_a$ of -5.6 corresponds to a driving force for deprotonation of the radical cation relative to that of the parent of 7.6 kcal/mol.

As in the case of the adenine system, the one–electron–oxidized molecule G$^{\bullet+}$ or G(-H)$^{\bullet}$ (≡ G(N1-H)$^{\bullet}$) can also be obtained by reaction in aqueous solution with the OH radical, via addition at C4[22] followed by elimination of H$_2$O (k = 5x10^3 s^{-1}).[23]

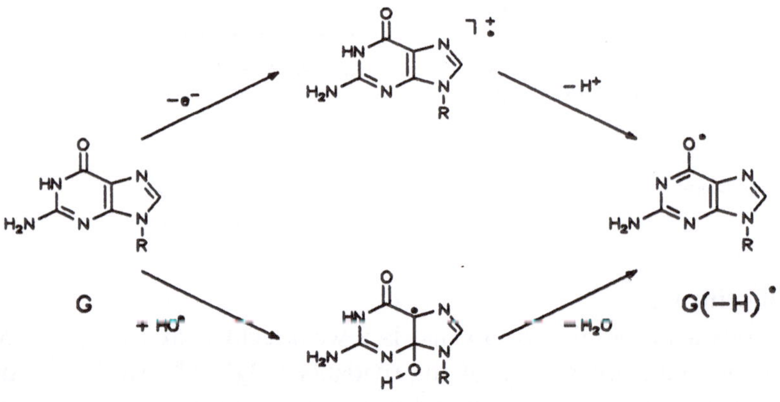

(4)

R = Me, (2'-deoxy)ribose(5-phosphate)

Thus, with the two purines, A and G, direct oxidation (ionization) and indirect oxidation (via the OH radical) lead to the same radicals. However, the OH radical does not only add at C4 (leading to one–electron oxidation of the purine) but also at C8, in which case opening of the imidazole ring may occur yielding radicals which are reducing, or, alternatively, oxidation of the radical to yield 8-hydroxyguanine derivatives.[1,23]

Concerning the deprotonated radical cation, G(-H)$^{\bullet}$, this species is strongly oxidizing.[22,24] Its reduction potential at pH 7, E^1_7, can be estimated to be ≥ 1.2 V/NHE.[25]

1.3 Deoxycytidine (C)

In aqueous solution, C can be oxidized rapidly only with $SO_4^{\bullet-}$. Not sufficiently reactive are $Br_2^{\bullet-}$, $N_3^{\bullet}$ or Tl(II). Ionization can also be achieved with 193 nm light.[9] The reaction of $SO_4^{\bullet-}$ with C has been suggested[7] to lead to an anilino–type radical formed by one–electron oxidation followed by deprotonation from the exocyclic nitrogen:

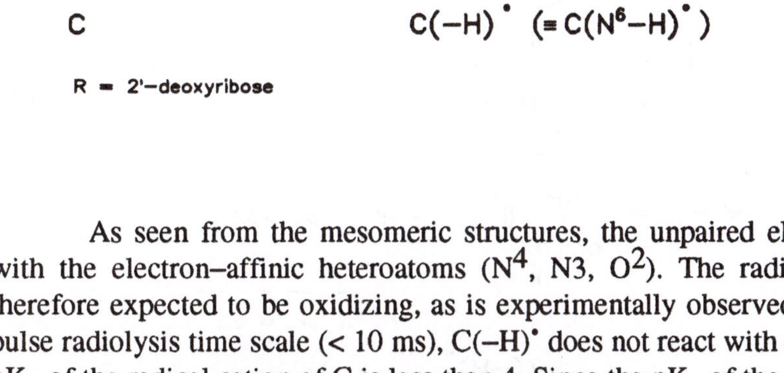

$$C \qquad\qquad C(-H)^{\bullet} \ (\equiv C(N^6-H)^{\bullet})$$

R = 2'–deoxyribose

(5)

As seen from the mesomeric structures, the unpaired electron resides with the electron–affinic heteroatoms (N^4, N3, O^2). The radical C(–H)$^{\bullet}$ is therefore expected to be oxidizing, as is experimentally observed.[7,26] On the pulse radiolysis time scale (< 10 ms), C(–H)$^{\bullet}$ does not react with oxygen.[7] The pK_a of the radical cation of C is less than 4. Since the pK_a of the parent, C, is > 13, ΔpK_a (C/C$^{\bullet+}$) $\leq$ -9, which means that the Brönsted acidity of C increases by ≥ 9 orders of magnitude on one–electron oxidation.

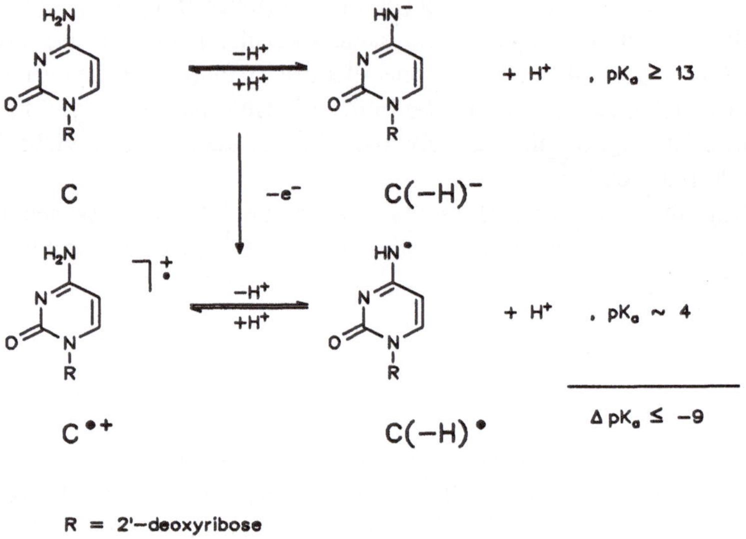

$$C \qquad C(-H)^-$$

(6)

+ H⁺ , pKₐ ≥ 13 → rendered: + H$^+$, $pK_a \geq 13$

$$\Delta pK_a \leq -9$$

R = 2'—deoxyribose

1.4 Thymidine (T).

Pulse radiolysis experiments with optical and conductance detection have shown that the radical cation T$^{\bullet+}$ reversibly deprotonates at N3 to give the N3—centered neutral radical, T(N3-H)$^{\bullet}$ ($\equiv$ T(-H)$^{\bullet}$). The pK_a of the radical cation is 3.6. Since the parent, T, has a pK_a of 9.9, it is evident that one—electron removal from the molecule leads to a drastic increase in its acidity.

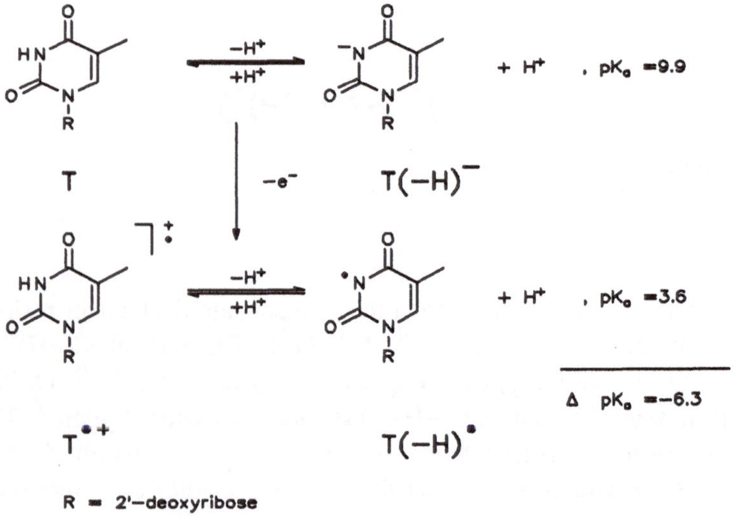

$$T \qquad T(-H)^-$$

(7)

+ H$^+$, $pK_a = 9.9$

+ H$^+$, $pK_a = 3.6$

$$\Delta \ pK_a = -6.3$$

R = 2'—deoxyribose

In addition to the reversible proton transfer from and to N3 (eq 7), there is an irreversible deprotonation from the methyl group at C5 and also a nucleophilic attack of a water molecule at C6 followed by deprotonation to give the (oxidizing) 6–hydroxy-5,6-dihydrothymin-5-yl radical (T(C6OH)˙), as shown in eq 8:[27]

$$T(C^5-H)˙ \qquad (8)$$

$$T(C6OH)˙$$

R = Me, 2'–deoxyribose

The deprotonation from the methyl group of T to give the allyl-type radical (eq 8a) is the dominant reaction of the radical cation in the solid state[4a,c,13,28,29] so there is again an independence of reactivity on the environment (crystalline/solution). However, it is worth mentioning that the deprotonation product from N3 of the radical cation of T, T(N1-H)˙ (see scheme 11), has so far not been seen in irradiated single crystals or matrices.[30]

2. Electron addition (One–electron reduction).

All the nucleic acid bases have a very high reactivity with the hydrated electron, e^-_{aq} (for a collection of the rate constants see[1,31]), a reaction in which the corresponding "electron adducts" are formed:

$$e^-_{aq} + base \rightarrow base˙^-$$
$$\text{electron adduct}$$

Due to the increase in electron density, the electron adducts are considerably stronger bases (proton acceptors) than their parental precursors. Examples for this phenomenon will be given in the following sections.

2.1 Thymidine (T)

On the basis of electron spin resonance[32] and pulse radiolysis with optical[33-35] and conductance[35] detection, the electron adduct of T is negatively charged (i.e. it is a radical anion) at pH values above ~ 7. At lower pH values, it protonates rapidly and reversibly at O^4 to form the neutral radical $T(O^4H)^{\cdot}$ (eq 9).[35] The pK_a has been determined by time–resolved optical spectroscopy (pulse radiolysis) to be 6.9.[34] If this value is compared with that for the parent (≈ -5)[36] it is evident that electron addition to T increases its basicity by ~ 12 orders of magnitude:

$$(9)$$

R = 2'–deoxyribose

One of the consequences of the increased electron density of the T system is that protonation at carbon becomes possible. This irreversible reaction takes place at C6 leading to the (oxidizing) 5,6–dihydrothymidine–5–yl radical,[32,35] abbreviated as $T(C6H)^{\cdot}$ (eq 10). The protonation on carbon is catalyzed by phosphate, the great efficiency of this reaction making it possible to observe the product radical by pulse radiolysis or in–situ–radiolysis ESR on the ~ ms timescale.[32,35] The radical $T(C6H)^{\cdot}$, often abbreviated as $TH^{\cdot}$ in the ESR literature, has been observed also on irradiation of single crystals of T systems[37] and of DNA at temperatures $\geq$ 77 K,[38] where it is the most prominent, if not the only, "final" radical product from the chain of events that start with the capture of the electron ejected by the ionizing radiation.[4,13,28]

$$T^{\bullet-} \qquad \xrightarrow{+H_2O} \qquad T(C6H)^{\bullet} \qquad + \ OH^- \qquad (10)$$

R=2'-deoxyribose(5'-phosphate)

It is noteworthy that in single crystals of thymidine (and of 1-methylthymine) at 8 K the electron adduct is protonated at O^4 (like it is in aqueous solution at pH ≤ 7).[37,39]

2.2 Deoxycytidine (C).

The radical anion $C^{\bullet-}$, formed by electron addition to C in neutral solution, is protonated by water[40] (eq 11) in less than 4 ns,[41] as shown by conductance techniques. This protonation does not lead to a loss of reducing ability[40] (the neutral radical is able to one–electron–reduce even very weak oxidants such as N–methylpyridine cations or even thymidine).[34] From this it is concluded that $C^{\bullet-}$ gets protonated on a hetero–atom (O^2 or N3, see eq 11), and not on carbon, in which case the radical would not be a good reductant (as concluded from the behavior[42] of the corresponding radical from uracil or thymine).

$$C^{\bullet-} \qquad C(N3H)^{\bullet} \qquad C(O^2H)^{\bullet} \qquad (11)$$

R = 2'-deoxyribose(5'-phosphate)

With conductance methods it was shown[40] that $C(N3H)^{\bullet}$ does not deprotonate up to pH 10.6. The optical absorption spectrum of the radical remains unaltered from pH 6 up to 13.[1,34] If it is assumed that the spectra of $C(N3H)^{\bullet}$ and $C^{\bullet-}$ are different (as they are for $T(O^4H)^{\bullet}/T^{\bullet-}$,[33-35] U(racil) $(O^4H)^{\bullet}/U^{\bullet}$,[34] and $A(NH)^{\bullet}/A^{\bullet-}$,[43] this observation means that the pK_a of $C(N3H)^{\bullet}$ is larger than 13. Comparison with the pK_a for the parent base (4.4) gives ΔpK_a $(C(N3H)^+/C(N3H)^{\bullet}) \geq 8.6$, i.e. the basicity of C is increased by electron addition by ≥ 8.6 orders of magnitude.

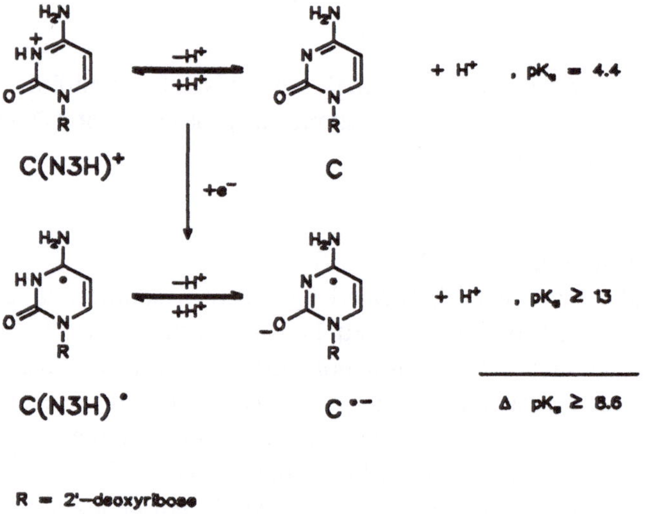

(12)

$R = 2'-deoxyribose$

2.3 Adenosine (A).

The adenosine electron adduct, $A^{\bullet-}$, has been shown by conductance to be rapidly protonated by water.[17,44] The initial protonation is on the nitrogens and is reversible with a pK_a of the N–protonated neutral radical $A(NH)^{\bullet}$ (this radical probably exists in aqueous solution as a mixture of the N1, N3, and N7 protonated isomers, of 12.1.[43,1] Compared with the pK_a of the protonated parent (3.5), this means an increase in basicity by 8.6 orders of magnitude $(\Delta pK_a$ $(A(NH)^+/A(NH)^{\bullet} = 8.6)$.

$$(13)$$

R = ribose

A$^{\bullet-}$ also reacts rapidly by protonation on <u>carbon</u> (k = 3.6 x 10^6 s^{-1}), and even the less electron–rich A(NH)$^{\bullet}$ still protonates on carbon with k = 1 x 10^4 s^{-1}. This rate can be considerably accelerated by phosphate.[43] Protonation occurs at C2 and C8,[43] giving the same radicals (A(C2H)$^{\bullet}$ and A(C8H)$^{\bullet}$, respectively) as those[4a,c,11,12,28a,b,45] observed in single crystals. Of the two adducts at carbon, A(C8H)$^{\bullet}$ is the thermodynamically more stable one, in the solid state[45] as well as in aqueous solution.[43]

2.4 Deoxyguanosine (G).

The behavior of the radical anion is similar to that of adenosine, i.e. that there is a rapid protonation by water on a heteroatom (probably at O^6) followed by a transformation (k $\geq$ 3x10^5 s^{-1}) leading to protonation on carbon, probably at C8.[24] This conclusion is based on the very weak reducing properties of the resulting radical(s). In acid solution, the C-protonated electron adduct undergoes a further protonation to give a radical cation G(H$_2$)$^{\bullet+}$, whose pK$_a$ is equal to 5.3.[24] Protonation of G$^{\bullet-}$ on O^6 has also been observed to take place in single crystals, and the production of the "H-adduct" to C8 of the guanine moiety has been interpreted[4c] as due to protonation of the radical anion at C8.

3. Proton–transfer between the bases of a pair ("Intra–pair" proton transfer).

As has been pointed out previously,[1,46] in DNA the stage is set in a perfect way for proton transfer to occur, i.e. along the preset channels of the hydrogen bonds between the bases. In principle, these proton transfers could be very fast, involving just one vibrational period. The adjustment of the system to changes in proton affinity due to ionization (electron loss) or electron gain can thus take place essentially instantaneously. An aspect of this concept is that, in order to understand the redox chemistry of a particular base in DNA, the properties and the behavior of its partner have to be considered as well.

In the following, the information given in sections 1 and 2 on the acid/base properties of the one–electron oxidized/reduced (2'-deoxy)nucleosides[47] is combined with the aim of obtaining a quantitative picture of redox–induced proton transfer, hopefully reflecting the situation in DNA.

3.1 Oxidation of bases in the base pairs.
3.1.1 The G–C pair.

Eq 14 describes what probably happens after an electron is removed from a guanine moiety. Deoxyguanosine is a weak acid ($pK_a = 9.4$). The radical cation, however, is a much stronger acid ($pK_a = 3.9$). If G is ionized, the equilibrium positions of one or of all of the three protons involved in the hydrogen bonds with the cytosine moiety (C) will therefore be shifted toward the cytosine with the result of an overall transfer of positive charge to that base, whereas the unpaired spin will of course remain on the oxidized base. Proton transfer thus leads to a separation of charge from spin: The spin residing on the damaged base and the charge on the complementary base in the other strand. For better "visibility", in eq 14 and the following ones, the proton moving across the strands is encircled.

The equilibrium for protonation of C or for deprotonation from $G^{\cdot+}$ can be easily calculated by combining the corresponding reactions, as follows:

$$G^{\cdot+} \rightleftharpoons G(-H)^{\cdot} + H^+ \qquad K = 10^{-3.9}$$

$$C + H^+ \rightleftharpoons C(H)^+ \qquad K = 10^{4.3}$$

$$G^{\cdot+} + C \rightleftharpoons G(-H)^{\cdot} + C(H)^+ \qquad K = 10^{0.4}$$

(14)

The result is that in DNA the radical cation of the G moiety should deprotonate by H^+ transfer to C. However, the equilibrium constant for this process is not very large, which means that G will also have some radical cation nature. The radical $G(-H)^{\bullet}$ has in fact been observed on oxidation of double-stranded DNA in aqueous solution at room temperature. In agreement with the above scheme (see also ref 5), it was found that $G^{\bullet+}$ deprotonates at N1,[48] i.e. that it is present as $G(N1-H)^{\bullet}$.

If ionization occurs at the cytosine end of the G-C base pair (eq 15), the ionization–induced driving force for deprotonation is considerably larger than in the case of ionization of G (since ΔpK_a $(C/C^{\bullet+}) \leq -10$ compared to ΔpK_a $(G/G^{\bullet+}) = -5.6$). On the other hand, G is a weaker <u>base</u> than C, so the tendency of G to <u>accept</u> a proton from $C^{\bullet+}$ is less than in the reverse case (eq 14). These opposing properties result in a situation where the proton remains essentially at the C moiety. In other words, the radical cation produced by ionization of C <u>remains a radical cation</u>, since it does not lose its charge by proton transfer to its base partner. This is thus different from the situation resulting from ionization of G, where the equilibrium between $G^{\bullet+}$ and $G(-H)^{\bullet}$, formed by proton transfer to C, is in favor of the neutral $G(-H)^{\bullet}$.

$$C^{\bullet+} \rightleftharpoons C(-H)^{\bullet} + H^+ \qquad K \sim 10^{-4}$$

$$G + H^+ \rightleftharpoons G(H)^+ \qquad K = 10^2$$

$$C^{\bullet+} + G \rightleftharpoons C(-H)^{\bullet} + G(H)^+ \qquad K \sim 10^{-2}$$

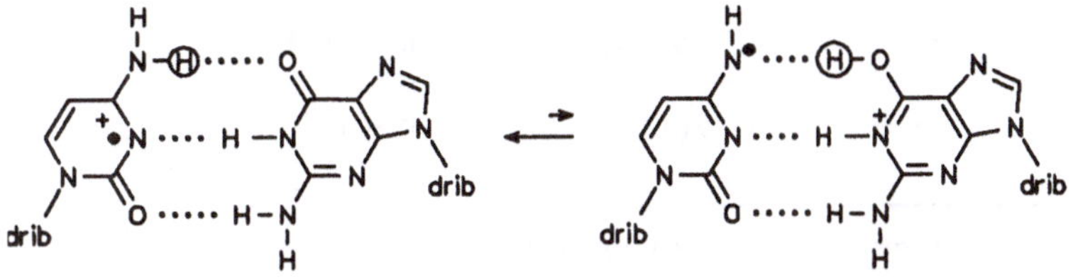

(15)

3.1.2 The A–T pair.

In the case of the A–T base pair, ionization of A leads to the strongly acidic species $A^{\bullet+}$, so one would expect this species to deprotonate. However, this is essentially prevented by the extremely low basicity of T (pK_a $(T(H)^+) =$

238

−5). Therefore, analogous to the case of $C^{\cdot +}$, it is concluded that the radical cation of A remains a cationic species.

$$A^{\cdot +} \quad \rightleftharpoons \quad A(-H)^{\cdot} + H^+ \qquad K \geq 10^{-1}$$

$$T + H^+ \quad \rightleftharpoons \quad TH^+ \qquad K = 10^{-5}$$

$$A^{\cdot +} + A \quad \rightleftharpoons \quad A(-H)^{\cdot} + TH^+ \qquad K \geq 10^{-6}$$

(16)

The situation is different in the A–T pair if it is T which is ionized. Combining pK_a $(T^{\cdot +}) = 3.6$ with pK_a $(A(H)^+) = 3.8$ yields the value of 1.6 $(\equiv 10^{0.2})$ for the proton transfer equilibrium from $T^{\cdot +}$ to A. It can therefore be concluded that the reactivity of one–electron oxidized T in DNA should be characterized by its two protonation states, $T^{\cdot +}$ and $T(N3–H)^{\cdot}$, being approximately equally important.

$$T^{\cdot +} \quad \rightleftharpoons \quad T(N3-H)^{\cdot} + H^+ \qquad K = 10^{-3.6}$$

$$A + H^+ \quad \rightleftharpoons \quad A(H)^+ \qquad K = 10^{3.8}$$

$$T^{\cdot +} + A \quad \rightleftharpoons \quad T(N3-H)^{\cdot} + A(H)^+ \qquad K = 10^{0.2}$$

(17)

3.2 Reduction of bases in the base pairs.

The rate constants for reaction (in aqueous solution) of e^-_{aq} with purines and pyrimidines are essentially the same.[1,31] On this basis it may be assumed that the chances of an electron reacting with a purine–pyrimidine base pair to end up on either are equal.

3.2.1 The A–T pair.

If the electron is picked up by adenine, the Brönsted basicity of this molecule increases by 8.6 orders of magnitude, as reflected by the pK_a for deprotonation from the protonated radical anion of 12.1. Thymine, the base partner of adenine, is a weak acid. Thus, it should be able to protonate strong bases. Combining pK_a (A(NH)$^\bullet$) = 12.1 with the pK_a for deprotonation of neutral thymidine (= 9.6) shows that the equilibrium for proton transfer between T and A$^{\bullet-}$ is such that A$^{\bullet-}$ is quantitatively protonated:

$$A^{\bullet-} + H^+ \longrightarrow A(NH)^\bullet \qquad K = 10^{12.1}$$

$$T \longrightarrow T(N3-H)^- + H^+ \qquad K = 10^{-9.9}$$

$$A^{\bullet-} + T \longrightarrow A(NH)^\bullet + T(N3-H)^- \qquad K = 10^{2.2}$$

(18)

This means that the chances for one–electron reduced adenine to remain a radical anion are low. Thermodynamically, the system is more stable if a proton is transferred to A$^{\bullet-}$, i.e. if the negative charge is on the thymine moiety (with the unpaired spin, of course, still on the adenine). It is evident that also in the case of reduction-induced proton transfer all proton transfer reactions that occur in a base pair lead to a separation of charge from spin such that charge and spin end up in opposite strands of the double helix.

If the electron is scavenged not by A but by T, the resulting radical anion remains an anion, in spite of the fact that the Brönsted basicity of T$^{\bullet-}$ is ≥ 12 orders of magnitude higher than that of T. The reason is that the base partner, A, is an extremely weak acid (pK_a (A) ≥ 14). The following equations describe the situation:

$$T^{\bullet -} + H^+ \quad \rightleftharpoons \quad T(O^4H)^{\bullet} \qquad K = 10^{8.9}$$

$$A \quad \rightleftharpoons \quad A(N^6-H)^- + H^+ \qquad K \leq 10^{-14}$$

$$T^{\bullet -} + A \quad \rightleftharpoons \quad T(O^4H)^{\bullet} + A(N^6-H)^- \qquad K \leq 10^{-7.1}$$

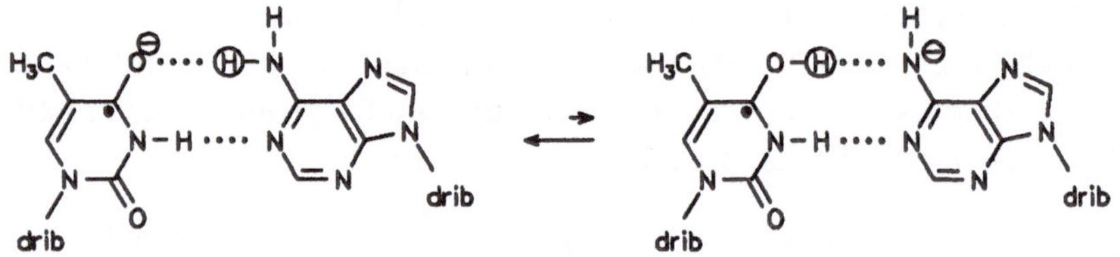

(19)

With the negative charge on T not neutralized by proton transfer from A, other protonating agents such as water get a chance to react with $T^{\bullet -}$. In DNA, it is the C5/C6 double bond of T which is exposed to the water molecules in the hydration shell of the helix. The consequences of this are discussed in section 5.

3.2.2 The G–C pair.

If the electron is scavenged by C, the Brönsted basicity of this molecule is further increased, as reported in section 2.2. Nature has coupled this powerful base with the strongest acid among the nucleic acid bases, namely G. Proton transfer from G to $C^{\bullet -}$ is therefore expected to be highly favored, and this is the case as seen from the following equations:

$$C^{\bullet -} + H^+ \quad \rightleftharpoons \quad C(H)^{\bullet} \qquad K \geq 10^{13}$$

$$G \quad \rightleftharpoons \quad G(-H)^- + H^+ \qquad K = 10^{-9.5}$$

$$C^{\bullet -} + G \quad \rightleftharpoons \quad C(H)^{\bullet} + G(-H)^- \qquad K \geq 10^{3.5}$$

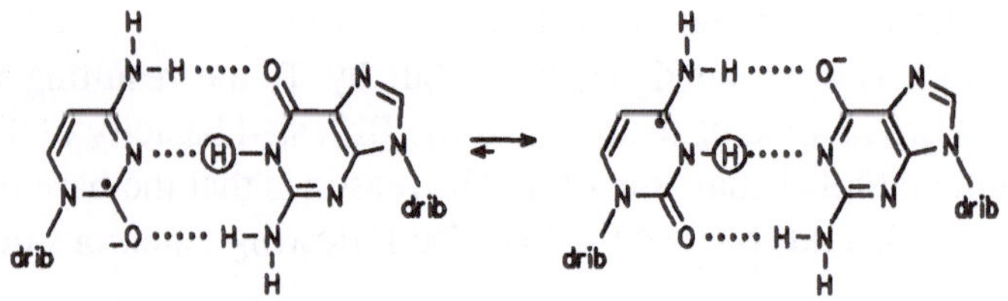

(20)

The conclusion is thus that, in contrast to the pyrimidine base T, in DNA the cytosine radical anion is <u>protonated</u> (by its complementary base, G) on a hetero atom, probably on N3, as shown above. One of the consequences is that the radical is thereby protected with respect to undergoing irreversible protonation on carbon (see section 5.2).

It has in fact been found experimentally, using ESR, that in DNA at low-temperature (4 - 77 K) the electron adduct of C is <u>protonated</u>,[49-51,52,53] probably at N3.[51] Furthermore, the abundance of this radical is much larger than that of the electron adduct of T, i.e. the electron prefers to localize at C rather than at T.[52,53] On the basis of the acidity/basicity concepts developed above (eq's 14 - 20), this preference for C is easy to understand: In the G-C pair protonation of the electron adduct of C is thermodynamically highly favored, which is much in contrast to the situation in the A-T pair if the electron is scavenged by T. If it is assumed that the electron affinity of C is probably less than that of T,[54] this is an example which shows that the thermodynamics of proton transfer may qualitatively change the energetics of electron transfer.

4. Involvement of carbon. The complementary base influences also the likelihood and the pattern of de/protonation from/on <u>carbon</u> of the radical cation/anion in a particular strand. Little is known about the details of these reactions. For general considerations, the reader is referred to ref. *46.*

Summary and conclusions

It has been demonstrated that protonations and deprotonations result from one-electron reduction or oxidation, respectively, of the DNA bases. These proton transfer processes (by which the charge of the molecule is altered) not only change the nature and thereby the reactivity of the base radicals, but they also have a strong effect on electron transfer processes via their "modulation" of the free energies of reaction. An example for this is the base cytosine whose "electron affinity under protonating conditions" is <u>higher</u>, whereas the gas phase electron affinity is probably <u>lower</u> than that of the other bases. In addition to the reversible proton transfer processes (between the hetero-atoms) there are the irreversible ones (involving carbon atoms). The former are more likely to occur in the G-C base pair, whereas the A-T pair is more prone to engage in the latter. Since the reversible proton transfers are rapid they determine the <u>initial</u> sites of electron and positive hole deposition (as determined at $\approx$ 4-10 K), whereas in the irreversible processes, although they are slow, the <u>final</u> sites of damage (as measured at $\geq$ 77 K) are fixed. Future work will have to focus on the detailed mechanisms of "damage migration" from the initial to the final sites.

242

References and Notes

[1] S. Steenken, (1989) Purine bases, nucleosides and nucleotides: Aqueous solution chemistry and transformation reactions of their radical cations and e⁻ and OH adducts. Chemical Reviews, **89**, 503–520.

[2] For information on the pattern of interaction of water molecules with the bases in DNA and oligonucleotides see a) B. Wolf and S. Hanlon, (1975) Structural transitions of deoxyribonucleic acid in aqueous electrolyte solutions. II. The role of hydration. Biochemistry, **14**, 1661-70. b) M.L. Kopka, A.V. Fratini, H.R. Drew and R.E. Dickerson, (1983) Ordered water structure around a β-DNA dodecamer. A quantitative study. Journal of Molecular Biology, **163**, 129-46. c) E. Westhof, T. Prange, B. Chevrier and D. Moras, (1985) Solvent distribution in crystals of B-oligomers and Z-oligomers. Biochimie, **67**, 811-17. d) R. Brandes, A. Rupprecht and D.R. Kearns, (1989) Interaction of water with oriented DNA in the A- and B-form conformations. Biophysical Journal, **56**, 683-91. e) S. Kuwabara, T. Umehara, S. Mashimo and S. Yagihara, (1988) Dynamics and structure of water bound to DNA. Journal of Physical Chemistry, **92**, 4839-41. For further information on the crystal structure of oligomers of DNA see A.V. Fratini, M.L. Kopka, H.R. Drew and R.E. Dickerson, (1982) Reversible bending and helix geometry in a B-DNA dodecamer: CGCGAATT^{Br}CGCG. Journal of Biological Chemistry, **257**, 14686. E. Westhof, P. Dumas and D. Moras, (1985) Crystallographic refinement of yeast aspartic-acid transfer-RNA. Journal of Molecular Biology, **184**, 119-45. O. Kennard, W.B.T. Cruse, J. Nachman, T. Prange, Z. Shakked and D. Rabinovi, (1986) Ordered water structure in an A-DNA octamer at 1.7 A resolution. Journal of Biomolecular Structure and Dynamics, **3**, 623-47.

[3] See D.W. Boerth and P.K. Bhowmik, (1989) Purine nucleotide cations. 2. Energetics and conformational effects on protonation–deprotonation of purine nucleoside. Journal of Physical Chemistry, **93**, 3327–34.

[4] For reviews see a) D.M. Close, W.H. Nelson, and E. Sagstuen, (1987) EPR and ENDOR study of γ-irradiated single crystals of purines at 4.2 K. In: J.A. Weil (ed.) Electronic Magnetic Resonance of the Solid State, Canadian Society of Chemistry (CSC Symposium Series, 1), Ottawa, 237-50. b) M.C.R. Symons, (1987) Application of electron spin resonance spectroscopy to the study of the effects of ionising radiation on DNA and DNA complexes. Journal of the Chemical Society, Faraday Transactions 1, **83**, 1-11. c) J. Hüttermann, (1991) Radical ions and their reactions in DNA and its constituents: Contributions of electron spin resonance spectroscopy. In: A. Lund and M. Shiotani (eds.) Radical Ionic Systems, Kluwer, Dordrecht, pp. 435-62.

5 It is, however, possible to oxidize A with Tl(II) (k = 1.3×10^8 $M^{-1}s^{-1}$ for dAMP at pH $\approx$ 7): a) S.V. Jovanovic and M.G. Simic, (1989) The DNA guanyl radical: Kinetics and mechanisms of generation and repair. Biochimica et Biophysica Acta, **1008**, 39–44. As with deoxycytidine and thymidine (see sections 1.3 and 1.4), the one–electron oxidant $Br_2^{\bullet-}$ (E^0 = 1.6 V/NHE) reacts with A with a rate constant < 10^7 M^{-1} s^{-1}, which means that it is not possible to produce, in dilute solutions, the one–electron oxidized species in times sufficiently short (i.e. < 1 ms) to allow their study before they undergo (undesired) radical–radical reactions.

6 L. Eberson, (1982) Electron-transfer reactions in organic chemistry. Advances in Physical Organic Chemistry, **18**, 79-185.

7 P. O'Neill and S.E. Davies, (1987) Pulse radiolytic study of the interaction of $SO_4^{\bullet-}$ with deoxynucleosides. Possible implications for direct energy deposition. International Journal of Radiation Biology, **52**, 577-87.

8 A.J.S.C. Vieira and S. Steenken, (1987) Pattern of OH radical reaction with N^6, N^6–dimethyladenosine. Production of three isomeric OH adducts and their dehydration and ring–opening reactions. Journal of the American Chemical Society, **109**, 7441–8.

9 L.P. Candeias and S. Steenken, (1992) Ionization of purine nucleosides and nucleotides and their components by 193 nm laser photolysis in aqueous solution: Model studies for oxidative damage of DNA. Journal of the American Chemical Society, **114**, 699-704.

10 A.J.S.C. Vieira and S. Steenken, (1987) Pattern of OH radical reaction with 6– and 9–substituted purines. Effect of substituents on the rates and activation parameters of the unimolecular transformation reactions of two isomeric OH adducts. Journal of Physical Chemistry, **91**, 4138–44.

11 L. Kar and W.A. Bernhard, (1983) Electron gain and electron loss radicals stabilized on the purine and pyrimidine of a cocrystal exhibiting base-base interstacking: ESR-ENDOR of X-irradiated adenosine:5-bromouracil. Radiation Research, **93**, 232-53.

12 D.M. Close and W.H. Nelson, (1989) ESR and ENDOR study of adenosine single crystals X–irradiated at 10 K. Radiation Research, **117**, 367–78.

13 E. Sagstuen, E.O.Hole, W.H. Nelson and D.M. Close, (1991) The effect of environment upon DNA free radicals. In: E. M. Fielden and P. O'Neill (eds.) The Early Effects of Radiation on DNA, Nato ARW Series, Vol. H **54**, Springer, pp. 215-30.

14 P. O'Neill, P.W. Chapman and D.G. Papworth, (1985) Repair of hydroxyl radical damage of dA by antioxidants. Life Chemistry Reports **3**, 62-9.

15 A.J.S.C. Vieira and S. Steenken, (1990) Pattern of OH radical reaction with adenine and its nucleosides and nucleotides. Characterization of two types of isomeric OH adduct and their unimolecular transformation reactions. Journal of the American Chemical Society, **112**, 6986–94.

[16] As in the case of $C^{\bullet+}$ (see section 1.3), this conclusion is of course based on the assumption that the optical absorption spectra of $A^{\bullet+}$ and $A(-H)^{\bullet}$ are sufficiently different (as they are in the case of G, see ref 5) to enable their distinction by optical detection.

[17] A. Hissung, C. von Sonntag, D. Veltwisch and K.-D. Asmus, (1981) The reactions of the 2'-deoxyadenosine electron adduct in aqueous solution. The effects of the radiosensitizer p-nitroacetophenone. A pulse spectroscopic and pulse conductometric study. International Journal of Radiation Biology, **39**, 63-71.

[18] Unless otherwise noted, the pK_a values of the parent "bases" refer mainly to the 2'-deoxynucleosides and are from a) P.O.P. Tso, (1974) Bases, nucleosides, and nucleotides. In: Basic Principles in Nucleic Acid Chemistry, P.O.P. Tso, ed., Vol. **1**, Academic, New York, pp 454-584.

[19] L.P. Candeias and S. Steenken, (1989) Structure and acid-base properties of one-electron-oxidized deoxyguanosine, guanosine, and 1-methylguanosine. Journal of the American Chemical Society, **111**, 1094-99.

[20] B. Rakvin, J.N. Herak, K. Voit and J. Hüttermann, (1987) Free radicals from single crystals of deoxyguanosine-5'-monophosphate (Na salt) irradiated at low temperatures. Radiation and Environmental Biophysics, **26**, 1-12.

[21] E.O.Hole, W.H. Nelson, D.M. Close and E. Sagstuen, (1987) ESR and ENDOR study of the guanine cation: Secondary product in 5'-dGMP. Journal of Chemical Physics, **86**, 5218-20.

[22] P. O'Neill, (1983) Pulse radiolytic study of the interaction of thiols and ascorbate with OH adducts of dGMP and dG: Implications for DNA repair processes. Radiation Research, **96**, 198-210.

[23] L.P. Candeias and S. Steenken, (1991) Transformation reactions of two isomeric OH-adducts of 2'-deoxyguanosine. In: E. M. Fielden and P. O'Neill (eds.) The Early Effects of Radiation on DNA, Nato ARW Series, Vol. H **54**, Springer, pp. 265-6.

[24] L.P Candeias, P. Wolf, P. O'Neill and S. Steenken, (1992) Reaction of hydrated electrons with guanine nucleosides: Fast protonation on carbon of the electron adduct. Journal of Physical Chemistry, **96**, 10302-10307.

[25] Based on the value published by S.V. Jovanovic and M.G. Simic, (1986) One-electron redox potentials of purines and pyrimidines. Journal of Physical Chemistry, **90**, 974-8, and using an improved number (1.08 V/NHE) (from G. Merenyi, J. Lind and X. Shen, (1988) Electron transfer from indoles, phenol, and sulfite (SO_3^{2-}) to chlorine dioxide ($ClO_2^{\bullet}$). Journal of Physical Chemistry, **92**, 134-7) for the potential of the reference compound.

[26] D.K. Hazra and S. Steenken, (1983) Pattern of OH radical addition to cytosine and 1-, 3-, 5-, and 6-substituted cytosines. Electron transfer and dehydration reactions of the OH adducts. Journal of the American Chemical Society, **105**, 4380-86.

[27] D.J. Deeble, M.N. Schuchmann, S. Steenken and C. von Sonntag, (1990) Direct evidence for the formation of thymine radical cations from the reaction of $SO_4^{•-}$ with thymine derivatives: A pulse radiolysis study with optical and conductance detection. Journal of Physical Chemistry, 94, 8186–92. J.R. Wagner, J.E. van Lier, C. Decarroz, M. Berger and J. Cadet, (1990) Photodynamic methods for oxy radical-induced DNA damage. Methods in Enzymology, 186, 502-11.

[28] See, e.g., a) W.A. Bernhard, (1981) Solid–state radiation chemistry of DNA: The bases. Advances Radiation Biology, 9, 199–280. b) J. Hüttermann, (1982) Solid state radiation chemistry of DNA and its constituents. Ultramicroscopy, 10, 25-40.

[29] J. Hüttermann, (1970) Electron-spin-resonance spectroscopy of radiation-induced free radicals in irradiated single crystals of thymine monohydrate. International Journal of Radiation Biology, 17, 249-59.

[30] The radical cations of uracil and thymine have been postulated to be formed in CF_3Cl: C.J. Rhodes, I.D. Podmore and M.C.R. Symons, (1988) Radical cations of N,N-dimethyluracil and N,N-dimethylthymine. Journal of Chemical Research (S), 120-1. However, on the basis of a comparison of the reported coupling constants with those[43a] of the neutral radicals (formed by deprotonation of the radical cations from N1) as determined in aqueous solution, the species in CF_3Cl may as well have been the neutral radicals. a) H.M. Novais and S. Steenken, (1987) Reactions of oxidizing radicals with 4,6–dihydroxypyrimidines as model compounds for uracil, thymine, and cytosine. Journal of Physical Chemistry, 91, 426–33.

[31] G.V. Buxton, C.L. Greenstock, W.P. Helman and A.B. Ross, (1988) Critical review of rate constants for reactions of hydrated electrons, hydrogen atoms and hydroxyl radicals ($OH^•/O^{•-}$) in aqueous solution. Journal of Physical and Chemical Reference Data, 17, 513-886.

[32] H.M. Novais and S. Steenken, (1986) ESR studies of electron and hydrogen adducts of thymine and uracil and their derivatives and of 4,6–dihydroxypyrimidines in aqueous solution. Comparison with data from solid state. The protonation at carbon of the electron adducts. Journal of the American Chemical Society, 108, 1–6.

[33] E. Hayon, (1969) Optical absorption spectra of ketyl radicals and radical anions of some pyrimidines. Journal of Chemical Physics, 51, 4881-92.

[34] S. Steenken, J.P. Telo, H.M. Novais and L.P. Candeias, (1992). One-electron potentials of pyrimidine bases, nucleosides and nucleotides in aqueous solution. Consequences for DNA redox chemistry. Journal of the American Chemical Society, 114, 4701-4709.

[35] D.J. Deeble, S. Das, C. von Sonntag, (1985) Uracil derivatives: Sites and kinetics of protonation of the radical anions and the UV spectra of the C(5) and C(6) H-atom adducts. Journal of Physical Chemistry, 89, 5784-8.

246

36 B. Garcia and J. Palacios, (1988) Protonation study of biological bases of DNA. Berichte der Bunsengesellschaft für Physikalische Chemie, **92**, 696-700.

37 E. Sagstuen, E.O. Hole, W.H. Nelson and D.M. Close, (1989) Structure of the primary reduction product of thymidine after X Irradiation at 10 K. Journal of Physical Chemistry, **93**, 5974–77.

38 A. Gräslund, A. Ehrenberg, A. Rupprecht and G. Ström, (1971) Ionic base radicals in γ–irradiated DNA. Biochimica et Biophysica Acta, **254**, 172–86. J. Hüttermann, K. Voit, H. Oloff, W. Köhnlein, A. Gräslund and A. Rupprecht, (1984) Specific formation of electron gain and loss centers in X–irradiated oriented fibers of DNA at low temperatures. Faraday Discussions of the Chemical Society, **78**, 135–49. I. Zell, J. Hüttermann, A. Gräslund, A. Rupprecht and W. Köhnlein, (1989) Free radicals in irradiated oriented DNA fibers: Results from B–form DNA and from deuterated DNA samples. Free Radical Research Communications, **6**, 105–6.

39 E.O. Hole, E. Sagstuen, W.H. Nelson and D.M. Close, (1991) Primary and secondary radicals in thymine derivatives: Solid state ESR/Endor study of 1-methylthymine and thymidine. In: E. M. Fielden and P. O'Neill (eds.) The Early Effects of Radiation on DNA, Nato ARW Series, Vol. H **54**, Springer, pp. 409-10.

40 A. Hissung and C. von Sonntag, (1979) The reaction of solvated electrons with cytosine, 5–methylcytosine and 2'–deoxycytidine in aqueous solutions. The reaction of the electron adduct intermediates with water, p–nitroacetophenone and oxygen. A pulse spectroscopic and pulse conductometric study. International Journal of Radiation Biology, **35**, 449–58.

41 K.J. Visscher, private communication.

42 M.N. Schuchmann, S. Steenken, J. Wroblewski and C. von Sonntag, (1984) Site of OH radical attack on dihydrouracil and some of its derivatives. International Journal of Radiation Biology, **46**, 225-32.

43 L.P. Candeias and S. Steenken, (1992) Electron adducts of adenine nucleosides and nucleotides in aqueous solution: Protonation at two carbon sites (C2 and C8) and intra- and intermolecular catalysis by phosphate. Journal of Physical Chemistry, **96**, 937-944.

44 a) K. J. Visscher, M. P. de Haas, H. Loman, B. Vojnovic, J.M. Warman, (1987) Fast protonation of adenosine and of its radical anion formed by hydrated electron attack; a nanosecond optical and dc-conductivity pulse radiolysis study. International Journal of Radiation Biology, **52**, 745. b) K. J. Visscher, M. Hom, H. Loman, H. J. W. Spoelder, and J. B. Verberne, (1988) Spectral and kinetic properties of intermediates induced by reaction of hydrated electrons with adenine, adenosine, adenylic acid and polyadenylic acid: A multicomponent analysis. Radiation Physics and Chemistry, **32**, 465-73.

45 H. Zehner, W. Flossmann and E. Westhof, (1976) Formation of H-addition radicals in adenine derivatives. Zeitschrift für Naturforschung, **31c** , 225 31. H.

Zehner, E. Westhof, W. Flossmann and A. Müller, (1977) Formation of H-addition radicals in adenine derivatives: Part II. Zeitschrift für Naturforschung, 32c, 1-10. E. Westhof, W. Flossmann, H. Zehner and A. Müller, (1978) Formation of H-adition radicals in some pyrimidine and purine derivatives. Faraday Discussions of the Chemical Society, 63, 248-54.

46 S. Steenken, (1992) Electron-transfer-induced acidity/basicity and reactivity changes of purine and pyrimidine bases. Consequences of redox processes for DNA bsase pairs. Free Radical Research Communications, 16, 349-379.

47 It is assumed and there is experimental evidence[48] that the redox–induced changes in acid/base properties of the nucleotides are not very different from those of the nucleosides.

48 K. Hildenbrand and D. Schulte–Frohlinde, (1990) ESR Spectra of radicals of single–stranded and double–stranded DNA in aqueous solution. Implications for •OH–induced strand breakage. Free Radical Research Communications, 11, 195–206.

49 M.C.R. Symons, (1990) ESR spectra for protonated thymine and cytidine radical anions: their relevance to irradiated DNA. International Journal of Radiation Biology, 58, 93–6.

50 W.A. Bernhard, (1991) Initial sites of one electron attachment in DNA. In: E. M. Fielden and P. O'Neill (eds.) The Early Effects of Radiation on DNA, Nato ARW Series, Vol. H 54, Springer, pp.141-54.

51 J. Barnes, W.A. Bernhard and K.P. Mercer, (1991) Distribution of electron trapping in DNA: Protonation of one-electron reduced cytosine. Radiation Research, 126, 104-7. J. Hüttermann, J. Ohlmann, H. Schaefer and W. Gatzweiler, (1991) The polymorphism of a cytosine anion studied by electron paramagnetic resonance spectroscopy. International Journal of Radiation Biology, 59, 1297-1311.

52 W.A. Bernhard, (1989) Sites of electron trapping in DNA as determined by ESR of one–electron–reduced oligonucleotides. Journal of Physical Chemistry, 93, 2187–9.

53 M.D. Sevilla, D. Becker, M. Yan and S.R. Summerfield, (1991) Relative abundance of primary ion radicals in γ–irradiated DNA: Cytosine vs thymine anions and guanine vs adenine cations. Journal Physical Chemistry, 95, 3409–15.

54 Experimentally determined electron affinity values for the nucleic acid bases do not seem to be available.

REDOX REGULATION OF THE HUMAN IMMUNODEFICIENCY VIRUS TYPE 1 (HIV-1).

J. PIETTE, C. SAPPEY, B. PIRET and S. LEGRAND-POELS

Laboratory of Fundamental Virology, Institute of Pathology B23,

University of Liège, B-4000 LIEGE, BELGIUM.

1. Introduction

One of the most intensive investigative efforts of cell and molecular biology is devoted to the elucidation of the molecular switches that regulate mammalian cell proliferation. The control of cell proliferation involves a variety of genetic and biochemical factors that still remains poorly understood. These factors are also important for the propagation of viruses and for their reactivation from a latent stage. In this paper, we will summarize informations concerning the propagation of the human immunodeficiency virus in relation with the redox status of the infected cell and we will emphasize the correlation between this propagation and the inducibility of a redox-controlled transcription factor (NF-κB).

2. HIV-1 infectious cycle

The natural history of HIV disease is best described as a continuous process in which immune dysfunction and the loss of CD4+ helper T-cells begins at the time of infection and progressively increases [1,2,3] leading finally to the wasting syndrome, opportunistic infections and malignancies that constitute clinically defined AIDS (the acquired immunodeficiency syndrome) [4,5]. Three consecutive stages can be viewed : a first acute stage corresponding to primary infection with HIV-1 (CDC stage I), a chronic stage representing a period of clinical latency (CDC stage II and III) and a crisis stage where the profound immunodeficiency is manifested by the development of opportunistic infections and other pathologic conditions (CDC stage IV) [see 4 for review].

249

B. Meunier (ed.), DNA and RNA Cleavers and Chemotherapy of Cancer and Viral Diseases, 249–268.
© 1996 Kluwer Academic Publishers. Printed in the Netherlands.

The symptoms of the primary infection consist of fever, headache, lymphadenopathy, myalgia [6,7], and in such patients have been found transiently high levels of infectious virus in the plasma with a significant percentage of infected peripheral blood lymphocytes (PBL) (up to 1 % of CD4$^+$ T-cells) [8]. In parrallel, the number of circulating CD4$^+$ falls acutely but transiently. One month after the infection, the number of infected CD4$^+$ found in the peripheral blood starts to decline and the apperance of an anti-viral humoral and cellular immune response correlates with the diminished level of HIV-1 replication [9].

When the symptoms of the primary infection disappear with the evidence of an anti-viral immune response, HIV-1 infected patients enter in a chronic, clinically asymptomatic state often called the latent period. This stage is estimated to be 8-11 years before entering the last stage [10,11]. However, AIDS can also occur in 2 to 4 months [12,13], whereas some infected individuals may remain asymptomatic for more than 13 years [14]. Although clinically latent, recent studies suggest that the virus continues to replicate at a low level throughout this period in which absolute levels of HIV RNA in the pheripheral blood are low but still demonstrable [15,16,17]. However, high levels of virus replication and of "trapped" extracellular virus (e.g., on follicular dendritic cells) are found within lymphoid organs [18,19,20].

The development of clinically significant immunodeficiency is presaged by evidence of increasing virus burden, a significant level of plasma viremia and elevated levels of HIV-1 DNA and RNA in the blood [22]. Increasing virus replication has also been correlated with accelerated depletion of CD4$^+$. In HIV-1 pathogenesis, an increased viral load correlates with CD4 lymphocyte depletion and disease progression [23-25], but until recently relatively little information was available on the kinetics of virus and CD4 turn over *in vivo* [26,27]. The treatment of AIDS patients with ABT-538, an inhibitor of the HIV-1 protease, causes plasma HIV-1 levels to decrease exponentially (mean half-life, 2.1 ± 0.4 days) and CD4 counts to rise substantially. Almost complete replacement of wild type virus in plasma by drug-resistant variants occurred after 14 days, indicating that HIV-1 viraemia is sustained primarily by a dynamic process involving continous rounds of *de novo* virus infection and replication with a rapid cell turnover. Although, it is emphasized that most virus in plasma derives from actively replicating short-lived population of cells, latently infected cells that become activated or chronically producing cells that generate proportionately less virus may nonetheless be important in HIV-1 pathogenesis. Several correlations suggest that disease progression is associated with a relative change in the proportion of cells carrying latent genomes and of those that actively replicate virus. First, the putative mechanisms for cellular latency have most readily observed in cross-sectional studies of patients with early, asymptomatic disease. Secondly, prospectives studies have revealed a time-dependent loss of cellular latency as disease progresses reflecting a shift away from latency even within the lymphoid compartments [28 for review]. Based on *in situ* analysis [21], latently infected cells far outnumber the actively replicating pool and the diversity of their constituent viral genomes represents a potentially important source of clinically relevant variants, including those conferring drug resistance. On the other hand, long-term survivor patients show low levels of HIV-1 in the presence of strong virus-specific immune responses combined with some degree of viral attenuation, thereby tipping the balance in favor of the infected host [29]. Overall, the viral burden in the plasma and peripheral blood mononuclear cells (PBMC) of long-term survivors is orders of magnitude lower than typically found in subjects with progressive disease. They also exhibit preserved lymphoid tissue with reduced formation of germinal centers and reduced HIV trapping, despite a low but persistent level of viral replication [30].

One important point which should be mentionned is the dependence of the first step of the HIV-1 replication cycle on the state of the T-lymphocyte. Incomplete reverse transcription may occur in unstimulated cells leading to an unstable partially reverse transcribed HIV-1 DNA intermediate [31]. This molecule is composed of partial minus strand DNA sequences including one Long Terminal Repeat (LTR) but no *gag* sequences. Recent data suggest that the efficiency of the completion of viral DNA production from these partial transcripts is very poor. Only 5 % of this intermediate can be rescued to produce virus 15 h after infection [32]. Thus, this intermediate may contribute to inefficient infection of resting T-cells but is unlikely to provide a state of true virus latency. A second report shows that unstimulated lymphocytes in cell culture fully transcribe HIV-1 but the proviral DNA does not integrate in the host genome [33]. Only after cellular stimulation is the proviral DNA able to integrate and productively express virion progeny. Unintegrated viral DNA is shown in PBL of certain HIV-1 infected individuals and stimulation of these cells with mitogens in cell culture leads to viral integration [34]. Similarly, HIV-1 proviruses integrated into the genome of cells of the monocyte-macrophage lineage, which are permissive for proviral synthesis and integration even in the absence of cellular proliferation, might also be capable of maintaining a latent state. Subsequent activation of these cells by antigens, cytokines, or other stimuli might then result in reactivation of a non-productive HIV-1 infection.

Some cell culture systems have been proposed to represent *in vitro* models for HIV-1 proviral latency. These are cultured cell lines which contain the viral genome, but produce only very low quantities of viral progeny. Two models are often used : ACH-2 cell line which is a derivative of the CEM T lymphocytic cell line which contains a single integrated provirus, and U1 cell line which is derived from the U397 monocyte/macrophage which contains two integrated proviruses. It has been shown that in ACH-2 cells, latency is not secondary to the general characteristics of the cellular system, but rather to the integration of the HIV-1 provirus in a site of the cellular genome which offers little support for transcription due to either an inhibitory effect of the chromatin structure in that region or to cellular sequences acting as a silencer on the HIV-1 promoter [35].

3. HIV-1 infection and cell redox status

Gluthatione (GSH), the main intracellular defence against oxidative stress, is decreased in plasma, lung epithelial fluid, erythrocytes and lymphocytes, including T cell subsets of individuals infected with HIV-1 [36 and 37 for review]. Several groups have demonstrated depleted GSH in the tissue of infected patients. Eck *et al.* [38], are first to show that HIV-infected persons have decreased concentrations of acide-soluble thiols (cysteine and GSH) in their plasma and in cell lysates of PBMC. These changes are found in symptom-free individuals, indicating that they are not the consequence of a wasting syndrome. The same patients also have increased plasma glutamate concentrations, which inhibit the cysteine uptake needed for GSH synthesis. Buhl *et al.* confirmed the lower plasma GSH concentrations and, in addition, showed that GSH is decreased in lung epthelial-lining fluid of 14 symptom-free patients [39]. This findings indicates that HIV infection is accompanied by a systemic defiency of extracellular GSH withn little change in the amount of its oxidized form (GSSG) [40]. This deficiency may potentiate HIV replication and accelerate disease progression, especially in individuals with increased concentrations of inflammatory cytokines because such cytokines stimulate HIV replication more efficiently in GSH-depleted cells [41].

GSH is a cysteine-containing tripeptide (γ-glutamyl-cysteinyl-glycine) that is found in eukaryotic cells at millimolar concentrations. Its cellular functions include aminoacid transport, acting as a cofactor in enzymatic reactions, and maintenance of protein sulphydryl redox status. Furthermore, GSH is a defence against electrophilic xenobiotics and intracellular oxidants (reactive oxygen species, ROS). Therefore, GSH is also a regulator of cellular redox potential. Synthesis and degradation of GSH are part of the γ-glutamyl cycle in which GSH is synthesized by the action of two ATP-consuming enzymes: γ-glutamylcysteine synthase and GSH synthase. Reduced GSH can be oxidized to GSH disulphide either non-enzymatically or through the activity of GSH peroxidase. Under normal conditions, GSH disulphide can be reduced by GSH disulphide reductase to regenerate GSH.

Adequate concentrations of GSH are required for mixed lymphocyte reactions [42], T-cell proliferation [42], T and B cell differenciation [43], cytotoxic T-cell activity, and natural killer cell activity [44]. Decreasing GSH by 10 to 40 % can inhibit completely T-cell activation *in vitro* [45]. Oxidative stress conditions stimulate overexpression of TNF-α, the IL-2 receptor, and inflammatory cytokines favoring HIV-1 replication in lymphocytes and mononuclear cells [41]. Thus, an intracellular GSH deficiency in lymphocytes has profound effects on immune functions emphasizing the importance of a correct intracellular redox balance to prevent the progression towards the final stage of the disease.

Other antioxidant molecules are also lowered in HIV-infected individuals either at the CDC II (mean CD4 count , 396/mm3) or IV stages (mean CD4 count, 56/mm3) [46,47]. It is particularly obvious for levels/or activities of superoxide dismutase (SOD) [48], catalase [49], vitamin E, retinol carotenoid and several trace elements like selenium and zinc [46,47] (Fig. 1). These authors have found a severe defect in carotenoids and β-carotene in the plasma of HIV-infected patients and the difference is particularly important at stage II. While the mechanism for this defect still remains unknown, the authors postulate that the degree of reduction in carotene levels is due to its antioxidant functions. This hypothesis is strengthened by an increased level of circulating lipd peroxidation byproducts in HIV patients, such as hydroperoxides and malondialdehyde (Fig.1). In addition, SOD activity is reported to be depressed in pheripheral blood cells isolated from children infected with HIV [48], and more importantly, the HIV Tat protein is capable to downregulate synthesis and overrides induction of the MnSOD which is induced by oxidative stress in cultured cells [50]. All these observations strongly suggest that oxidative stress conditions found in HIV-infected patients is not merely an epiphenomenon but a central event of the HIV pathogenesis. All these findings support a model of disease pathogenesis which has been proposed by Fuchs *et al.* [51]. In this model, HIV and opportunistic infections directly or indirectly promote oxidative stress. The pro-oxidative conditions cause activation of free-radical-producing immune cells, enhancement of viral replication, and weakening of the antioxidant defense system. This cycle becomes autocatalytic and facilitates the disease progression.

The restoration of a well-balanced redox status in HIV-infected individual will decrease the generation of ROS, inhibit HIV stimulation by inflammatory cytokines and partly decrease virus production and therefore may be desirable for therapy of HIV in AIDS during the asymptomatic stage as well as later stages. Drugs such as vitamin C or penicillamine alleviate oxidative stress by neutralizing oxidants but do not directly increase GSH levels. In contrast, nontoxic GSH prodrugs such as *N*-acetyl-L-cysteine (NAC) and L-2-oxothiazolidine-4-carboxylate (OTC) are readily taken up into cells and converted into cysteine by *N*-acetylase and 5-oxoprolinase, respectively. Normal levels of GSH can then be regenerated from newly converted cysteine, the limiting precursor in GSH biosynthesis. Both NAC and OTC have been shown to replenish GSH levels in mice [52] and to enhance both murine and human T cell activation and mitogenesis [53,54]. Although NAC is widely used to restore GSH, some questions have been raised about the potential availability of orally administered NAC in HIV-infected patients. Orally administered NAC is rapidly converted to cysteine, cystine, or GSH in the gut and the liver before reaching the circulation [55].

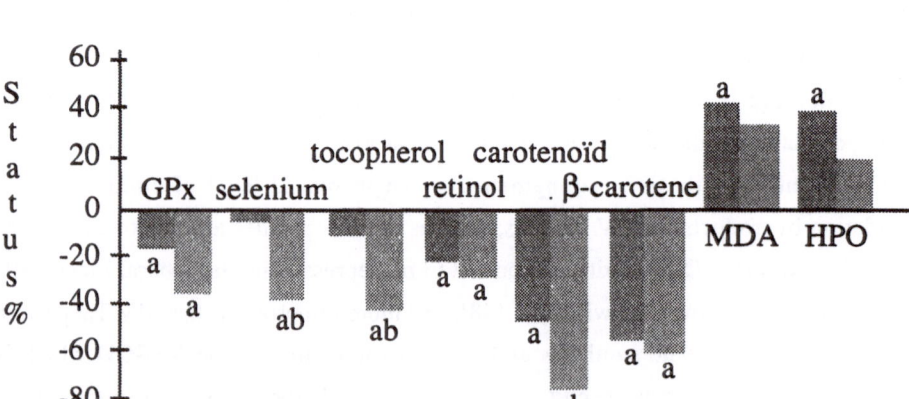

a : significatively different from the controls
b : significatively different between CDC II & CDC IV

Figure 1. Relative importance of antioxidant deficiency and peroxide formation expressed as percentage of normal value found in uninfected patients.

Low blood levels of NAC are actually indicative of its efficient conversion to functional compounds such as GSH [56]. The failure of a single dose of NAC to increase intracellular GSH could be due to the turn over of GSH in lymphocytes is rather slow, such that a longer exposure to high extracellular concentrations of NAC would be required to result in a measurable increase in GSH synthesis. While supplementation of AIDS patients with 1.8 g/day of NAC for 2 weeks increases the plasma concentration of cysteine, it does not significantly increase GSH levels in plasma and PBMC [57]. The failure of sulphydryl supplementation to increase GSH suggests that the low concentrations of GSH are also the consequence of a decreased rate of GSH synthesis in HIV patients.

4. Oxidative stress is a part of a general mechanism of HIV-1 activation

To determine the eventual role of oxidative stress conditions in HIV-1 activation, two cell lines supporting a latent HIV-1 infection (the promonocytic U1 and the lymphocytic ACH-2 cell lines) have been subjected to stress conditions by addition of H_2O_2 [58]. These cell lines have been used as model systems to explore HIV-1 post-integration latency in cell culture [59,60]. In the base line unstimulated state, these cells express the multiply-spliced HIV-RNA and some singly-spliced HIV-1 RNA but an extremely low level of the full-length unspliced RNA [61]. Upon stimulation, this pattern undergoes a switch to the synthesis of unspliced transcripts with a concomitant upregulation of total viral RNA transcription [61]. Thus, these cells express a specific RNA pattern analogous to the early stage and appear to be blocked in progressing to the late stage of productive infection unless they are stimulated [62,63].

A similar situation is recorded in numerous asymptomatic patients, in which by *in vitro* reverse transcription coupled to PCR, it has been shown that the ratios between multiply-spliced to unspliced HIV-1 RNA are dramatically higher in these patients than in patients with AIDS [64]. Over 70 % of PBMC samples from asymptomatic individuals reveal a viral RNA pattern with blocked early-stage latency whereas over 80 % of PMBC samples from patients with AIDS exhibit a productive pattern. It can be concluded that the asymptomatic HIV-1 infection in human could be characterized by a majority of HIV-1-infected PBMCs which have HIV-1-specific RNA expression patterns consistent with a blocked early-stage latency.

U1 and ACH-2 cell lines have been treated with increasing concentrations of H_2O_2 to characterize their susceptibility to these stress conditions [58]. These cells are rather sensitive to H_2O_2 and small increases in concentration lead to a proportional decrease in viability. The viability begins to significantly drop at a H_2O_2 concentration higher than 200 µM and leads to a very low surviving fraction above 2 mM in the case of U1 cells. The lymphocytic cell line ACH-2 appears to be far more sentive to H_2O_2 than U1; a significant lethal effect is already observed at a very low H_2O_2 concentration such as 50 µM.

To estimate the effect of H_2O_2 on HIV-1 reactivation in these latently infected cells two different techniques can be used : (i) the measurement of a reverse transcriptase (RT) activity in cell supernatants and (ii) the detection of intracellular viral antigen. An increasing RT activity is determined in supernatant fluid after H_2O_2 exposure and this phenomenon turned out to be transient with a maximun arising between 48 and 72 h. In addition, a dose-response curve is obtained (Fig. 2) and again low H_2O_2 concentrations are needed to transiently produce HIV particles in cell supernatants. With the ACH-2 cells, the optimal concentration of H_2O_2 leading to the higher HIV-1 reactivation is much lower than observed with U1 cells, and corresponds to a range of H_2O_2 concentrations observed in physiological conditions. An increase of intracellular HIV-1 proteins synthesis can be evidenced by an immunofluorescence (IF) assay [58]. Treatment of U1 cells with H_2O_2 between 0.5 and 1 mM leads to an increase in the frequencies of IF positive cells with an optimum 48 h after the oxidative stress.

A.

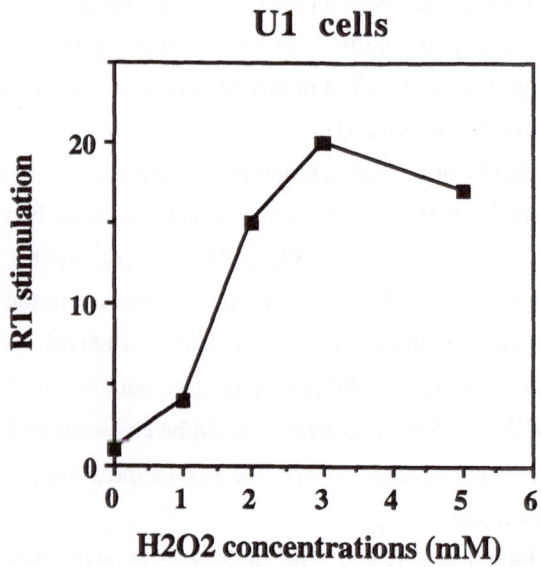

B.

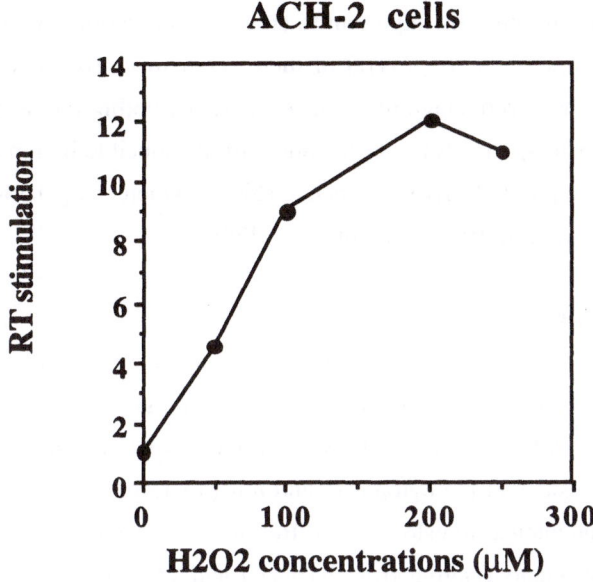

Figure 2. HIV-1 reactivation by H_2O_2 from two non-productively infected cell lines (A: U1 cells, B: ACH-2 cells).

These data demonstrated for the first time that HIV-1 can be activated to replicate in non productively infected cells just by displacing the intracellular redox balance toward a pro-oxidant state. Antioxidants like NAC, OTC [see 36, for review] or selenium supplementation in order to increase the gluthation peroxidase activity [65], or iron chelation [66] can all counteract this activation mediated by an oxidative stress. Because HIV-1 gene activation in these pro-oxidant conditions is under the control of the HIV-1 long terminal repeat, these data point out the importance of host transcriptional factors in the control of HIV gene expression.

5. Importance of host factors in HIV-1 activation

Numerous papers have demonstrated the key role of several cellular factors and of the HIV-1 LTR in the passage between a latently infected cell and a virus producing cell. In many respects, the HIV-1 LTR functions like the promoter of a T-cell activation gene [67]. Transcriptional induction driven by the LTR is critically dependent upon the presence of two juxtaposed κB enhancer elements that specifically engage members of the NF-κB / Rel family transcription factors [68]. Additional DNA-binding proteins have been identified which bind to the downstream LTR region like the LBP, UBP-1 and CTF/NFI factors or to regions upstream the transcription initiation site like Sp1, USF, NFAT-1 and AP-1 sites [69]. Therefore, it is important to consider that various stimuli or cofactors favoring the induction of one or several of these regulatory proteins can promote the switching towards the severe stage of the disease by disrupting the latent stage of HIV-1. Among these factors, NF-κB has been shown to be inducible by tumor necrosis factor α (TNF-α) and by phorbol myristate acetate (PMA) which can generate intracellular oxidants and by the extracellular addition of H_2O_2 [70].

6. Structure and function of NF-κB

NF-κB and the other member of the Rel family of transcriptional activator proteins are a focal point for understanding how extracellular signals induce the expression of specific sets of genes in higher eukaryotes [for reviews, 71-73]. Unlike most transcriptional factors, this family of proteins resides in the cytoplasm and must therefore translocate into the nucleus to function. The nuclear translocation of Rel proteins is induced by an extraordinarily large number of agents ranging from bacterial and viral pathogens to immune and inflammatory cytokines to a variety of agents that damage cells. Remarkably, an even larger number of genes appear to be targets for the activation by Rel proteins.

6.1 THE REL AND IκB FAMILIES

The Rel protein family has been divided into two groups based on differences in their structures, functions, and modes of synthesis. The first group consists of p50 (NF-kB1) and p52 (NF-kB2), which are synthesized as precursor proteins of 105 and 100 kDa respectively (Fig.3). The mature proteins, which are generated by proteolytic processing, have a so-called Rel homology domain that includes DNA-binding and dimerization domains and a nuclear localization signal. The mature proteins form functional Rel dimers with other members of the family, while dimers containing the unprocessed proteins remain sequestred in the cytoplasm. The second group of Rel proteins, which includes p65 (RelA), Rel (c-Rel), RelB, and the drosophila Rel proteins dorsal and Dif, are not synthesized as precursors (Fig.3). In addition to the Rel homology domain, they possess one or more transcriptional activation domains. Members of both groups of Rel proteins can form homo- or hetero-dimers; e.g., NF-κB is the classical p50-p65 heterodimer which binds to the 5'-GGGANNYYCCC-3' consensus sequence.

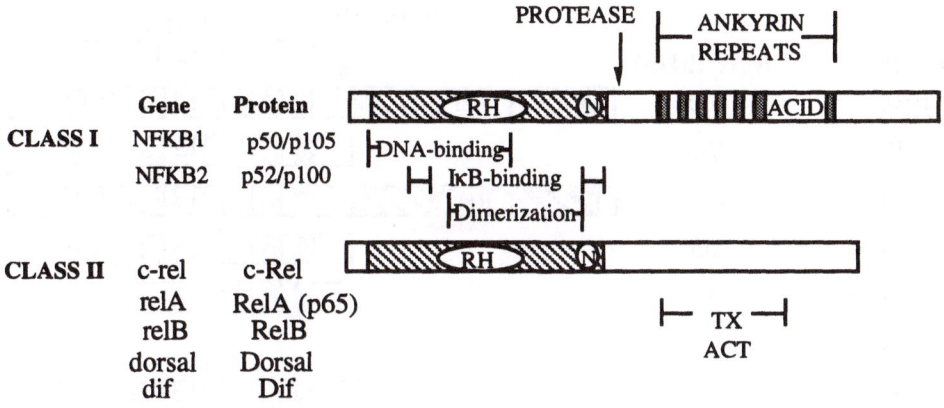

Figure 3. The class I and class II members of the Rel protein family

Two types of Rel protein complexes are found in the cytoplasm prior to induction. The first type consists of Rel homo- or hetero-dimers (e.g. p50 and p65) bound to a member of the IκB family of inhibitor proteins (IκB-α, IκB-β, IκB-γ, Bcl-3 and the drosophila protein cactus; Fig. 4). Members of this family share a characteristic ankyrin repeat motif that is required for their interactions with Rel proteins and a C-terminal PEST sequence thought to be involved in protein degradation [74]. The second type of complex consists of a heterodimer formed between a mature Rel protein (e.g., p65) and an unprocessed Rel protein precursor (e.g., p105). An induction signal leads to the phosphorylation of both IκB and p105 [75]. This phosphorylation is the signal for IκB degradation and p105 processing, both of which generate active Rel dimeric complexes that translocate to the nucleus and activate genes containing Rel protein-binding sites (κB sites).

Several different transduction pathways have been implicated in NF-κB activation, all of which culminate in the phosphorylation of serine residues 32 and 36 of the amino-terminal part of IκB [76]. Unfortunately, none of these pathways have been fully elucidated, and the a so-called IκB kinase has yet to be definitely identified. The sphingomyelinase/ceramide and the Raf pathways have been implicated in TNF-α induction of NF-κB [see 72, for review] while induction by anti-CD28 antibodies involves the rapamycine sensitive p70 S6 kinase pathway [77].

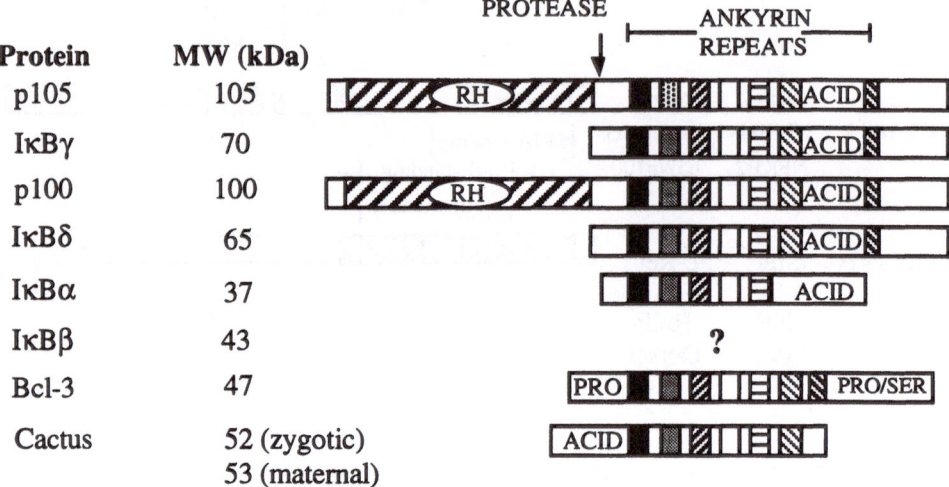

Figure 4. The member of the IκB family.

The common step in all these induction pathways is the involvement of ROS [70]. This conclusion is based largely on the inhibition of NF-κB activation by a series of antioxidants. Antioxidants have been reported to block NF-κB activation in many instances, although the extent of this block appears to vary depending on cell and signal. Inhibitory antioxidants with diverse chemical properties include NAC, dithiocarbamates, vitamin E derivatives, and various metal chelators. Support for the involvement of ROS as common messenger also derives from evidences showing elevated cellular level of ROS in response to TNF-α, IL-1, PMA, LPS, UV light and gamma irradiation. In addition, direct addition of hydrogen peroxide or other well-known ROS generator like photosensitizer also induce nuclear translocation of NF-κB demonstrating that ROS can specifically function in the NF-κB signaling cascade.

The data summarized above demonstrate the interplay between HIV-1, NF-κB and the establishment of oxidative stress conditions in T-lymphocytes. Recently, the role of NF-κB-dependent signals in activating the transcriptional activity of the HIV-1 LTR has been analyzed in human CD4+ cells isolated from peripheral blood and in transformed lymphoblastoid T cell line [78]. These authors show that unstimulated CD4+ cells offer a cellular environment of very low permissivity to HIV-1 LTR functioning. This is in sharp contrast to the high spontaneous LTR activity observed in lymphoblastoid T cells, where LTR activity is essentially dependent on the κB-responsive elements. Due to the low basal LTR activity in resting T cells, NF-κB-dependent transactivation is a *sine qua non* event for induction of the HIV LTR reinforcing the role of prooxidant conditions in favouring NF-κB induction and HIV-1 replication.

7. Acknowledgments

J.P. is Research Director from the Belgian Fund for Scientific Research (NFSR, Brussels, Belgiun), BP is a grantee from the Belgian FRIA and SLP was funded by the NFSR-AIDS research fund. This research program was granted by the Belgian NFSR and the Belgian National Lottery.

8. References

1.   Clerici, M., Stocks, N.I., Zajac, R.A., Boswell, R.N., Lucey, D.R., Via, C.S., and. Shearer, G.M (1989) Detection of three distinct patterns of T helper cell dysfunction in asymptomatic, human immunodeficiency virus-seropositive patients, *J. Clin. Invest* **84**,1892.

2.   Cuthbert, R.J.G., Ludham, C.A., Tucker, J., Steel, C.M., Beatson, D., Rebus, S., and Peutherer, J.F. (1990) Five year of prospective study of HIV infection in the Edinburgh haemophiliac cohort, *Br. Med. J.* **301**,956.

3.   Philips, A.N., Lee, C.A., Elford, J., Janossy, G., Timms, A., Bofill, M., and Kernoff, P.B.A. (1991) Serial CD4 lymphocyte counts and development of AIDS, *Lancet* **337**, 389.

4.   Levy, J.A. (1993) Pathogenesis of human immunodeficiency virus infection, *Microbiol. Rev.* **57**, 183.

5.   Kaslow,R.A., Duquesnoy, R., Van Raden, M., Kingsley, L., Marrari, M., Friedman,. Su, H., Saah, A.J. S, Detels, R., Phair, J., and Rinaldo, C. (1990) A1, Cw7, B8, DR3 HLA antigen combination associated with rapid decline of T-helper lymphocytes in HIV-1 infection, *Lancet* **335**, 927.

6.   Franzetti, F., Cavalli, G.,. Foppa, C.U, Amprimo, M.C., Gaido, P., and Lazarin, A. (1988) Raised serum β-2 microglobulin levels in different stages of human immunodeficiency virus infection, *J. Clin. Lab. Immunol.* **27**, 133.

7.   Gaines, H., Von Sydow, M., Pehrson, P.O., and Lundbergh, P.(1988) Clinical picture of primary HIV infection presenting as a glandular-fever-like illness, *Br. Med. J.* **297**, 1363.

8.   Lang,W., Perkins, H., Anderson, R.E., Royce, R., Jewell, N., and Winkelstein, W. (1989) Patterns of T lymphocyte changes with immunodeficiency virus

infection : from seroconversion to the development of AIDS, *J. AIDS* **2**, 23.

9. Allain, J.P., laurin, Y., Paul, D.A., Verroust, F., Leuther, M., et al. (1987) Long-term evaluation of HIV antigen and antibodies to p24 and gp41 in patients with haemophilia, *New Engl. J. Med.* **317**, 1114.

10. Bachetti, P. and Moss, A.R. (1989) Incubation period of AIDS in San Francisco, *Nature* **338**, 251.

11. Downs, A.M., Ancelle-Park, R.A., Costagliola, D., Rigaut, J.P., and Brunet, J.B. (1991) Transfusion-associated AIDS cases in Europe : estimation of the incubation period distribution and prediction of future cases, *J. AIDS* **4**, 805.

12. Isaksson, B., Albert, J., Chiodi, F., Furucrona, A., Krook, A., and Putkonen, P. (1988) AIDS two months after primary human immunodeficiency virus infection, *J. Infect. Dis.* **158**, 866.

13. Mc Lean, K.A., Holmes, D.A., Evans, B.A., Mac Alpine, L., Thorp, R., Parry, J.V., and Glaser, M.G. (1990) Rapid clinical and laboratory progression of HIV infection, *AIDS* **4**, 369.

14. Lifson, A.R., Butchbinder, S.P., Sheppard, H.W., Mawle, A.C., Wilber, J.C., Stanley, M., Hart, C.E., Hessol, N.A., and Holmberg, S.D. (1991) Long-term human immunodeficiency virus infection in asymptomatic homosexual and bisexual men with normal CD4+ lymphocyte counts : immunologic and virologic characteristic, *J. Infect. Dis.* **163**, 959.

15. Coombs, R.W., Collier, A.C., Allain, J.P., Nikora, Y., Leuther, M., Gjerset, G.F., and Corey, L. (1989) Plasma viremia in human immunodeficiency virus, *New Engl. J. Med.* **321**, 1626.

16. Simmonds, P., Balfe, P., Peutherer, J.F., Ludlam, C.A., Bishop, J.O., and Leigh Brown, A.J. (1990) Human immunodeficiency virus-infected individuals contain provirus in small numbers of peripheral mononuclear cells and at low copy numbers, *J. Virol.* **64**, 864.

17. Piatak, M., Saag, M.S., Yang, L.C., Clark, S.J., Kappes, J.C., Luk, K.-C., Hahn, B.H., Shaw, G.M., and Lifson, J.D. (1993) High levels of HIV-1 in plasma during all stages of infection determined by competitive PCR, *Science* **259**, 1749.

18. Fox, C.H., Tenner-Rcz, K., Racz, P., Firpo, A., Pizzo, P.A. and Fauci, A.S. (1991) Lymphoid germinal centers are reservoirs of human immunodeficiency virus type 1 RNA, *J. Infect. Dis.* **164**, 1051.

19. Pantaleo, G., Graziosi, C., Butini, L., Pizzo, P.A., Schnittman, S.M., Kottler, D.P., and Fauci, A.S. (1991) Lymphoid organs function as major reservoirs for human immunodeficiency virus, *Proc. Natl. Acad. Sci. USA* **88**, 9838.

20. Pantaleo, G., Graziosi, C., Demarest, J.F., Butini, L., Montroni, M., Fox, C.H., Orenstein, J.M., Kottler, D.P., and Fauci, A.S. (1993) HIV infection is active and progressive in lymphoid tissue during the clinically latent stage of disease, *Nature* **362**, 355.

21. Embretson, J., Zupancic, M., Ribas, J.L., Burke, A., Racz, P., Tenner-Racz, K., and Haase, A.T. (1993) Massive covert infect of helper T lymphocytes and macrophages by HIV during the incubation period of AIDS, *Nature* **362**, 359.

22. Saag, M.S., Crain, M.J., Decker, W.D., Campbell-Hill, S., Robinson, S., Brown, W.E., Leuther, M., Wihtley, R.J., Hahn, B.H., and Shaw, G.M. (1991) High level of viremia in adults and children infected with human immunodeficiency virus: relation to disease stage and CD4+ lymphocyte levels, *J. Infect. Dis.* **164**, 72.

23. Weiss, R.A. (1993) How does HIV causes AIDS, *Science* **260**, 1272.

24. Ho, D.D., Moudgil, T., and Alam, M. (1989) Quantitation of human immunodeficiency virus type 1 in the blood of infected persons, *New England J. Med.* **321**, 1621.

25. Simmonds, P., Balfe, P., Peutherer, J.F., Ludlam, C.A., Bishop, J.O., and Leigh Brown, A.J. (1990) Human immunodeficiency virus-infected individuals contain provirus in small numbers of peripheral mononuclear cells and at low copy numbers, *J. Virol.* **64**, 864.

26. Ho, D.D., Neumann, A.U., Perelson, A.S., Chen, W., Leonard, J.M., and Markovitz, M. (1995) Rapid turnover of plasma virions and CD4 lymphocytes in HIV-1 infection, *Nature* **373**, 123.

27. Wei, X., Ghosh, S.K.,. Taylor, M.E, Jonhson, V.A., Emini, E.A., Deutsch, P., Lifson, J.D., Bonhoeffer, S., Nowak, M.A., Hahn, B. H., Saag, M.S., and Shaw, G.M. (1995) Viral dynamics in human immunodeficiency virus type 1 infection, *Nature* **373**, 117.

28. McCune, J.M. (1995) Viral latency in HIV disease, *Cell* **82**, 183.

29. Cao, Y., Qin, L., Zhang, L., Safrit, J., and Ho, D.D. (1995) Virologic and immunologic characterization of long-term survivors of human immunodeficiency virus type 1 infection, *New England J. Med.* **332**, 201.

30. Pantaleo, G., Menzo, S., Vaccareza, M., Graziosi, C., Cohen, O.J., Demarest,

J.F., Montefiori, D., J.M. Orenstein, C. Fox, Schrager, L.K., Margolick, J.B., Bucbinder, S., Giorgi, J.V., and Fauci, A.S. (1995) Studies in subjects with long-term non-progressive human immunodeficiency virus infection, *New England J. Med.* **332**, 209.

31. Zack,J.A., Arrigo, S.J., Weitaman, S.R.,. Go, A.S, Haislip, A., and Chen, I.S.Y. (1990) HIV-1 entry into quiescent lymphocytes : molecular analysis reveals a labile, latent viral structure, *Cell* **61**, 213.

32. Zack, J.A., Haislip, A.M., Krogstad, P., and Chen, I.S.Y. (1992) Incompletely reverse transcribed HIV-1 genomes in quiescent cells can function as intermediate in retroviral life cycle, *J. Virol.* **66**, 1717.

33. Stevenson,M., Stanwick, T.L., Dempsey, M.P., and Lamonica, C.A. (1990) HIV-1 replication is controlled at the level of T-cell activation and proviral integration, *EMBO J.* **9**, 1551.

34. Bukrinsky, M.I., Stanwick, T.L., Dempsey, M.P., and Stevenson, M. (1991) Quiescent T lymphocytes as an inducible virus reservoir in HIV-1 infection, *Science* **254**, 423.

35. Winslow, B.J., Pommerantz, R.J., Bagasra, O., and Trono, D. (1993) HIV-1 latency due to the site of proviral integration, *Virology* **196**, 849 (1993).

36. Staal, F.J.T., Ela, S.W., Roederer, M., Anderson, M.T., Herzenberg, L.A., and Herzenberg,L.A. (1992) Glutathione deficiency and human immunodeficiency virus infection, *Lancet* **339**, 909.

37. Pace, G.W. and Leaf, C.D. (1995) The role of oxidative stress in HIV disease (1995) *Free Radical Biol. Med.* **19**, 523.

38. Eck, H-P., Gmünder, H., Hartman, M., Petzold, D., Daniel, V., and Dröge, W. (1989) Low concentrations of acid-soluble thiols (cysteine) in the blood plasma of HIV-1 infected patients, *Biol. Chem Hoppe Seyler* **370**, 101.

39. Buhl, R., Jaffe, H.A., Holroyd, K.J., Wells, F.B., Mastrangelli, A., Saltini, C., Cantin, A.M., and Crystal, R.G. (1989) Systemic glutathione deficiency in symptom-free HIV-seropositive individuals, *Lancet* **ii**, 1294.

40. Buhl, R. (1994) Imbalance between oxidants and antioxidants in the lungs of HIV-seropositive individuals, *Chem.-Biol. Interact.* **91**, 147.

41. Poli, G. and Fauci, A.S. (1992) The effects of cytokines and pharmacological agents on chronic HIV infection, *AIDS Res. Human Retroviruses* **8**, 191.

42. Meister, A. (1988) Glutathione metabolism and its selective modification, *J.*

Biol. Chem. **263**, 17205.

43.   Hamilos, D.L. and Wedner, H.J. (1985) The role of glutathione in lymphocyte activation, I. comparison of inhibitory effects of buthionine sulfoximine and 2-cyclohexene-1-one by nuclear size transformation, *J. Immunol.* **135**, 2740.

44.   Dröge, W., Pottmeyer-Gerber, C., Schmidt, H., and Nick, S. (1986) Glutathione augments the activation of cytotoxic T lymphocytes in vivo, *Immunobiology* **172**, 151.

45.   Suthanthiran, M., Anderson, M.E., Sharma, V.K., and Meister, A. (1990) Glutathione regulate activation-dependent DNA synthesis in highly purified normal human T lymphocytes stimulated *via* the CD2 and CD3 antigens, *Proc. Natl. Acad. Sci USA* **87**, 3343.

46.   Favier, A., Sappey, C., Leclerc, P., Faure, P., and Micoud, M. (1994) Antioxidant status and lipid peroxidation in patients infected with HIV, *Chem-Biol. Interact.* **91**, 165.

47.   Sappey, C., Leclercq, P., Coudray, C., Faure, P., Micoud, M., and Favier, A. (1994) Vitamin, trace elements, and peroxide status in HIV seropositive patients: asymptomatic patients present a severe β-carotene deficiency, *Clin. Chem. Acta* **230**, 35.

48.   Polyakov, V.M., Shepelev, A.P., Kokovkina, O.E., and Vtornikova, I.V. (1994) Superoxide anion (O2-) production and enzymatic disbalance in pheripheral blood cells isolated from HIV-infected children, *Free Radical Biol. Med.* **16**, 15.

49.   Leff, J.A., Oppegard, M.A., Curiel, T.J., Brown, K.S., Schooley, R.T. and Repine, J.E. (1992) Progressive increases in serum catalase activity in advancing human immunodeficiency virus infection, *Free Radical Biol. Med.* **13**, 143.

50.   Flores, S.C., Marecki, J.C., Harper, K.P., Bose, S.K., Nelson, S.K., and MC Cord, J.M. (1993) The HIV-1 Tat protein represses Mn superoxide dismutase expression in HeLa cells, *Proc. Natl. Acad. Sci. USA* **90**, 1.

51.   Fuchs, J., Milbradt, R., Ochensdorf, F., Rubsamen-Waigmann, H., and Schofer, H. (1991) Oxidative imbalance in HIV infected patients, *Medicine Hypoth.* **7**, 60.

52.   Meister, A. (1981) Metabolism and functions of glutathione, *Trends Biol. Sci.* **6**, 231.

53.   Fidelius, R.K. and Tsan, M.F. (1986) Enhancement of intracellular glutathione promotes lymphocyte activation by mitogen *Cell Immunol.* **97**, 155.

54.   Eylar, E., Quinones, C, Molina, C., Baez, I., and. Mercado, C (1993) *N*-acetyl-

L-cysteine enhances T cell functions and T cell growth in culture, *Int. Immunol.* **5**, 97.

55. Holdiness, M.R. (1991) Clinical pharmacokinetics of *N*-acetyl-L-cysteine, *Clin Pharmacokinet.* **20**, 123.

56. DeQuay, B., Malinverni, R., and Lauterburg, B.H. (1992) Glutathione depletion in HIV-infected patients: role of cysteine deficiency and effect of oral *N*-acetyl-L-cysteine, *AIDS* **6**, 815.

57. Witschi, A., Junker, E., Schranz, C.,. Speck, R.F, and Lauterburg, B.H. (1995) Supplementation of *N*-acetyl-L-cysteine fails to increase glutathione in lymphocytes and plasma of patients with AIDS, *AIDS Res.* **11**, 141.

58. Legrand-Poels, S., Vaira, D., Pincemail, J., Van de Vorst, A., and Piette, J. (1990) Activation of human immunodeficiency virus type 1 by oxidative stress, *AIDS Res.* **6**, 1389.

59.. Folks, T.M., Justement, J., Kinter, A., Schnittman, S., Orenstein, J., Poli, G., and A.S. Fauci (1988) Characterization of a promonocyte clone chronically infected with HIV and inducible by PMA, *J. Immunol.* **140**, 1117.

60. Clouse, K.A., Powell, D., Washington, I., Poli, G., Strebel, K., Farrar, W., Barstad, D., Kovacs, J., Fauci, A.S., and Folks, T.M. (1989) Monokine regulation of HIV-1 expression in a chronically infected human T cell clone, *J. Immunol.* **142**, 431.

61. Pommerantz, R.J., Trono, D., Feinberg, M.B., and Baltimore, D. (1991) Cells non productively infected with HIV-1 exhibit an aberrant pattern of viral RNA expression : a molecular model for latency, *Cell* **6**, 11271.

62. Saksela, K., Stevens, C., Rubinstein, R., and Baltimore, D. (1994) Human immunodeficiency type 1 mRNA expression in pheripheral blood cells predicts disease progression independently of the numbers of CD4+ lymphocytes, *Proc. Natl. Acad. Sci. USA* **91**, 1104.

63. Michael, N.L., Mo, T., Merzouki, A., O'Shaughnessy, M., Oster, C., Burke, D.S., Redfield, R.R., Birx, D.L., and Cassol, S.A. (1995) Human immunodeficiency virus type 1 cellular RNA load and splicing patterns predict disease progression in a longitudinal cohort, *J. Virol.* **69**, 1868.

64. Pomerantz, R.J., Bagasra, O., and Baltimore, D. (1992) Cellular latency of immunodeficiency virus type 1, *Current Opinion Immunol.* **4**, 475.

65. Sappey, C., Legrand-Poels, S., Best-Belpomme, M., Favier, A., Rentier, B., and Piette, J. (1994) Stimulation of glutathione peroxidase activity decreases HIV

type 1 activation after oxidative stress, *AIDS Res.* **10**, 1451.

66.   Sappey, C., Boelaert, J.R., Legrand-Poels, S., Forceille, C., Favier, A., and Piette, J. (1995) Iron chelation decreases NF-κB and HIV-1 activation due to oxidative stress, *AIDS Res.* **11**, 1149.

67.   Molitor, J.A., Walker, W.A., Doerre, S., Ballard, D.W., and Greene, W.C. (1990) NF-κB: a family of inducible and differentially expressed enhancer binding protein in T cells, *Proc. Natl. Acad. Sci. USA* **87**, 10028.

68.   Nolan, G.P., Ghosh, S., Liou, H-C., Tempst, P., and Baltimore, D. (1991) DNA binding and I-κB inhibition of the clonad p65 subunit of NF-κB, a Rel-related polypeptide, *Cell* **64**, 961.

69.   Folks,T.M., Justement, J., Kinter, A., Schnittman, S., Orenstein, J., Poli, G., and A.S. Fauci, (1988) Characterization of a promonocyte clone chronically infected with HIV and inducible by PMA, *J. Immunol.* **140**, 1117.

70.   Schreck, R., Rieber, P., and Baeuerle, P.A. (1991) Reactive oxygen intermediates as apparently widely used messengers in the activation of the NF-κB transcription factor and HIV-1, *EMBO J.* **10**, 2247.

71.   Baeuerle, P.A. and Henkel, T. (1994) Function and activation of NF-κB in the immune system, *Annu. Rev. Immunol.* **12**, 141.

72.   Siebelnist, U., Franzoso, G., and. Brown, K (1994) Structure, regulation and function of NF-κB, *Annu. Rev. Cell. Biol.* **10**, 405).

73.   Grilli, M., Chiu, J.J.-S. and Lenardo, M.J. (1993) NF-κB and rel-participants in a multiform transcriptional regulatory system, *Int Rev Cytol.* **143**, 1.

74.   Beg, A.A. and Baldwin, A.S. (1993) The IκB proteins: multifunctional regulators of Rel/NF-κB transcription factors, *Genes Dev.* **7**, 2064.

75.   Mercurio, F., DiDonato, J.A., Rosette, C. and Karin, M. (1993) p105 and p98 precursor proteins play an active role in NF-κB-mediated signal transduction, *Genes and Dev.* **7**, 705.

76.   Traenckner, E.-B., Pahl, H.L., Henkel, T., Schmidt, K.N., Wilk, S. and Baeuerle, P.A. (1995) Phosphorylation of human IκB-α on serines 32 and 36 controls IκB-α proteolysis and NF-κB activation in response to diverse stimuli, *EMBO J.* **14**, 2876.

77.   Lai, J.H. and Tan, T.H. (1994) CD28 signaling causes a sustained down-regulation of IκB-α which can be prevented by the immunosuppressant Rapamycine, *J. Biol. Chem.* **269**, 30077.

78.   Alcami, , de Lera, T.L., Folgueira, L., Pedraza, M-A., Jacqué, J.M., Bachellerie, F., Noriega, A.R., Hay, R.T., Harrich, D., Gaynor, R.B., Virelizier, J.L., and Arenzana-Seisdedos, F. (1995) Absolute dependence on κB responsive elements for initiation and Tat-mediated amplification of HIV transcription in blood CD4 T lymphocytes, *EMBO J.* **14**, 1552.

RNA Cleavage by RNase H

RIBONUCLEASE H : FROM ENZYMES TO ANTISENSE EFFECTS OF OLIGONUCLEOTIDES

J.J. TOULMÉ, C. BOIZIAU, B. LARROUY, P. FRANK[a], S. ALBERT[b] AND R. AHMADI.
INSERM U386, Laboratoire de Biophysique Moléculaire, Université de Bordeaux II, 146 rue Léo Saignat, 33076, Bordeaux, France. Present address : a) Institute for Tumor Biology and Cancer Research, Vienna, Austria b) Université de Perpignan, France

Ribonucleases H are enzymes that specifically cleave the RNA strand of double-stranded RNA-DNA hybrids. They are found in every cell from bacteria to higher eukaryotes and are involved in the development of retroviruses. Discovered in 1969 by Stein and Hausen [1], RNase H was recently brought to the front of the stage for two reasons : first, as a key activity for maturing the genome of the Human Immunodeficiency Virus, RNase H constitutes a valid target for designing antiviral drugs [2]; second, RNase H is likely to play a role in the effects mediated by antisense oligonucleotides which are promising potentially new therapeutic agents [3]. Several reviews [2, 4, 5] and a book [6] were recently devoted to ribonucleases H. This paper will focus on particular aspects dealing with the activity of eukaryotic RNase H on heteroduplexes formed by RNA and synthetic chemically-modified oligonucleotides designed to regulate translation or reverse transcription. We will first briefly summarize the main properties of RNases H and recall what is known on RNA-DNA hybrids.

1. Ribonucleases H

The knowledge on these enzymes is quite paradoxical. Numerous informations have been accumulated on their biochemistry ; the crystal structure of an *E. coli* enzyme and of the RNase H domain of the human immunodeficiency virus-1 (HIV-1) reverse transcriptase (RT) has been solved. But, surprisingly,

B. Meunier (ed.), DNA and RNA Cleavers and Chemotherapy of Cancer and Viral Diseases, 271–288.
© *1996 Kluwer Academic Publishers. Printed in the Netherlands.*

except for retroviral enzymes, very little is known about the biological role of RNases H.

1.1. PROPERTIES OF RNASE H

In *E. coli* two enzymes with RNase H activity (HI and HII) have been discovered. HI from *E. coli*, a 18 kDa protein, is by far the most studied RNase H [7]. It was proposed [8], mainly from genetic studies, that RNase HI i) removes Okazaki fragments which prime the replication of the lagging DNA strand, ii) is involved in the initiation of replication of Col E1-type plasmids and iii) ensures the initiation of replication of the *E. coli* chromosome from *oriC*.

In higher eukaryotes, the situation is more obscure. Two classes, termed HI and HII, have been identified [9]. The bovine enzymes are the best characterized [10, 11], but the human ones have been recently purified [12, 13]. Class I enzymes which constitute the major RNase H activity in the cell have a high molecular weight and are insensitive to SH-blocking agents like N-ethylmaleimide (NEM). Class II RNases H have a low molecular weight and are sensitive to NEM (Table 1). Another major difference between the two classes is the cation which is required for activity : whereas RNase HI is active in the presence of either magnesium or manganese, RNase HII is activated only by Mg^{2+} and inhibited by Mn^{2+} ions. In addition, due to their different iso-electric point (pI), RNase HI binds to anion exchangers but not to cation exchangers ; the reverse is true for HII. This property can be used to separate the two enzymes in one step of chromatography [13]. RNases HI are likely involved in DNA replication as their activity is increased in parallel with DNA synthesis [14]. It has been suggested that RNase HII is linked to transcription as its activity increases with RNA synthesis.

	Mw (kDa)	Cation	Inhibition by NEM or Mn^{2+}	pI	Amount in cells
RNase HI	60-90	Mn^{2+}(0,5mM) Mg^{2+}(10mM)	No	5	80%
RNase HII	30-45	Mg^{2+}(10mM)	Yes	7	20%

Table 1 : Properties of class I and II RNases H from eukaryotic cells.

Reverse transcriptase is a multifunctional enzyme which possesses RNase H activity required at several steps of the conversion of the single-stranded viral RNA in double-stranded DNA, prior to the integration of the viral genome in the host genome. Retroviral RNase H is involved i) in the degradation of the retroviral strand, ii) in the production of the plus-strand primer, and iii) in the removal of tRNA and plus-strand primers [15]. In the case of HIV, the reverse transcriptase is composed of two subunits, p66 and p51, the small subunit being produced by proteolytic cleavage of the carboxy-terminal part of the large one. The 15 kDa RNase H domain is located at the carboxy terminus of the p66 domain and is therefore not present in the p51 subunit.

1.2. STRUCTURE OF *E. COLI* AND HIV RNASES H

A high resolution structure was obtained for both *E. coli* and HIV-1 RNases H. The bacterial enzyme is made of a single polypeptide, 155 amino acids long, folded into five α-helices and a β-sheet, the latter forms with two of the α-helices a large cleft, which likely represents a binding site for the RNA-DNA substrate [16, 17]. Interestingly, the three carboxylic acid residues, Asp-10, Glu-48 and Asp-70, involved in the catalytic activity are located in this area, along with four other important residues. Katayanagi *et al.* [16] showed that Mg^{2+} binds close to the carboxylic acid triad. On one side of the β-sheet is located a short region, about 18 amino acids long, known as "handle" or "basic protrusion" as it contains six Lys and Arg residues important for substrate binding.

The comparison of the primary structure of *E. coli* RNase HI and of the RNase H domain of the HIV-1 reverse transcriptase reveals that seven amino acids are conserved, including the three carboxylic acid residues crucial for catalytic activity. The three dimensional structure of the retroviral RNase H domain [18] is strikingly similar to that of the bacterial enzyme. One major difference is the absence of the handle region in the retroviral enzyme. This might explain in part the inability of the isolated RNase H domain of HIV to bind to RNA-DNA hybrids.

All RNases H produce 5'-phosphate and 3'-hydroxyl termini but their mechanism of action is still controversial. On the one hand, based on the *E. coli* RNase H structure, a carboxylate-hydroxyl relay mechanism similar to that of bovine DNase I has been proposed [19]. On the other hand, a two metal ion mechanism has been suggested for the HIV enzyme, as two Mn^{2+} ions, approximatively 4Å apart have been identified in the crystal structure [18], a situation similar to that of the 3' -> 5' exonuclease domain of the *E. coli* DNA Pol I. Due to the similar structure of HIV-1 and *E. coli* RNases H, it is very unlikely that these enzymes use two different mechanisms to achieve the same reaction.

2. RNA-DNA hybrids

RNase H selectively cleaves RNA-DNA hybrids and has no activity on double-stranded RNA. How does RNase H recognize RNA-DNA heteroduplexes ? Because of their biological importance RNA-DNA hybrids have been the subject of structural studies on crystals and in solution. In double-stranded DNA, the strands generally exhibit a C2'-*endo* sugar pucker which is typical of the B form, whereas in RNA the ribose ring adopts a C3'-*endo* pucker resulting in A-type helices.

The situation for RNA-DNA hybrid is still controversial : the proposed structures include A type [20], B type [21] or heteronomous duplexes in which the nucleotides of the RNA strand have C3'-*endo* sugar pucker (A form) but that of the DNA strand are close to the C2'-*endo* domain (B form) [22]. It is generally considered that enzymes that do not display sequence specificity interact with the minor groove of double-stranded nucleic acids. The minor groove of RNA-DNA hybrids is narrower and deeper than in double helical A-RNA. Recently, B. Reid and co-workers concluded that whereas the RNA strand was close to A form, the DNA strand was neither A nor B but assumed an intermediate conformation, the sugar pucker being O4'-*endo* [23]. These authors derived a model for such an RNA-DNA hybrid structure bound to the E. coli RNase H. In this model a better contact is obtained between the carboxylic acids involved in the catalytic activity and the substrate than in the one previously published by Nakamura *et al.* [19] that assumed an A form for the hybrid. It is likely that the local structure will vary with the hybrid sequence. This might explain the RNase H resistance of a few segments like the polypurine tract generating the primer for plus-strand synthesis in reverse transcription of retroviral genomes [15].

The presence of mixed backbones (chimeric molecules ; see 4.2 below) or the use of chemically-modified analogues for antisense studies may be another source of alterated conformation of the heteroduplex. For instance, it has been shown that a model for Okazaki fragments made of a double-stranded decanucleotide comprising three ribo-residues at the 5' end of one strand adopted an A-type conformation [20]. Therefore the RNA stretch drove the overall conformation of the hybrid. Very little information is available for heteroduplexes formed with chemically-modified oligomers although this is of prime importance for antisense studies. It is likely that the structure of such RNA-modified oligonucleotide hybrids is changed compared to regular ones as many of them are not substrates for RNase H. A study of a series of oligonucleotides containing various 2'-O-modified adenosine residues showed that the duplexes they formed with RNA exhibited CD spectra close to that of RNA-DNA hybrids [24].

3. RNase H contribution to antisense effects

Antisense oligonucleotides are synthetic oligomers complementary to a portion of a target RNA : pre-RNA, mRNA or viral RNA. The formation of the RNA-oligonucleotide hybrid can potentially interfere with mRNA maturation, translation or with viral development. Therefore antisense oligonucleotides constitute a way to specifically modulate the expression of one selected gene, as the association between the antisense and the target sequences is driven by the formation of nucleic acid base pairs. For chemical reasons synthetic antisense sequences were prepared as oligodeoxynucleotides : DNA is easier and less expensive than RNA to synthesize in amounts large enough to perform experiments with live biological material. Numerous successful reports have been published on the use of oligonucleotides to artificially regulate gene expression in cell-free extracts, in micro-injected or in cultured cells and, more recently, *in vivo* (see [3, 25] for reviews). Clinical trials are underway for several diseases : infection by HIV and cytomegalovirus, acute and chronic leukemias, restenosis.

In this chapter we will focus on the role played by RNase H in the effects induced by antisense oligonucleotides. Two mechanisms have been described for the inhibition of translation by antisense oligonucleotides. In the first one the oligomer constitutes a physical blockade to the reading of the information borne by the target RNA [26]. This mechanism is restricted to target sequences located in the 5' leader region. In other words, the direct competition between the antisense oligonucleotide and the translation machinery can prevent the initiation but not the elongation step. The second mechanism is operative whatever the location of the target along the message : the RNA-oligonucleotide hybrid is recognized as a substrate by RNase H which cleaves the portion of the RNA bound to the antisense oligomer. The subsequent RNA fragments are generally trimmed by exonucleases ; this prevents the direct observation of the breakdown products, the net result being a decrease of the target RNA. The involvement of RNase H in antisense oligonucleotide-mediated effects has been clearly established in wheat germ extracts, used for cell-free translation assays, and in *Xenopus* oocytes [27, 28, 29]. Several examples suggest that RNase H is also involved in antisense effects in mammalian cells in culture [30, 31] but the situation is not clear yet and may vary from one cell type to another. This has been reviewed in details recently [32]. The same two mechanisms, *i.e.* physical blockade and RNase H-induced degradation of the RNA, are responsible for the inhibition of reverse transcription on an RNA template by antisense oligonucleotides [33, 34].

Antisense oligonucleotides used with cultured cells are generally not unmodified phosphodiesters (PO) due to their susceptibility to the nucleases present in the cell and in the growth medium. Numerous nuclease-resistant derivatives have been prepared and evaluated for antisense purposes [35]. The most frequently used are phosphorothioate (PS) and methylphosphonate (MP) oligonucleotides in which a non-bridging oxygen atom of the phosphodiester bond is substituted by a sulfur atom and by a methyl group, respectively. 2'-O-alkyl (2'-O) and alpha (α) analogues are also nuclease-resistant compounds. Interestingly, only phosphorothioate derivatives elicit RNase H activity.

One of the key feature of the antisense strategy is the selectivity of the effect. One expects that the inhibition will be restricted to the gene of interest if the antisense oligomer binds exclusively to the target RNA and if the complementary sequence is unique in the all genome. Statistically, uniqueness is achieved with oligomers 11-15 nucleotides long in the case of higher eukaryotes [36]. However, RNase H can cleave mismatched RNA-DNA hybrids and, therefore, induce the inhibition of non-target genes as far as they contain sequences partly homologous to the target.

4. RNA-chimeric oligonucleotide duplexes

It has long been reported that RNase H can cleave RNA sequences partially complementary to an antisense oligonucleotide [27]. It is important to study the activity of various RNases H of interest on non-perfect hybrids to determine which mismatches are tolerated by the enzyme. This is a key point for the use of antisense oligonucleotides against dysfunctional genes resulting from a mutation. It is also important to design oligomers able to optimize the RNase H contribution. RNase H activity is useful *per se* as, in combination with a nuclease resistant oligomer, a catalytic antisense effect will result : once the bound RNA has been cleaved, the antisense oligonucleotide is free to bind to a second one. Multiple rounds of degradation can be achieved with a single oligonucleotide molecule. This has been reported for a PS 17-mer targeted to the rabbit β-globin mRNA in microinjected *Xenopus* oocytes [37]. But the optimal design of antisense sequences should reduce the non-specificity derived from RNase H activity on RNA partly homologous to the target. This can be achieved using sandwich or chimeric oligomers, as described below (Section 4.2).

4.1. RNASE H CLEAVAGE OF MISMATCHED HYBRIDS

Oligonucleotides are able to hybridize with non complementary sequences and form heteroduplexes with one or more mismatches. The stability of the hybrids will depend on the number, the nature and the distribution of the mismatches. In addition, the neighbouring sequence will play a role : for example, in DNA, G.A mismatches are as stable as Watson-Crick duplexes in the context YGAR but are destabilizing in the context YAGR [38]. Very little information is available for thermodynamic stability of mismatched RNA-DNA hybrids. Free energy increments have been evaluated for point mutations in RNA-phosphodiester and phosphorothioate oligomers. From experiments with one PS sequence it appears that C.C, C.U and C.A mismatches are the most destabilizing [39]. It is worth mentioning that lengthening the antisense sequence will decrease the specificity : due to their high energy of interaction, long ODNs may tolerate many mismatches.

Oligonucleotides which can direct RNase H have been reported to show non-specific toxic effects in *Xenopus* embryos and in cultured cells. In some cases, the authors suggested that this toxicity was related to the destruction of non-target RNAs by RNase H [40, 41]. This has been demonstrated recently in cell-free systems : in wheat germ extract, RNase H was shown to be responsible for the non-specific inhibition of translation of rabbit β-globin mRNA by an anti-α-globin 15 mer [42]. Ten sites in the β-globin message exhibited more than 60% complementarity with this antisense oligonucleotide termed ODN1. Northern blot analysis of β-globin after *in vitro* translation in the presence of ODN1 clearly showed that three of these partially complementary regions could hybridize with the antisense oligomer, leading to the cleavage of β-globin RNA and consequently to the non-specific inhibition of the β-globin synthesis.

As one of the ultimate goals of the antisense strategy is the design of new therapeutic agents, we decided to study the effect of human RNase HII that we recently purified [13] on the mismatched hybrids formed by ODN1 and the rabbit β-globin mRNA. We evaluated the activity of this enzyme on two synthetic RNA-PO oligonucleotide hybrids, each having three mismatched bases. As shown in figure 1a, human RNase HII is able to cleave the non-perfect hybrids. From the patterns shown in figure 1b, it can be seen that cleavage profiles with mismatched hybrids are different from the one obtained with the perfect one. The mismatches G.A (ARN 250-ODN1) and U.C and G.A (ARN 390-ODN1) do not prevent RNase H activity ; on the contrary, the main sites of cleavage are located just next to these mismatched pairs. Interestingly, these mismatches are among the most destabilizing for RNA-DNA hybrids [39], but they are not recognized by RNase A [43]. Therefore, RNase H is able to accomodate mismatches in the hybrid duplex, and may exhibit increased sensitivity at these positions. Whether these are particular cases or reflect a general situation remains to be demonstrated. It is also likely that reactivity will vary with neighbouring bases.

a)   b)

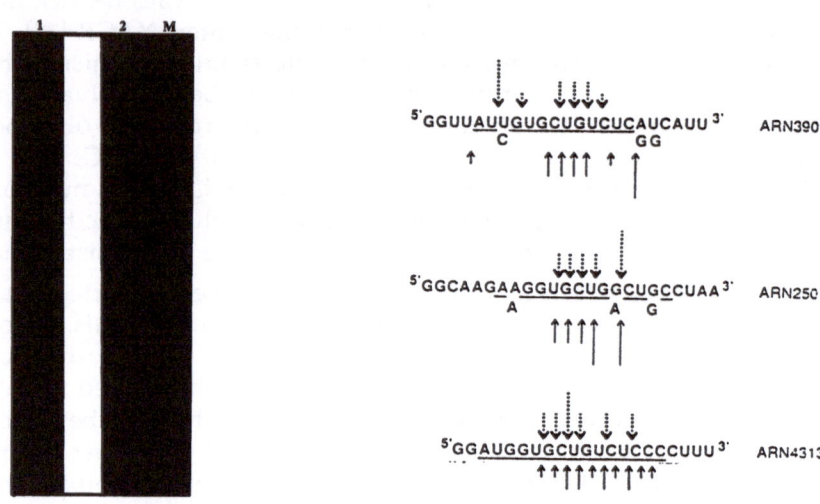

Figure 1 : Cleavage of RNA/oligonucleotide hybrids by human RNase HII.

a- Autoradiograph of RNA-oligonucleotide hybrids cleaved by human RNase HII. [^{32}P] 5' end-labeled RNA390 (the sequence is presented in panel b) was incubated with RNaseHII either in the absence (lane 1) or in the presence (lane 2) of ODN1. Lane M corresponds to the alkaline hydrolysis of [^{32}P]- labelled RNA. Experimental conditions : 0.1 µM RNA was mixed with 1 µM ODN1 in the presence of 0.3 units of human RNase HII, prepared as described by Frank *et al.* [13]. Incubation was performed for 30 min at 37°C in 30 µM Tris-HCl buffer pH 7.8 containing 50 mM $(NH_4)_2SO_4$, 20 mM $MgCl_2$ and 0.01% β-mercapto ethanol. Electrophoresis was run on a 20% polyacrylamide 7M urea gel.

b- Location of cleavage sites induced by human RNase HII on a perfect (ARN 4313) and on two mismatched hybrids (ARN 390 and ARN 250) formed with ODN1. Only RNA sequences are shown ; the underlined nucleotides correspond to the region paired with ODN1. The shifted nucleotides are the mismatched ones. The arrows indicate both the location and the relative intensity of the cleavage sites deduced from densitometric analysis of autoradiographs similar to the one shown in panel a ; long, medium and short tailed arrows correspond to 25-50%, 5-25% and 0-5% cleavage, respectively. Solid and dotted arrows refer to 3' and 5' end-labeled RNA, respectively.

4.2. CLEAVAGE OF RNA-CHIMERIC OLIGONUCLEOTIDE HYBRIDS BY RNASE H

How to concile within a single antisense sequence the ability to elicit RNase H activity and the limited cleavage at given position(s) of the RNA-DNA substrate ? The first demonstration of RNase H and oligonucleotide acting as a sequence-specific RNase was established by Inoue and coworkers in 1987 : a 9-mer analogue composed of a central stretch of four or five unmodified deoxynucleotides, surrounded with 2'-O-methyl modified residues, induced RNase H-mediated cleavage at a single site [44]. This strategy to design "restriction RNases" has been extended to the antisense field to yield molecules of high selectivity. So-called sandwich or chimeric oligomers are made of a stretch of nucleotides that elicit RNase H activity (PO or PS) flanked by two blocks of derivatives that do not (2'-O, MP or α). In addition these oligomers are resistant to exonucleases which are mainly responsible for the degradation of oligonucleotides in cell growth medium and inside the cell. Another type of oligonucleotide made of conjugated deoxyribo (D) and ribo (R) nucleotides giving rise to DRD chimeras has been used to elucidate the role of RNase H in the cell. Some results obtained with these sandwich oligomers are summarized below for three different types of RNase H.

4.2.1. *Cleavage by the E. coli RNase H*
The first question concerns the width of the PO (or PS) window that allows RNase H activity. The result of several studies is that four consecutive PO or PS nucleotides are enough to direct the *E. coli* RNase H activity, whatever the chemical modifications in the flanks, MP [45], 2'-O [44] or α [46]. But oligonucleotides with alternating PO and MP linkages do not form substrates, in contrast to molecules with phosphorothioate linkages either alternating with PO [47] or capping the oligonucleotide [48]. Presently, it is not fully understood which parameters determine the RNase H to cleave such modified hybrids : affinity of the enzyme for oligonucleotide/RNA duplexes, conformation of the hybrids, position of the catalytic center with respect to the window. It should be noted that modified oligomers display quite different properties which can perturb the binding and/or the activity of the enzyme on the hybrid : MP are neutral analogues, α bind in the reverse orientation compared to the other derivatives, 2'-O have a substituted 2' position.

DRD/D hybrids were synthesized as mimics of Okasaki fragments, to check if the *E. coli* RNase HI could be implicated in the removal of these fragments, during replication. A window of either 13, 21 or 31 ribonucleotides upstream of a deoxy region led to several cleavage sites, without any preference for the 3' side of the RNA region [49], in contrast to what is observed with mammalian RNases H I (see below).

4.2.2. *Cleavage by human RNases H*

As mentioned in part 1, two distinct enzymes showing RNase H activity have been identified : HI, which is the most abundant and displays a nuclear localization, and the minor activity HII. For cleavage analysis, hybrids were incubated either with purified enzymes or in a nuclear extract (which is supposed to contain mainly RNase HI).

Human RNase HI : Phosphorothioate oligonucleotides can promote the action of human RNase HI, although with less efficiency than the PO homologues [50]. Chimeric oligonucleotides are encountering an increasing interest, because of their long life time in cells and culture media. Although RNase HI has a large size compared to the *E. coli* enzyme (89 kDa and 18 kDa, respectively), only five PO or PS nucleotides are required to direct cleavage by the human enzyme, compared to four with the bacterial enzyme : a window of five PO, surrounded with MP stretches [50], or of five PS surrounded with 2'-O nucleotides [51] elicited RNA degradation in HeLa cell nuclear extracts.

DNA duplexes with one to several ribonucleotides (DRD/D hybrids) incubated in the presence of the purified human RNase HI were cleaved specifically one nucleotide upstream of the R-D junction, leaving the 3' fragment of the DRD strand with one ribonucleotide at the 5' end [12]. Thus, this enzyme could remove ribonucleotides misincorporated in the nascent DNA strand, or Okazaki fragments with the additional help of a 5' RNase. This behaviour could be associated with class I RNases H, as the same property was reported for the bovine RNase HI [49].

Human RNase HII : This enzyme has been purified in our laboratory from placenta [13]. In contrast to RNase HI, we could not detect cleavage of a DRD/D hybrid, with a one or two ribonucleotide region (unpublished results). We analysed the cleavage of an oligoribonucleotide bound to a series of αβ chimeric 17mers or to the parent oligomers, either unmodified (β17) or fully modified (α17). As seen in figure 2, the RNA target incubated in the presence of β17 and of RNase HII was cleaved at several positions, indicating a non-sequence specific activity of the enzyme. However, more efficient cleavage was observed in the purine rich part of the RNA. As previously reported [13] the cleavage pattern was shifted toward the 5' end of the oligonucleotide, indicating a side position of the catalytic center with respect to the binding domain of the enzyme. When the RNA was hybridized with α2β15, a chimeric oligonucleotide with two nucleotides in the alpha configuration, the cleavage pattern was shifted by two nucleotides (Fig. 2). Extending the α motif up to 7 nucleotides leads to a further shift of the cleavage pattern.

In both cases the most 5' cleavage site (with respect to the RNA strand) is about 3 nucleotides away from the αβ boundary. This indicates a preferential binding of the enzyme on the RNA/β part of the hybrid. No cleavage was seen with the α12β5 chimera, suggesting that a window of 5 β residues is not wide enough to allow binding or/and cleavage of the hybrid.

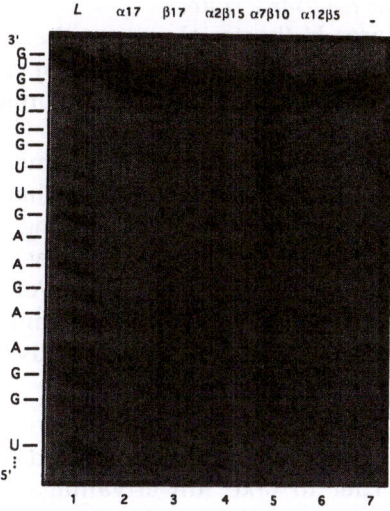

Figure 2 : Cleavage of RNA/chimeric αβ hybrids by human RNase HII.

[32]P-5'end labelled RNA was incubated alone (lane 7) or in the presence of the oligonucleotide indicated at the top of lanes 2-6. The sequence of β17 is 3'AC ACC TTC TTC AAC CAC. Due to its reverse polarity of binding to the complementary sequence, α17 has the reverse orientation (5'AC ACC TTC TTC AAC CAC). αβ chimeras have 5' ends identical to the parent compounds and therefore show a 3'-3 junction 2, 7 or 12 nucleotides from the α 5'end. Lane L corresponds to size markers generated by alkaline hydrolysis of the labelled RNA. Experimental conditions : 50 nM RNA was incubated with 0.1 u of RNase HII for 30' at 37°C in the presence of 100 nM oligonucleotide, in a 30 mM Tris- HCl buffer pH 8.5, containing 75 mM $(NH_4)_2 SO_4$, 20 mM $MgCl_2$ and 0,01% βmercaptoethanol. Samples were run on a 20% polyacrylamide gel containing 7 M urea.

4.2.3. RNase H from reverse transcriptase

In contrast to other RNases H (bacterial, mammalian ...), this enzyme is able to degrade an RNA/DNA hybrid with a certain specificity, dictated by the distance to the 3' OH extremity of the DNA oligonucleotide, (except when the 5' end of the RNA strand is recessed [52]). This is related to the distance, about 18-19 nucleotides between the DNA polymerase and the RNase H active sites [52, 53, 54]. It has been demonstrated, that after the first cleavage occuring 18-19 nt downstream, the subsequent binding of RT molecules induces other cleavage sites due to the 3' → 5' exonuclease activity, giving 10-14 nucleotide long RNA fragments [53]. Moreover, the location of the oligonucleotide binding site on the target is important : depending on wether it is internal or at the RNA 5' end, the remaining RNA fragments are 8-10 or 12-15 nucleotides long, respectively [52].

As aforementionned, antisense oligonucleotides can prevent reverse transcription by physical arrest or RNase H-induced degradation of the RNA template [33, 55]. In order to take full advantage of RNase H mediated-cleavage and nuclease resistance, chimeric oligonucleotides are potentially interesting molecules. Indeed, we recently reported that the $\alpha\beta$ chimeras described in figure 2 were able to prevent cDNA synthesis by the reverse transcriptase from Moloney murine leukemia virus (MMLV), mainly through the cleavage of the RNA template, with an efficiency that decreased with the size of the β stretch [56]. It should be noted that, at variance from human RNase HII (figure 2), MMLV RNase H was able to cleave RNA/$\alpha12\beta5$ hybrids.

We also investigated the properties of 2'-O.PO.2'-O sandwich oligonucleotides. 2'-O methyl oligoribonucleotides display a higher affinity for RNA than unmodified RNA or DNA. Such sandwich oligomers were able to inhibit cDNA synthesis, essentially via the RNase H-dependent mechanism only when adjacent to the primer. Surprisingly, they were not able to physically compete with the binding to (or the scanning of) the template by the RT whereas fully modified 2'-O-methyl oligonucleotides were inhibitory, obviously through an RNase H-independent mechanism [34]. This might mean that the PO window was not appropriately located with respect to the catalytic center when the RT is positionned on the RNA/oligonucleotide hybrid. The presence of an adjacent (unmodified) oligonucleotide primer gives the RT some more latitude.

4.2.4. Increased specificity for RNase H-mediated cleavage of RNA -sandwich oligonucleotide hybrids

Antisense oligonucleotides with an unmodified PO window inserted between two flanks that do not allow RNase H activity were developped to restrict cleavage to a part of the hybrid whereas the nucleic acid bases from both unmodified and chemically-modified moieties participate in the recognition of the target. An increased specificity was expected from such composite oligomers. Indeed MP.PO.MP sandwich oligonucleotides were shown to induce a more specific cleavage leading to a selective inhibition of translation of α-globin mRNA in wheat germ extract whereas the unmodified homologues displayed a non specific effect [42]. Similar compounds also restored specific cleavage *in vitro* with purified *E. coli* RNase H [45]. This was mainly ascribed to the reduced affinity that these MP.PO.MP oligomers display for the complementary RNA sequence compared to the parent PO oligomer. This was supported by the observation that 2'-O.PO.2'-O analogues of the methylphosphonate chimera targeted to the α-globin mRNA were non-specific inhibitors of *in vitro* translation, like the unmodified sequence [57]. As pointed out previously, 2'-O-methyl oligoribonucleotide form more stable hybrids with RNA than do regular phosphodiester oligomers. Therefore, for antisense oligonucleotides acting in an RNase H context the concept of "the stronger the better" is not valid. These results have been recently extended to experiments with cells in culture : Tidd and co-workers were able to achieve single base discrimination with MP.PO.MP chimeric oligonucleotides [58, 59].

4.2.5 RNase H cleavage of RNA-antisense oligonucleotide hybrids in eukaryotic cells.

Until recently only indirect indications suggested that RNase H played a role in antisense oligonucleotide-mediated inhibition of gene expression in intact cells [30, 31]. A direct demonstration of the involvement of RNase H was brought by Tidd and co-workers [59, 60] for myelogeneous leukemia KYO1 cells in culture, incubated with either PS or MP.PO.MP antisense oligonucleotides. Generally, RNA fragments generated by RNase H cannot be detected in cells due to their fast degradation by RNases. The site specific degradation of RNA induced by antisense oligonucleotides in cultured cells was demonstrated using reverse ligation-mediated PCR (RL-PCR) [61]. This procedure involves three successive steps : i) ligation of an RNA linker to all 5' phosphate groups, ii) reverse transcription using a gene specific primer and, iii) PCR amplification of the cDNA using a linker-specific primer on the 5' side and a gene-specific primer nested inside the DNA fragment generated by the reverse transcription primer on the 3' side. Intact RNA should not produce any amplified product because it does not contain 5' phosphate group. In contrast, 3' fragments of mRNA cleaved by RNase H will give rise to amplification products whose size should correspond to the distance between the 3' PCR primer and the antisense oligonucleotide binding site.

It was shown by this method that chimeric MP.PO.MP oligonucleotides were able to achieve single base discrimination in streptolysin O-permeabilized cells [59, 60]. Therefore, compounds with altered backbone could allow to modulate the affinity of the oligonucleotide for its target, leading to the saturation of the target sequence under concentration conditions which lead to limited amounts of undesired complexes at non target sites. Chimeras combined with RNase H could constitute an antisense system of high efficiency and selectivity for manipulating gene expression.

6. Acknowledgements

We thank C. Cazenave for the critical reading of the manuscript. We are grateful to B. Rayner and J.L. Imbach (Montpellier) for a generous gift of αβ chimeric oligonucleotides. This work was supported in part by the *Conseil Régional d'Aquitaine* and by the *Association pour la Recherche contre le Cancer*. B.L. is the recipient of a DRET fellowship.

7. References

1- Stein, H. and Hausen, P. (1969) Enzyme from calf thymus degrading the RNA moiety of RNA-DNA hybrids : effect on DNA-dependent RNA polymerase. *Science* 166, 393-395.

2- Hostomsky, Z., Hostomska, Z. and Matthews, D. A. (1993) "Ribonucleases H."in Nucleases. pp 341-376, Cold Spring Harbor Laboratory Press;

3- Hélène, C. and Toulmé, J. J. (1990) Specific regulation of gene expression by antisense, sense and antigene nucleic acids. *Biochim. Biophys. Acta* 1049, 99-125.

4- Crouch, R. J. (1990) Ribonuclease H : from discovery to 3D structure. *The New biologist* 2, 771-777.

5- Wintersberger, U. (1990) Ribonucleases H of retroviral and cellular origin. *Pharmacol. Ther.* 48, 259-280.

6- Crouch, R. J. and Toulmé, J. J. (1996) Ribonucleases H. In press.

7- Kanaya, S. and Crouch, R. J. (1983) DNA sequence of the gene coding for Escherichia coli ribonuclease H. *J. Biol. Chem.* 258, 1276-1281.

8- Kogoma, T. (1986) RNase H defective mutants of Escherichia coli. *J. Bacteriol.* 166, 361-363.

9- Büsen, W. and Frank, P. (1996) "Bovine ribonucleases H."in Ribonucleases H. Crouch, R. J. and Toulmé, J. J. (ed.), pp In press, INSERM, John Libbey ; Paris; Paris.

10- Büsen, W. (1980) Purification, subunit structure, and serological analysis of calf thymus ribonuclease HI. *J. Biol. Chem.* 255, 9434-9443.

11- Vonwirth, H., Frank, P. and Büsen, W. (1989) Serological analysis and characterization of calf thymus ribonuclease HIIb. *Eur. J. Biochem.* 184, 321-329.

12- Eder, P. S. and Walder, J. A. (1991) Ribonuclease-H from K562 human erythroleukemia cells - purification, characterization, and substrate specificity. *J. Biol. Chem.* 266, 6472-6479.

13- Frank, P., Albert, S., Cazenave, C. and Toulmé, J. J. (1994) Purification of human RNase HII. *Nucleic Acids Res.* 22, 5247-5254.

14- Büsen, W., Peters, J. H. and Hausen, P. (1977) Ribonuclease H levels during the response of bovine lymphocytes to Concanavalin A. *Eur. J. Biochem.* 74, 203-208.

15- Champoux, J. J. (1993) "Roles of RNase H in reverse transcription."in Reverse transcriptase. Skalka, A. M. and Goff, S. P. (ed.), pp 103-117, Cold Spring Harbor Laboratory Press; Cold Spring Harbor.

16- Katayanagi, K., Miyagawa, M., Matsuhima, M., Ishikawa, M., Kanaya, S., Matsuzaki, T. and Morikawa, K. (1990) Three-dimensional structure of ribonuclease H from E. coli. *Nature* 347, 306-309.

17- Yang, W., Hendrickson, W. A., Crouch, R. J. and Satow, Y. (1990) Structure of ribonuclease H phased at 2 Å resolution by MAD analysis of the selenomethionyl protein. *Science* 249, 1398-1405.

18- Davies, J. F., Hostomska, Z., Hostomsky, Z., Jordan, S. and Mathews, D. A. (1991) Crystal structure of the RNase H domain of HIV-1 reverse trranscriptase. *Science* 252, 88-95.

19- Nakamura, H., Oda, Y., Iwai, S., Inoue, H., Ohtsuka, E., Kanaya, S., Kimura, S., Katsuda, C., Katayanagi, K., Morikawa, K., Miyashiro, H. and Ikehara, M. (1991) How does RNase H recognize a DNA.RNA hybrid ? *PNAS* 88, 11535-11539.

20- Egli, M., Usman, N., Zhang, S. and Rich, A. (1992) Crystal structure of an Okazaki fragment at 2-Å resolution. *Proc. Natl. Acad. Sci. USA* 89, 534-538.

21- Gupta, G., Sarma, M. H. and Sarma, R. H. (1985) Secondary structure of the hybrid poly(rA).Poly(dT) in solution. *J. Mol. Biol.* 186, 463-469.

22- Arnott, S., Chandrasekaran, R., Millane, R. P. and Park, H. S. (1986) DNA-RNA hybrid secondary structures. *J Mol Biol* 188, 631-640.

23- Fedoroff, O. Y., Salazar, M. and Reid, B. R. (1993) Structure of a DNA:RNA Hybrid Duplex - Why RNase-H Does Not Cleave Pure RNA. *J. Mol. Biol.* 233, 509-523.

24- Lesnik, E. A., Guinosso, C. J., Kawasaki, A. M., Sasmor, H., Zounes, M., Cummins, L. L., Ecker, D. J., Cook, P. D. and Freier, S. M. (1993) Oligodeoxynucleotides containing 2'-O-modified adenosine - synthesis and effects on stability of DNA:RNA duplexes. *Biochemistry* 32, 7832-7838.

25- Milligan, J. F., Matteucci, M. D. and Martin, J. C. (1993) Current concepts in antisense drug design. *J. Med. Chem.* 36, 1923-1937.

26- Boiziau, C., Kurfurst, R., Cazenave, C., Roig, V., Thuong, N. T. and Toulmé, J. J. (1991) Inhibition of translation initiation by antisense oligonucleotides via an RNase-H independent mechanism. *Nucleic Acids Res.* 19, 1113-1119.

286

27- Cazenave, C., Loreau, N., Thuong, N. T., Toulmé, J. J. and Hélène, C. (1987) Enzymatic amplification of translation inhibition of rabbit β-globin mRNA mediated by anti-messenger oligodeoxynucleotides covalently linked to intercalating agents. *Nucleic Acids Res.* 15, 4717-4736.

28- Minshull, J. and Hunt, T. (1986) The use of single-stranded DNA and RNase-H to promote quantitative 'hybrid arrest of translation' of mRNA/DNA hybrids in reticulocyte lysate cell-free translations. *Nucleic Acids Res.* 14, 6433-6451.

29- Walder, R. Y. and Walder, J. A. (1988) Role of RNase-H in hybrid-arrested translation by antisense oligonucleotides. *Proc. Natl. Acad. Sci. USA* 85, 5011-5015.

30- Chiang, M. Y., Chan, H., Zounes, M. A., Freier, S. M., Lima, W. F. and Bennett, C. F. (1991) Antisense oligonucleotides inhibit intercellular adhesion molecule-1 expression by two distinct mechanisms. *J. Biol. Chem.* 266, 18162-18171.

31- Bonham, M. A., Brown, S., Boyd, A. L., Brown, P. H., Bruckenstein, D. A., Hanvey, J. C., Thomson, S. A., Pipe, A., Hassman, F., Bisi, J. E., Froehler, B. C., Matteucci, M. D., Wagner, R. W., Noble, S. A. and Babiss, L. E. (1995) An assessment of the antisense properties of RNase H-competent and steric-blocking oligomers. *Nucleic Acids Res.* 23, 1197-1203.

32- Toulmé, J. J. and Tidd, D. (1996) "Role of ribonuclease H in antisense oligonucleotide-mediated effects."in Ribonucleases H. Crouch, R. J. and Toulmé, J. J. (ed.), pp In press, John Libbey; Paris.

33- Boiziau, C., Thuong, N. T. and Toulmé, J. J. (1992) Mechanisms of the inhibition of reverse transcription by antisense oligonucleotides. *Proc. Natl. Acad. Sci. USA* 89, 768-772.

34- Boiziau, C., Larrouy, B., Sproat, B. and Toulmé, J. J. (1995) Antisense 2'-O-alkyl oligoribonucleotides are efficient inhibitors of reverse transcription. *Nucleic Acids Res.* 23, 64-71.

35- Goodchild, J. (1990) Conjugates of oligonucleotides and modified oligonucleotides: a review of their synthesis and properties. *Bioconjugate Chem.* 1, 165-187.

36- Hélène, C. and Toulmé, J. J. (1989) "Control of gene expression by oligodeoxynucleotides covalently linked to intercalating agents and nucleic acid cleaving reagents."in Oligodeoxynucleotides: antisense inhibitors of gene expression. Cohen, J. S. (ed.), pp 137-172, Macmillan Press; London.

37- Cazenave, C., Stein, C. A., Loreau, N., Thuong, N. T., Neckers, L. M., Subasinghe, C., Hélène, C., Cohen, J. S. and Toulmé, J. J. (1989) Comparative inhibition of rabbit globin mRNA translation by modified antisense oligonucleotides. *Nucleic Acids Res.* 17, 4255-4273.

38- Ebel, S., Lane, A. N. and Brown, T. (1992) Very stable mismatch duplexes : structural and thermodynamic studies on tandem G . A mismatches in DNA. *Biochemistry* 31, 12083-1208.

39- Freier, S. (1993) "Hybridization : considerations affecting antisense drugs."in Antisense Research and Applications. Lebleu, B. and Crooke, S. T. (ed.), pp 67-82, CRC; Boca Raton.

40- Smith, R. C., Bement, W. M., Dersch, M. A., Dworkin-Rastl, E., Dworkin, M. B. and Capco, D. G. (1990) Non-specific effects of oligodeoxynucleotide injection in xenopus oocytes - A reevaluation of previous D7 messenger RNA ablation experiments. *Development* 110, 769-779.

41- Jessus, C., Cazenave, C., Ozon, R. and Hélène, C. (1988) Specific inhibition of endogenous ß-tubulin synthesis in *Xenopus* oocytes by anti-messenger oligodeoxynucleotides. *Nucleic Acids Res.* 16, 2225-2233.

42- Larrouy, B., Blonski, C., Boiziau, C., Stuer, M., Moreau, S., Shire, D. and Toulmé, J. J. (1992) RNase-H mediated inhibition of translation by antisense oligodeoxynucleotides : a way to improve specificity. *Gene* 121, 189-194.

43- Myers, R. M., Larin, Z. and Maniatis, T. (1985) Detection of single base substitutions by ribonuclease cleavage at mismatches in RNA:DNA duplexes. *Science* 230, 1242-1246.

44- Inoue, H., Hayase, Y., Imura, A., Iwai, S., Miura, K. and Ohtsuka, E. (1987) Synthesis and hybridization studies of two complementary nona(2'-O-methyl)-ribonucleotides. *Nucleic Acids Res.* 15, 6131-6148.

45- Giles, R. V. and Tidd, D. M. (1992) Increased specificity for antisense oligodeoxynucleotide targeting of RNA cleavage by RNase-H using chimeric methylphosphonodiester/phosphodiester structures. *Nucleic Acids Res.* 20, 763-770.

46- Gottikh, M., Bertrand, J. R., Baud-Demattei, M. V., Lescot, E., Giorgio-Renault, S., Shabarova, Z. and Malvy, C. (1994) αβ Chimeric antisense oligonucleotides: synthesis and nuclease resistance in biological media. *Antisense Res. Develop.* 4, 251-258.

47- Furdon, P. J., Dominski, Z. and Kole, R. (1989) RNase-H cleavage of RNA hybridized to oligonucleotides containing methylphosphonate, phosphorothioate and phosphodiester bonds. *Nucleic Acids Res.* 17, 9193-9204.

48- Hoke, G. D., Draper, K., Freier, S. M., Gonzalez, C., Driver, V. B., Zounes, M. C. and Ecker, D. J. (1991) Effects of phosphorothioate capping on antisense oligonucleotide stability, hybridization and antiviral efficacy versus herpes simplex virus infection. *Nucleic Acids Res.* 19, 5743-5748.

49- Huang, L., Kim, Y., Turchi, J. J. and Bambara, R. A. (1994) Structure-specific cleavage of the RNA primer from Okazaki fragments by calf thymus RNase HI. *J. Biol. Chem.* 269, 25922-25927.

50- Agrawal, S., Mayrand, S. H., Zamecnik, P. C. and Pederson, T. (1990) Site-specific excision from RNA by RNase-H and mixed-phosphate-backbone oligodeoxynucleotides. *Proc. Natl. Acad. Sci. USA* 87, 1401-1405.

51- Monia, B. P., Lesnik, E. A., Gonzalez, C., Lima, W. F., Mcgee, D., Guinosso, C. J., Kawasaki, A. M., Cook, P. D. and Freier, S. M. (1993) Evaluation of 2'-

288

modified oligonucleotides containing 2'-deoxy gaps as antisense inhibitors of gene expression. *J. Biol. Chem.* 268, 14514-14522.

52- Destefano, J. J., Mallaber, L. M., Fay, P. J. and Bambara, R. A. (1993) Determinants of the RNase-H cleavage specificity of human immunodeficiency virus reverse transcriptase. *Nucleic Acids Res.* 21, 4330-4338.

53- Gopalakrishnan, V., Peliska, J. A. and Benkovic, S. J. (1992) Human immunodeficiency virus type-1 reverse transcriptase - spatial and temporal relationship between the polymerase and RNase-H activities. *Proc. Natl. Acad. Sci. USA* 89, 10763-10767.

54- Kohlstaedt, L. A., Wang, J., Friedman, J. M., Rice, P. A. and Steitz, T. A. (1992 Jun 26) Crystal structure at 3.5 Å resolution of HIV-1 reverse transcriptase complexed with an inhibitor. *Science* 256, 1783-1790.

55- Bordier, B., Hélène, C., Barr, P. J., Litvak, S. and Sarih-Cottin, L. (1992) In vitro effect of antisense oligonucleotides on human immunodeficiency virus type-1 reverse transcription. *Nucleic Acids Res.* 20, 5999-6006.

56- Boiziau, C., Debart, F., Rayner, B., Imbach, J. L. and Toulmé, J. J. (1995) Chimeric alpha-beta oligonucleotides as antisense inhibitors of reverse transcription. *FEBS Letters* 361, 41-45.

57- Larrouy, B., Boiziau, C., Sproat, B. and Toulmé, J. J. (1995) High affinity or selectivity : a dilemma to design antisense oligodeoxynucleotides acting in RNaseH-containing media. *Nucleic Acids Res.* in press.

58- Giles, R. V., Spiller, D. G. and Tidd, D. M. (1993) Chimeric oligodeoxynucleotide analogues - enhanced cell uptake of structures which direct ribonuclease-H with high specificity. *Anti-Cancer Drug Des.* 8, 33-51.

59- Giles, R. V., Ruddell, C. J., Spiller, D. G., Green, J. A. and Tidd, D. M. (1995) Single base discrimination for ribonuclease H-dependent antisense effects within intact human leukaemia cells. *Nucleic Acids Res.* 23, 954-961.

60- Giles, R. V., Spiller, D. G. and Tidd, D. M. (1995) Detection of ribonuclease H-generated mRNA fragments in human leukemia cells following reversible membrane permeabilization in the presence of antisense oligodeoxynucleotides. *Antisense Res. Dev.* 5, 23-31.

61- Bertrand, E., Fromontracine, M., Pictet, R. and Grange, T. (1993) Visualization of the interaction of a regulatory protein with RNA in vivo. *Proc. Natl. Acad. Sci. USA* 90, 3496-3500.

Hydrolysis of RNA by Ribozymes and Metal Complexes

STRUCTURE, ACTIVITY AND APPLICATION OF CHEMICALLY MODIFIED HAMMERHEAD RIBOZYMES

F. ECKSTEIN

Max-Planck-Institut für experimentelle Medizin, Hermann-Rein-Str. 3, D-37075 Göttingen; Germany

The hammerhead ribozyme is a small catalytically active RNA.[1] The structural motif of this ribozyme, indicated by the boxed nucleotides in the figure, is found in nature in some plant viroids where it is responsible for self-cleavage of the viroid RNA, in an intramolecular reaction. However, this motif can also be engaged in intermolecular reactions as indicated in the figure. This possibility makes the hammerhead rbozyme extremely interesting for the sequence-specific cleavage of mRNA for the inhibition of gene expression. This ribozyme can cleaves behind certain nucleotide triplets, such as GUC, by formation of an oligoribonucleotide terminating in a nucleoside $2'$, $3'$-cyclic phosphate and an oligoribonucleotide terminating in a $5'$-hydroxyl group. This reaction proceeds with inversion of configuration at phosphorus. Thus, the mechanism of the reaction is nominally identical to that of RNaseA.[2,3] The kinetics of this reaction have been studied in detail.[4]

We have engaged in studies on the structure - function relationship of this ribozyme and on developing it for the inhibition of gene expression, mainly by chemical modification. Early attempts were directed at trying to understand the function of the invariant nucleotides in the core region by replacement by analogues such as inosine, xanthosine, isoguanosine and 2-aminopurine nucleoside.

B. Meunier (ed.), DNA and RNA Cleavers and Chemotherapy of Cancer and Viral Diseases, 291–294.
© *1996 Kluwer Academic Publishers. Printed in the Netherlands.*

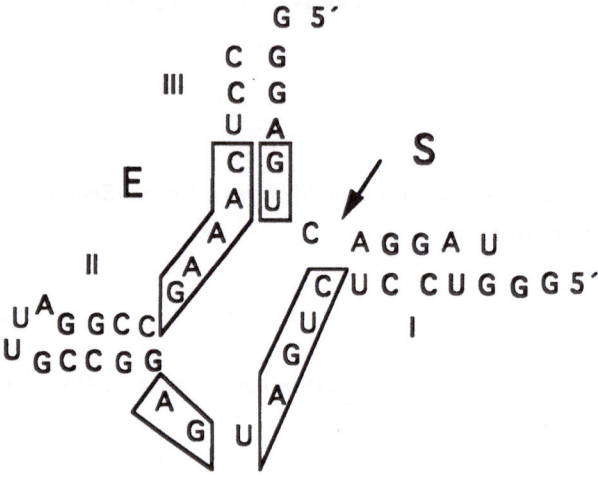

Basically all these changes abolished catalytic activity of the ribozyme and thus no useful information on structure-function relationship could be obtained.[5] More recently we have tried to obtain information on the three-dimensional arrangement of the 3 helices by fluorescence resonance energy transfer (FRET) measurements.[6] This study resulted in a Y-shaped model where helices I and II are very close and helix III positioned as a continuation of helix II. This structure is amazingly similar to that obtained by x-ray diffraction studies of crystals.[7,8] However, there are small differences, mainly in the positioning of helx I. To decide which of the two structures is kinetically competent we chemically crosslinked helix I with helix II with a disulfide linker which spans approximately 15 Å.[9] Two nucleotides were chosen in each structure which are 12 Å apart in the FRET model but 30 Å apart in the x-ray model and *vice versa* thus facilitating the distinction between the two models. The crosslinks were introduced *via* the 2´-amino groups of the two nucleotides. The crosslinked ribozyme based on the x-ray structure showed full catalytic activity whereas that based on the FRET study was inactive. The consistencies but also discrepancies between the x-ray structures and the chemical modification data are discussed by Tuschl et al.[10]

Ribozymes in which all the pyrimidine nucleosides are replaced by the corresponding 2´-fluoro- and 2´-aminonucleosides show no decrease in catalytic efficiency but are considerably stabilized against degradation by nucleases present in serum.[11,12] Additional stabilisation can be achieved by the incorporation of phosphorothioates to block 3´-5´-exonucleases. This observation lends itself to investigate the use of ribozymes to inhibit gene expression by exogenous delivery.[13] This method of application requires resistance of the ribozyme against nucleases present in the serum, a problem which we have solved with these 2´-modifications. *In vitro* transcripts of certain regions of the HIV genome of approximately 1000 nucleotides can be cleaved by such ribozymes. Cleavage of such long RNAs is several hundred fold slower than that of short substrates of 12 nucleotide length. This is due to a shift in the rate determining step from the chemisal step in the latter to complex formation between substrate and ribozyme in the latter. Fortunately, there are RNA-binding proteins in the cell which facilitate this complex formationd and which accelerate the cleavage approximately 50-fold, even with the chemically modified ribozymes.[14] Experiments to find out whether HIV replication can be inhibited of directing ribozymes to the LTR and *gag* region of the virus are under way in collaboration with Dr. Hunsmann´s group from the Primatenzentrum in Göttingen and with that of Dr. Sczakiel at the Deutsches Krebsforschungszentrum in Heidelberg.

Others have shown that this chemically modified ribozyme can inhibit the expression of the protein responsible for multiple drug resistance in cell culture.[15]

References

1. Symons, R. (1992) Small catalytic RNAs, *Ann. Rev. Biochem.* **61**, 641-671.
2. van Tol, H., Buzayan, J., Feldstein, P. A., Eckstein, F. and Bruening, G. (1990) Two autolytic processing reactions of a satellite RNA proceed with inversion of configuration, *Nucleic Acids Res.* 18, 1971-1975.
3. Slim, G. and Gait, M. J. (1991) Configurationally defined phosphorothioate-containing oligoribonucleotides in the study of the mechanism of cleavage of hammerhead ribozymes, *Nucleic Acids Res.* **19**, 1183-1188.
4. Hertel, K, Herschlag, D. and Uhlenbeck, O. C (1994) A kinetic and thermodynamic framework for the hammerhead ribozyme reaction, *Biochemistry* **33**, 3374-3385.

294

5. Ng, M. M. P., Benseler, F., Tuschl, T. and Eckstein, F. (1994) Isoguanosine substitution of conserved adenosines in the hammerhead ribozyme, *Biochemistry* **33**, 12119 and ref. cited therein.

6. Tuschl, T., Gohlke, C., Jovin, T. M., Westhof, E. and Eckstein, F. (1994) A three-dimensional model for the hammerhead ribozyme based on fluorescence measurments, *Science* **266**, 785-789.

7. Plej, H. W., Flaherty, K. M. and McKay. D. (1994) Three-dimensional structure of a hammerhead ribozyme, *Nature* **372**, 68-74.

8. Scott, W. G., Finch, J. T. and Klug, A. (1995) The crystal structure of an all RNA hammerhead ribozyme: A proposed mechanism for RNA catalytic clevage, *Cell* **81**, 991-1002.

9. Sigurdsson, S. Th., Tuschl, T. and Eckstein, F. (1995) Probing RNA tertiary structure: Interhelical crosslinking of the hammerhead ribozyme, *RNA*, in press

10. Tuschl, T., Thomson, J. B. and Eckstein, F. (1995) RNA cleavage by small catalytic RNAs, *Curr. Opinion Struct. Biol.* **5**, 296-302

11. Pieken, W. A., Olsen, D. B., Benseler, F., Aurup, H. and Eckstein, F. (1991) Kinetic characterisation of ribonuclease-resistant 2´-modified hammerhead ribozymes, *Science* **253**, 314-317.

12. Heidenreich O., Benseler, F., Fahrenholz, A. and Eckstein, F. (1994) Hammerhead ribozyme-mediated cleavage of the long terminal repeat RNA of human immunodeficiency virus type 1, *J. Biol. Chem.* **269**, 2131-2138.

13. Marschall, P., Thomson, J. B. and Eckstein, F. (1994) Inhibition of gene expression with ribozymes, *Cell. Mol. Neurobiol.* **14**, 523-538.

14. Heidenreich, O., Kang, S.-H., Brown, D. A., Xu, X., Swiderski, P., Rossi, J., Eckstein, F. and Nerenberg, M. (1995) Ribozyme-mediated RNA degradation in nuclei suspension, *Nucleic Acids Res.* **23**, 2223-2228.

15. Kiehntopf, M., Brach, M. A., Licht, T., Karawajew, L., Petschauer, S., Kirchning, C. and Herrmann, F. (1994) Ribozyme-mediated cleavage of the *MDR-1* transcript restores chemosensitivity in previously resistant cancer cells, *EMBO J.* **13**, 4645 4652.

THE HAIRPIN RIBOZYME

J.A. GRASBY, K.J. YOUNG, F. GILL, J.S. VYLE

Krebs Institute,
Department of Chemistry,
University of Sheffield,
Sheffield S3 7HF

1. Introduction

Ribozymes are RNA molecules which catalyse the formation and cleavage of phosphodiester bonds. A family of small ribozymes, which includes the hammerhead and hairpin ribozymes, catalyse similar reactions of RNA molecules with high specificity.[1] These catalytic RNA molecules provide an essential self-cleavage step during the replication of a number of satellite viruses mainly associated with plants. Currently, there is considerable interest in the structure, function and mechanism of catalytic RNA molecules because of their potential as chemotherapeutic agents and their role in a postulated RNA world. This review is concerned mainly with the hairpin ribozyme but where appropriate the properties of this enzyme are contrasted with the well-studied hammerhead variant.

2. Structural Requirements of the Cleavage Reaction

2.1 ESSENTIAL NUCLEOTIDES AND SECONDARY STRUCTURE ELEMENTS

The hairpin ribozyme motif has been reported to be involved in the self-cleavage step required by three plant satellite viruses; the minus strand of tobacco ring spot virus, arabis mosaic virus and chicory yellow mottle virus.[2-4] Due to the small amount of the phylogenetic data, a minimal catalytic motif has been determined in some elegant *in vitro* selection studies performed by Burke and co-workers.[5, 6] The essential nucleotide sequences and secondary structure elements of the hairpin ribozyme are shown in figure 1 and consist of two looped regions (A and B) containing the required nucleotides which are separated by two Watson-Crick base-paired helical regions (2 and 3) with a further two

B. Meunier (ed.), DNA and RNA Cleavers and Chemotherapy of Cancer and Viral Diseases, 295–306.
© *1996 Kluwer Academic Publishers. Printed in the Netherlands.*

helices (1 and 4) extending from the loops. Loops C and D may be varied in size or dispensed with. A recent review describes the *in vitro* selection experiments in detail.[7] The numbering system used in figure 1 has been employed throughout this review. The essential nucleotides, in loops A and B, have recently been confirmed by mutagenesis data.[8]

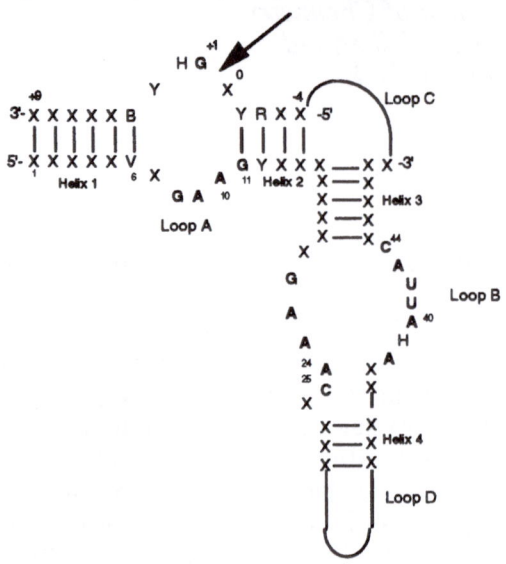

Figure 1. The Hairpin Ribozyme. Essential Residues are shown in bold. X is any nucleotide; B is U, C or G; H is A, U, or C; R is A or G; V is A, C or G; Y is U or C.

2.2 CLEAVAGE IN *TRANS*

Although *in vivo* the hairpin catalysed cleavage reaction and those of other small ribozymes occur in an intramolecular fashion, the RNA may be divided into two or more substrate and enzyme parts to provide ribozymes capable of *trans* cleavage. In these arrangements of two or more oligonucleotides the ribozyme is a true catalyst and capable of turnover. Many of these *trans* cleaving constructs obey Michaelis-Menten kinetics. The cleavage reaction requires the presence of the sequence 5'-RYX*GHYB-3' (where * denotes the site of cleavage) in the substrate. Providing these demands are met the ribozyme may be directed to cleave any RNA molecule. The applications of the hairpin ribozyme as a potential regulator of gene expression have been reviewed by Hampel and co-workers.[9]

Several constructions of hairpin ribozymes capable of supporting intermolecular cleavage have been reported. They differ in the site(s) of strand scission that have been employed in the creation of ribozyme and substrate. Common arrangements use substrate strands complementary to the ribozyme in helix 1 and 2 and thus include the site of cleavage as shown in figure 2a.[10] The minimal length for such a substrate is 14 bases; forming 4 base pairs in helix 2 and 6 base pairs in helix 1. Shorter substrates are cleaved at very much slower rates due to decreased affinity of the substrate. Extension of helix 1 eventually causes product release to become rate limiting for the enzyme catalysed reaction.

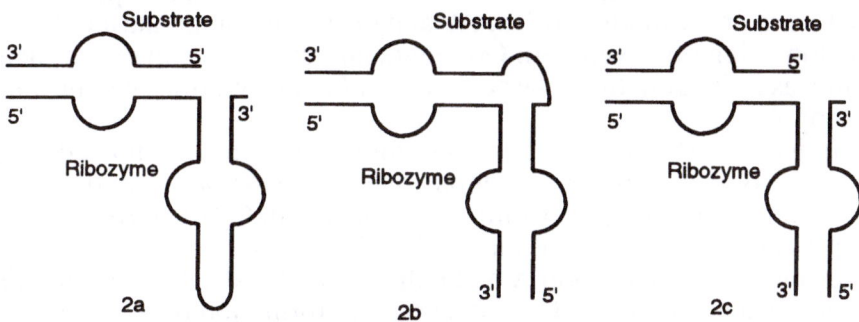

Figure 2. Alternative arrangements of the Hairpin Ribozyme capable of supporting catalysis in *trans*.

An alternative arrangement for *trans* cleaving ribozymes joins helix 2 and 3 through a linker loop C (typically pentacytidine) thus creating a substrate complementary to the ribozyme in all four helical regions (figure 2b).[11] This ribozyme arrangement particularly favours observation of the reverse ligation reaction (see section 3.2). Finally, a ribozyme arrangement with a similar substrate to that described in figure 2a but with the enzyme part divided into two shorter oligonucleotides (figure 2c) has been described by Burke and co-workers.[10] This ribozyme has been utilised in functional group mutation studies as the shorter oligoribonucleotides facilitate chemical synthesis approaches.

2.3 STUDIES ON THE GROUND STATE STRUCTURE OF THE ENZYME

2.3.1 *Cross-linking Experiments*

Favre and co-workers have studied the conformation of loop A by replacing the conserved guanosine of the substrate G+1 and residue X0 with 2'-deoxy-6-thioinosine.[12] For these experiments the ribozyme (type 2a) was supplied with a substrate which included the relevant modification and consisted of entirely 2'-deoxynucleosides. An apparent

K_d of approximately 5μM was observed for these modified substrates. These binding constants are very much in excess of the K_m values normally observed for similar sized cleavable substrates in the hairpin reaction. It should also be noted that replacement of G+1 with inosine has been shown to produce a non-cleavable substrate. Nevertheless, cross-linking was observed when the photoactivatible base was included at position +1 to all residues in loop A. The most efficient reaction occurred with V6. In contrast, inclusion of 2'-deoxy-6-thioinosine at position X0 resulted in cross-linking to G8, A9 and G11 with the greatest extent of reaction being observed with G8. Favre and colleagues interpret their results as indicating that a large number of conformations can be adopted by loop A.

Induced cross-linking of an unmodified ribozyme upon irradiation with UV light at 254nm has demonstrated that U41, an essential nucleoside within loop B, forms cross-links with G21 and /or A22 in both the presence and absence of magnesium ions.[13] A truncated form of the ribozyme which lacks loop A and helices 1 and 2 also behaves in this manner when irradiated. This suggests that formation of the loop B structure is independent of the substrate binding domain. Similarities between the UV cross-linking behaviour of loop B and other RNA molecules have led Burke and co-workers to propose a model for this region of the ribozyme.

2.3.2 *Chemical Mapping Studies*

Modification by chemical mapping studies is a well established technique for the investigation of RNA structure. This technique relies on various chemical reagents reacting with the functional groups of nucleobases, the reactivity being masked in cases where the nucleic acid structure involves these groups in hydrogen bonding. Studies of this nature have confirmed the secondary structure of the hairpin ribozyme as predicted from computer assisted secondary structure prediction and *in vitro* selection studies.[14] These results support the hypothesis that non Watson-Crick base pairs are formed in loop B prior to substrate addition and in the absence of magnesium. Interestingly, addition of magnesium ions alters the reactivity pattern; bases 21 to 26 of loop B are protected but increased sensitivity of bases 6 to 11 in loop A is observed. It is therefore proposed that magnesium may be essential to the stabilisation of tertiary contacts between loops A and B.

2.4 FUNCTIONAL GROUP REQUIREMENTS FOR THE CLEAVAGE REACTION

An alternative strategy for the elucidation of the structural requirements of the hairpin ribozyme involves the creation of mutant ribozymes and substrates followed by an investigation of the catalytic efficiency of these modified oligonucleotides in the cleavage reaction. Whilst chemical modification studies and cross-linking experiments on the enzyme-substrate complex reveal the ground state structure, those involving mutant ribozymes probe the functional groups essential for catalysis to take place. Typically ribozyme constructions of type 2c are used in these studies in order to limit the amount of chemical synthesis required.

2.4.1 *Modified Nucleoside Studies*

Chemical synthesis of oligoribonucleotides has facilitated studies which investigate the importance of specific functional groups of the essential nucleosides in ribozyme catalysis. Nucleoside analogues, which represent functional group deletions or subtle modifications, are site specifically incorporated into oligoribonucleotides in place of their essential natural counterparts. Investigations of this type have been extensively utilised in the study of the hammerhead ribozyme and more recently in the case of the hairpin.

Perhaps the simplest modification of a ribonucleoside involves its replacement with a 2'-deoxyribonucleoside. The importance of the 2'-hydroxyl function of essential residues in the substrate has been studied.[15] As expected from the proposed mechanism of the cleavage reaction (see section 3), substitution of a deoxynucleosides immediately 5'- to the cleaved phosphodiester bond inhibits the reaction completely. Interestingly, although an oligoribonucleotide containing this single deoxynucleoside substitution at the cleavage site is an inhibitor of the ribozyme, the K_i was found to be much greater than would be expected from the Michaelis constant of the hairpin reaction. This has led to the suggestion that the 2'-hydroxyl group at position 0 in the substrate is also involved in some interaction with the ribozyme in addition to being mechanistically essential. Other substitutions of deoxynucleosides within the hairpin substrate reduce the rate of reaction dramatically increasing K_m and decreasing k_{cat}. Burke and co-workers were unable to find any evidence that an entirely deoxynucleoside substrate is bound by the enzyme. This is in contrast to the case of the hammerhead ribozyme where a deoxynucleoside substrate is efficiently bound by the enzyme and has been used in crystallographic investigations of its structure.[16] It therefore appears that hydroxyl groups within the substrate strand may be involved in interactions with the ribozyme during both binding and catalysis.

Investigation of the requirement for 2'-hydroxyl functions within the ribozyme strands by 2'-deoxynucleoside and 2'-*O*-methylnucleoside substitution have revealed four sites (A10, G11, A24 and C25) in which this modification is strongly inhibitory to catalysis.[17] In the case of ribozyme modification the Michaelis constant remains unchanged and thus the changes in catalytic efficiency are manifested in k_{cat}. Increasing the concentration of magnesium ions rescued the inhibition found upon deoxynucleoside substitution at positions G11 and A24.

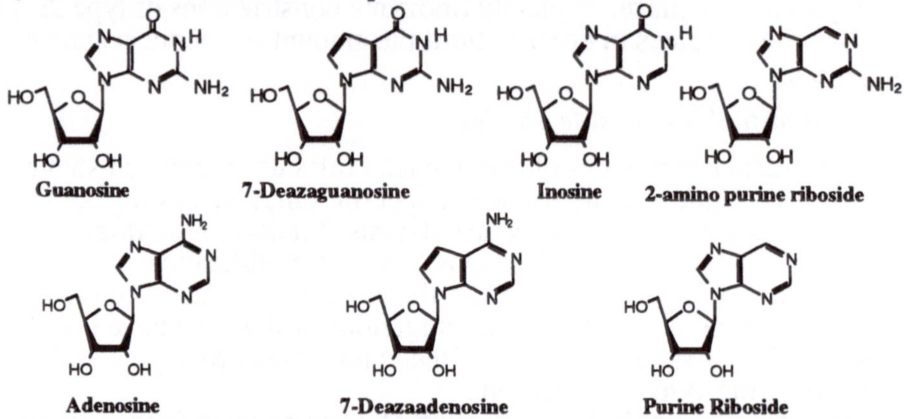

Figure 3. The structures of nucleoside analogues which have been used in identification of the important functional groups in hairpin ribozyme catalysis.

Burke and colleagues have studied the role of the essential guanosine residue, G+1, 3'- to of the cleaved phosphodiester bond.[18] Removal of the exocyclic amino group of this guanosine by substitution with inosine completely inhibits the reaction, whereas removal of the O-6 and N-1 proton by replacement with 2-amino purine riboside (see figure 3) only modestly affects K_m. It has been postulated that the amino function of G+1 is involved in the chemical reaction acting either as a general base and deprotonating the 2'-hydroxyl prior to reaction or by hydrogen bonding to one of the non-bridging phosphoryl oxygens of the scissile phosphodiester bond. However, the former suggestion appears unlikely in view of the known preference of guanosine for protonation at N-7 and N-3.

The role of the exocyclic amino groups and N-7 positions of essential adenosine and guanosine residues in catalysis have been investigated by replacement of these nucleosides with purine riboside, inosine, 7-deazaadenosine and 7-deazaguanosine respectively (see figure 3).[19] Dramatic decreases in the rate of reactions for all substitutions in loop A were observed. These decreases were largely observed in reduction of k_{cat} but replacement of guanosine G+1 in the substrate strand by 7-deazaguanosine resulted in a ribozyme with impaired k_{cat} and an increased K_m. Similar experiments in loop B revealed that the amino groups of A24, A40 and A43 and the N-7 of A40 and A24 could be deleted with only minimal effects on catalytic activity. Substitutions in all other essential purines in loop B resulted in decreased catalytic efficiency largely due to reductions in k_{cat}. However, removal of the N-7 of G21 by 7-deazaguanosine substitution also increased K_m six-fold. Investigations of the magnesium ion dependence of the cleavage reaction revealed that magnesium binding was impaired by 7-deazaguanosine substitutions at G+1 and 7-deazaadenosine replacement at A43 and A9. It is possible that these N-7 functions may be involved in magnesium binding. Alternatively removal of the N-7 functions of these nucleotides may cause a structural change which decreases the affinity for magnesium at a more distant site.

2.4.2 *Modified Phosphodiester Linkages*

The introduction of a phosphorothioate linkage in place of an internucleoside phosphodiester bond renders the phosphorus centre chiral and localises the charge associated with the diester on the sulphur atom. Due to the resistance of phosphorothioate internucleoside linkages to the action of enzymes which metabolise nucleic acids *in vivo*, there is considerable interest in the introduction of this modification into RNA to be used in chemotherapeutic applications. Consequently the effects of phosphorothioate substitution in the hairpin ribozyme have been investigated.[20] In this study the phosphorothioate linkage was introduced by transcribing the enzyme with nucleoside α-thiotriphosphates and so multiple incorporation of all sites 5'-of the corresponding nucleoside in the *Rp* configuration resulted. It was found that the ribozyme was essentially unaffected by phosphorothioate substitution 5' to all guanosines, cytidines and uridines, but that substitution adjacent to adenosine residues A7, A9 and A10 interfere with cleavage to a much greater extent.

2.4.3 *Modified Loop Size Studies*

Ohtsuka and co-workers have investigated replacement of the bulged loop C between helix 2 and 3 in a ribozyme of type 2b with a non-nucleotidic linker.[21] The linker consisted of multiple units of 1,3-propanediol phosphate. Loops containing more than three linker units were reasonable substrates for the ribozyme, but the rate of cleavage increased when more units were included.

3. The Cleavage/ Ligation Reaction

The cleavage reaction exhibited by small ribozymes (hammerhead, hairpin, hepatitis delta virus) results in the generation of oligoribonucleotides with a 2',3'-cyclic phosphate and a 5'-hydroxyl termini respectively. The stereochemical course of the hammerhead and hairpin reactions have been studied and found to proceed with inversion of configuration at phosphorus as shown in figure 4.[22-25] The simplest explanation of these results is that the reaction proceeds with in-line attack generating a pentacoordinate intermediate. 2'-O-Methyl nucleoside or 2'-deoxynucleoside substitution of the residue 3'- to the scissile phosphodiester bond completely inhibits the hammerhead and hairpin catalysed reactions.

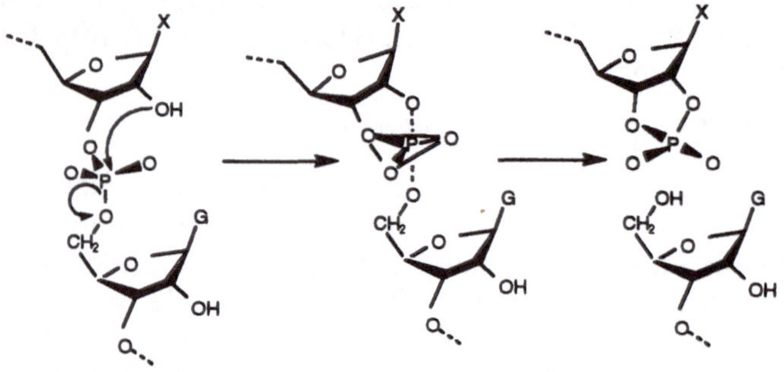

Figure 4 The mechanism of hairpin ribozyme catalysis consistent with the stereochemical course of the reaction.

3.1 THE ROLE OF METAL IONS IN HAIRPIN RIBOZYME REACTION

3.1.1 *Ionic Requirements of the Hairpin Ribozyme*

The ionic requirements of substrate binding and catalysis by the hairpin ribozyme have been studied.[10] The formation of a ribozyme substrate complex is not dependent on the presence of divalent metal ions. Mg^{2+}, Sr^{2+} and Ca^{2+} may all act as co-factors in the hairpin reaction, the maximal rate of reaction being observed with Mg^{2+}. The presence of Mn^{2+}, Co^{2+} and Zn^{2+} do not promote the reaction but cleavage is observed with these metal ions when spermidine is included in the reaction buffer. Low concentrations of spermidine also reduce the requirement for Mg^{2+} suggesting the binding of multiple divalent metal ions may be required for catalysis and that the polyamine may mimic the affect of one of these ions. In the presence of divalent metal ion chelaters (i.e. the absence of divalent metal ions) but the presence of spermidine or spermine, slow rates of reaction of hairpin and hammerhead catalysed reactions have been observed. In contrast to the hammerhead ribozyme, monovalent metal ions are inhibitory of the hairpin ribozyme reaction.

There has been much speculation about the role of divalent metal ions in the hammerhead and hairpin cleavage reactions. Based on a limited correlation between the ability of metal ions to promote the hammerhead cleavage reaction and the pK_a of the metal bound hydroxyl, it has been proposed that the initial deprotonation of the 2'-hydroxyl required for reaction is provided by a solvated divalent metal ion hydroxide.[26] Current evidence suggests that this relationship does not hold in the case of the hairpin reaction. [10] Furthermore, the possibility of a direct or water-mediated interaction of a metal ion stabilising the 5'-oxyanion leaving group has been discussed and is supported by theoretical calculations.[27] Since the stereochemical course of the reaction requires the attacking 2'-oxyanion at the opposite side to the leaving 5'-oxyanion, catalysis of the 2'-OH deprotonation and stabilisation of the leaving group could not be performed by the same metal ion in the absence of a drastic rearrangement of that ion. There is, therefore, the possibility of one or maybe two metal ions acting as general basic/acidic catalysts in ribozyme mediated reactions although there is no direct evidence to support this supposition. Finally, there remains the strong probability of a structural role for magnesium within ribozymes.

3.1.2 *Cleavage of a Phosphorothioate Internucleoside Bond*

Substitution of the proR_p oxygen of the scissile phosphodiester bond by sulphur in the hammerhead reaction severely reduces the rate of reaction observed in the presence of magnesium.[22, 23, 25] Manganese ions can rescue this observed decrease in rate.[22, 25] This ability has been cited as evidence that the divalent metal ion is bound directly to this oxygen, thus stabilising the transition state of the reaction. No such effects are observed upon substitution of the proS_p oxygen in the hammerhead cleavage site. In contrast to the hammerhead, no decrease in the rate of reaction are observed upon sulphur substitution of either oxygen of the cleaved phosphodiester bond in the hairpin-catalysed reaction. (Young, Gill and Grasby, unpublished results) This strongly suggests that the hairpin and hammerhead ribozyme cleavage reactions have different mechanisms.

3.2 THE REVERSE REACTION

In contrast to the hammerhead ribozyme, which catalyses the reverse reaction only very slowly, certain arrangements of the hairpin ribozyme such as those shown in figure 2b support catalysis of the formation of phosphodiester bonds.[11] The implication of this result is that the effective concentration of cleaved products in this ribozyme is higher than that in the hammerhead case, suggesting that the hairpin ribozyme may have more rigid geometry than the hammerhead. Feldstein and

Bruening have studied the ligation reaction by measuring the equilibrium concentration of a circular product formed from the ligation of a hairpin products contained within a single RNA strand. It was concluded that the bulged loop C between helix 2 and helix 3 is required for circularisation to occur. Since reduction of the loop size to one nucleotide, which would be expected to force helix 2 and 3 in a coaxial arrangement, inhibits the reaction completely this strongly argues for a parallel type of arrangement of helix 2 and 3 and suggests the possibility of tertiary contacts to the essential residues of loops A and B. To describe such an tertiary structure arrangement Feldstein and Bruening have coined the phrase "paper-clip ribozyme". These conclusions are also in accord with the studies of Komatsu *et al* [21] which define a minimal helix 2 and 3 linker length for the cleavage reaction to proceed.

Ohtsuka and co-workers have exploited the reverse reaction of the hairpin ribozyme to catalyse the formation of cross-ligated products.[28]

4. Summary and Future Prospects

Comparisons of the efficiency of the hairpin ribozyme with other small ribozymes (hammerhead and hepatitis delta virus), in *in vitro* and *in vivo* assays of the ability to undergo self-splicing has revealed very few differences between the three ribozymes. It thus appears likely that each ribozyme has considerable potential as possible chemotherapeutic agents. Further experiments interrogating both the structure and mechanism of the hairpin ribozyme should further knowledge about RNA structure and catalysis. Current evidence, which suggests the hairpin ribozyme reaction may mechanistically differ from that of other members of this family, is particularly intriguing. A further understanding of these features may assist in the more long term in the *de novo* design of novel nucleic acid based chemotherapeutic agents. Finally, a better understanding of catalysis and the rate limiting factors of the forward and back reactions via pre-steady state kinetic analysis may lead to the design of improved ribozymes for *in vivo* applications.

5. References

1. Symons, R. H. (1994) Ribozymes *Current Opinion In Structural Biology* **4**, 322-330.
2. Rubino, L., Tousignant, M. E., Steger, G. & Kaper, J. M. (1990) Nucleotide Sequence and Structural Analysis of two Satellite RNAs Associated with Chicory Yellow Mottle Virus *Journal of General Virlogy* **71**, 1897-1903.
3. Haseloff, J. & Gerlach, W. L. (1989) Sequences Required For Self-Catalysed Cleavage of the Satellite RNA of Tobacco Ringspot *Gene* **82**, 43-52.

4. Feldstein, P. A., Buzayan, J. M. & Bruening, G. (1989) Two Sequences Participating in the Autocatalytic Processing of Satellite Tobacco Ringspot Virus Complementary RNA *Gene* **82,** 53-61.
5. Joseph, S., Berzal-Herranz, A., Chowrira, B. M., Butcher, S. E. & Burke, J. M. (1993) Substrate Selection-Rules For the Hairpin Ribozyme Determined By *In Vitro* Selection, Mutation, and Analysis Of Mismatched Substrates *Genes & Development* **7,** 130-138.
6. Berzal-Herranz, A., Joseph, S., Chowrira, B. M., Butcher, S. E. & Burke, J. M. (1993) Essential Nucleotide-Sequences and Secondary Structure Elements Of the Hairpin Ribozyme *Embo Journal* **12,** 2567-2574.
7. Burke, J. M. (1994) The Hairpin Ribozyme in Eckstein, F. and Lilley D.M.J. (eds)*Nucleic Acids in Molecular Biology* Springer-Verlag, Berlin 105-118.
8. Anderson, P., Monforte, J., Tritz, R., Nesbitt, S., Hearst, J. & Hampel, A. (1994) Mutagenesis Of the Hairpin Ribozyme *Nucleic Acids Research* **22,** 1096-1100.
9. Hampel, A., Nesbitt, S., Tritz, R. & Altshuler, M. (1993) The Hairpin Ribozyme *Methods: A Companion to Methods in Enzymology* **5,** 37-42.
10. Chowrira, B. M., Berzal-Herranz, A. & Burke, J. M. (1993) Ionic Requirements For RNA-Binding, Cleavage, and Ligation By the Hairpin Ribozyme *Biochemistry* **32,** 1088-1095.
11. Feldstein, P. A. & Bruening, G. (1993) Catalytically Active Geometry In the Reversible Circularization Of Mini-monomer RNAs Derived From the Complementary Strand Of Tobacco Ringspot Virus Satellite RNA *Nucleic Acids Research* **21,** 1991-1998.
12. Vitorino Dos Santos, D., Fourrey, J.-L. & Favre, A. (1993) Flexibility of the Bulge Formed Between A Hairpin Ribozyme and Deoxy-Substrate Analogues *Biochemical and Biophysical Research Communications* **190,** 377-385.
13. Butcher, S. E. & Burke, J. M. (1994) A Photo-Cross-Linkable Tertiary Structure Motif Found In Functionally Distinct RNA Molecules Is Essential For Catalytic Function Of the Hairpin Ribozyme *Biochemistry* **33,** 992-999.
14. Butcher, S. E. & Burke, J. M. (1994) Structure-Mapping Of the Hairpin Ribozyme - Magnesium-Dependent Folding and Evidence For Tertiary Interactions Within the Ribozyme- Substrate Complex *Journal Of Molecular Biology* **244,** 52-63.
15. Chowrira, B. M. & Burke, J. M. (1991) Binding and Cleavage of Nucleic Acids by the Hairpin Ribozyme *Biochemistry* **30,** 8518-8522.
16. Pley, H. W., Flaherty, K. M. & McKay, D. B. (1994) 3-Dimensional Structure Of a Hammerhead Ribozyme *Nature* **372,** 68-74.
17. Chowrira, B. M., Berzal-Herranz, A., Keller, C. F. & Burke, J. M. (1993) 4 Ribose 2'-Hydroxyl Groups Essential For Catalytic Function Of the Hairpin Ribozyme *Journal Of Biological Chemistry* **268,** 19458-19462.
18. Chowrira, B. M., Berzal-Herranz, A. & Burke, J. M. (1991) Novel Guanosine Requirement for Catalysis by the Hairpin Ribozyme *Nature* **354,** 320-322.

306

19. Grasby, J. A., Mersmann, K., Singh, M. & Gait, M. J. (1995) Purine Functional-Groups In Essential Residues Of the Hairpin Ribozyme Required For Catalytic Cleavage Of RNA *Biochemistry* **34,** 4068-4076.
20. Chowrira, B. M. & Burke, J. M. (1992) Extensive Phosphorothioate Substitution Yields Highly Active and Nuclease-Resistant Hairpin Ribozymes *Nucleic Acids Research* **20,** 2835-2840.
21. Komatsu, Y., Koizumi, M., Nakamura, H. & Ohtsuka, E. (1994) Loop-Size Variation to Probe a Bent Structure Of a Hairpin Ribozyme *Journal Of the American Chemical Society* **116,** 3692-3696.
22. Dahm, S. C. & Uhlenbeck, O. C. (1991) Role of Divalent Metal Ions in the Hammerhead RNA Cleavage Reaction *Biochemistry* **30,** 9464-9469.
23. Koizumi, M. & Ohtsuka, E. (1991) Effects of Phosphorothioate and 2-Amino Groups in Hammerhead Ribozymes on Cleavage Rates and Mg2+ Binding *Biochemistry* **30,** 5145-5150.
24. van Tol, H., Buzayan, J. M., Feldstein, P. A., Eckstein, F. & Breuning, G. (1990) Two Autocatalytic Processing Reactions of a Satellite RNA Proceed with Inversion of Configuration *Nucleic Acids Research* **18,** 1971-1975.
25. Slim, G. & Gait, M. J. (1991) Configurationally Defined Phosphorothioate-Containing Oligoribonucleotides in the Study of the Mechanism of Cleavage of the Hammerhead Ribozyme *Nucleic Acids Research* **19,** 1183-1188.
26. Dahm, S. C., Derrick, W. B. & Uhlenbeck, O. C. (1993) Evidence for the Role of Solvated Metal Hydroxide in the Hammerhead Cleavage Mechanism *Biochemistry* **32,** 13040-13045.
27. Taira, K., Uebayasi, M., Maeda, H. & Furukawa, K. (1990) Energetics of RNA Cleavage: Implications for the Mechanism of Action of Ribozymes *Protein Engineering* **3,** 691-701.
28. Komatsu, Y., Koizumi, M., Sekiguchi, A. & Ohtsuka, E. (1993) Cross-Ligation and Exchange-Reactions Catalysed By Hairpin Ribozymes *Nucleic Acids Research* **21,** 185-190.

Sequence-Specific Cleavage of RNA Using Lanthanide Complexes Linked to Oligonucleotides

Robert Häner,* Jonathan Hall, Dieter Hüsken and Heinz E. Moser
Central Research Laboratories, Ciba
R-1060.1.30, CH-4002 Basle, Switzerland.

Abstract: The sequence-specific cleavage of synthetic RNA is described using macrocyclic lanthanide complexes covalently linked to oligodeoxyribonucleotides. The terpyridine derived metal complexes are completely stable towards chemical degradation under the experimental conditions (pH 7.4, 37°C). The cleavage efficiency of the conjugates is dependent on the type of lanthanide metal used. Whereas lanthanum(III) complexes are rather weak cleavers, almost complete cleavage of the RNA target is effected within 16 h using europium(III) complexes. Preferred sites of cleavage are NA-sites. Transesterification of the RNA backbone, which readily takes place in single stranded regions, is much more difficult to achieve in a duplex. Strand scission, however, can be induced within a duplex region by the presence of unpaired (bulged) RNA residues. Cleavage takes place at or near the bulge.

Introduction

The sequence-specific cleavage of ribonucleic acid (RNA) is currently an area of intensive investigation [1]. Metal complexes are among the best synthetic catalysts for the cleavage of RNA [2-7]. Covalent attachment of suitable metal complexes to oligonucleotides allows the cleavage of complementary single-stranded RNAs in a

---

* To whom correspondence should be addressed

B. Meunier (ed.), DNA and RNA Cleavers and Chemotherapy of Cancer and Viral Diseases, 307–320.
© 1996 *Kluwer Academic Publishers. Printed in the Netherlands.*

sequence-specific manner [8-11]. Artificial nucleases of this kind may be of importance for improving the efficiency of antisense oligonucleotides [12-13] and may find application as chemical probes for structural and functional studies of RNA [14].

The ability of rare earth metal salts to accelerate phosphodiester hydrolysis is well documented in literature [15-17]. Macromonocyclic [2a-c] and bicyclic [7] lanthanide complexes have been reported to be efficient promoters of RNA cleavage. Here we describe the covalent attachment of terpyridine derived lanthanide complexes to oligodeoxynucleotides and their use for the cleavage of complementary RNA at physiological pH and temperature.

Macrocyclic Lanthanide Complexes

Lanthanide complexes 1 and 2 (see Figure 1) have been synthesized in analogy to the known nickel (II) complex [18]. They are held together by stable hydrazone bonds, which provide the required chemical stability against hydrolytic decomposition of the complexes in aqueous solution. Selected lanthanide complexes prepared in this way are shown in Figure 1.

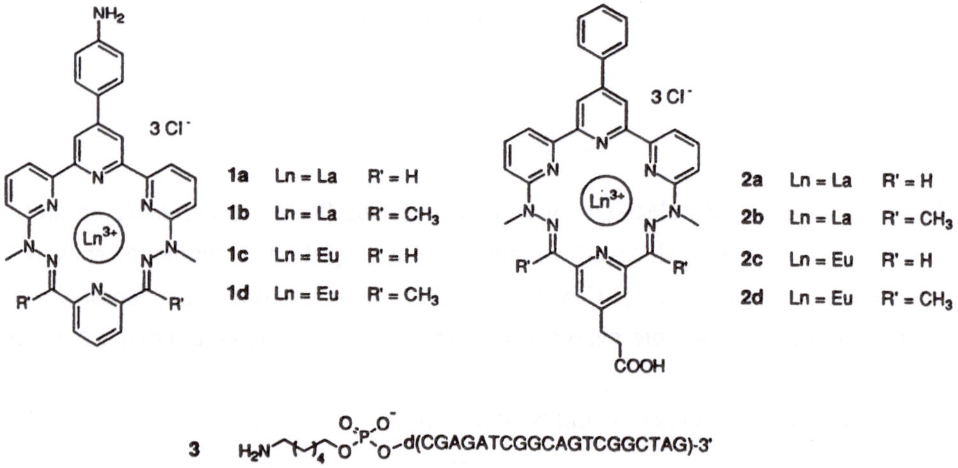

Figure 1. Terpyridine derived macrocyclic lanthanum(III) and europium(III) complexes (1 and 2) and amino-oligodeoxynucleotide 3 used for preparation of the conjugates.

Conjugation of Lanthanide Complexes to Oligonucleotides

Covalent attachment of complexes **1d** and **2c** to amino-oligodeoxynucleotide **3** yielded conjugates **4** and **5**, which are complementary to the 5'-end of the target oligoribonucleotides used throughout these studies. The metal complexes were linked to the 5'-hexylamino-modified oligodeoxynucleotide **3** *via* the intermediate isothiocyanate and *N*-hydroxysuccinimide ester derivatives, respectively (Figure 2). Conjugates **4** and **5** were purified by standard reverse phase HPLC and characterised by matrix assisted laser desorption ionisation mass spectrometry, capillary gel electrophoresis, UV/VIS spectroscopy and polyacrylamide gel electrophoresis analysis.

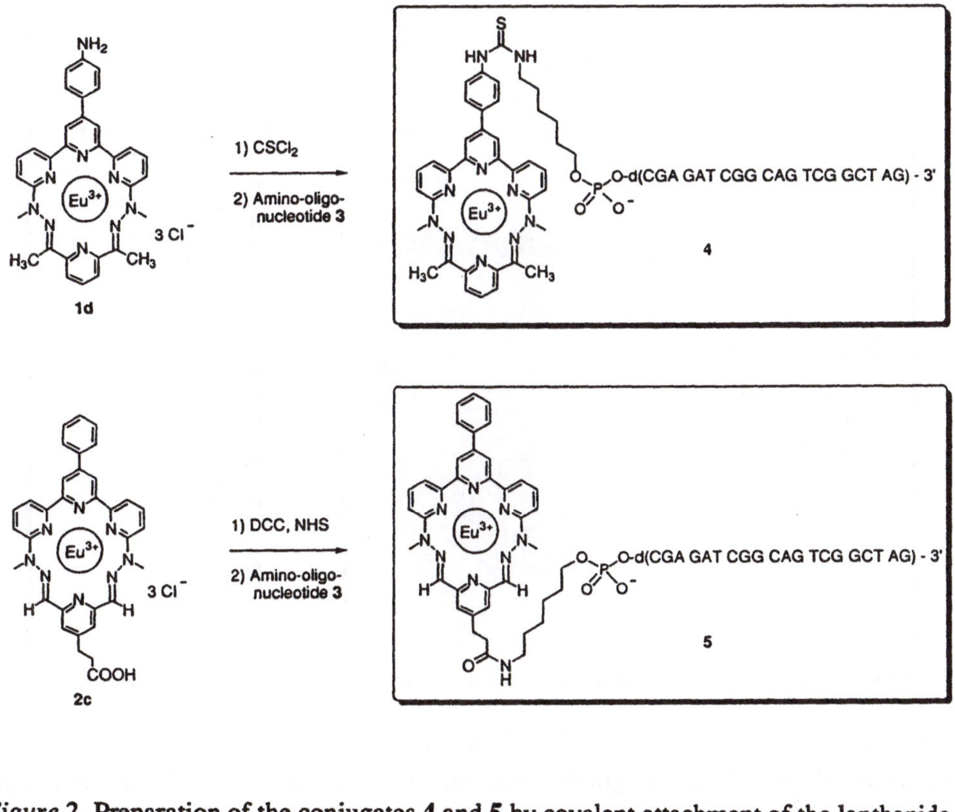

Figure 2. Preparation of the conjugates **4** and **5** by covalent attachment of the lanthanide complexes **1d** and **2c** to amino-oligodeoxynucleotide **3** *via* their intermediate isothiocyanate and *N*-hydroxy succinimide ester, respectively.

310

Cleavage of Complementary Single Stranded RNA

Cleavage experiments were carried out using the 29-mer oligoribonucleotide **6** as target and the products were analysed on a 12% denaturing polyacrylamide gel (see Figure 3). Incubation of the target RNA with the europium conjugate **4** (1μM, lane 6) resulted in scission of **6** in the single strand region, where the metal complex is expected to be approximately located. A second cleavage site is observed at the UAC-site at the 3'-end of the target RNA. This may arise from folding back of the negatively charged phosphate backbone of the target RNA onto the cationic metal complex.

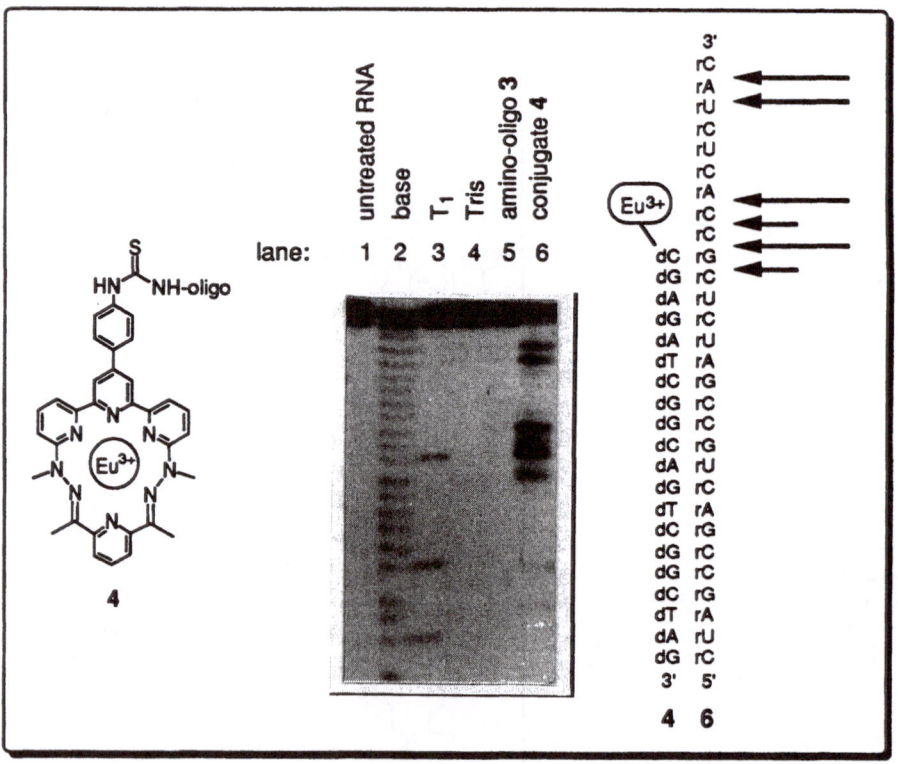

Figure 3. Autoradiograph of a 12 % denaturing polyacrylamide gel obtained after treatment of the 29-mer oligoribonucleotide **6** (^{33}P-labelled at the 5'-end) with conjugate **4**. Lanes 1 and 4: RNA before and after incubation in the reaction buffer; lane 2: alkaline hydrolysis (40 mM NaOH, 70°C, 20 min); lane 3: RNase T1 partial digestion; lanes 5 and 6: incubation (16 h, pH 7.4, 37°C) with amino-oligodeoxynucleotide **3** (lane 5) or conjugate **4** (lane 6).

Treatment of the target with the amino-oligodeoxynucleotide **3** did not result in any cleavage (lane 5) and treatment with **3** in combination with the unconjugated metal complex **1d** (400 µM) led to random cleavage of the target preferentially in the single stranded region (data not shown). Neither the efficiency nor the cleavage pattern was affected by the presence of 5 mM EDTA (data not shown), a condition which completely inhibits RNA cleavage by free lanthanide(III) salts. No cleavage was observed if non-complementary RNA was incubated with conjugate **4**.

The cleavage pattern obtained with the present lanthanide complexes is dependent on the target sequence. Although cleavage occurs at all four bases, we observe a preference for NA-sites, in particluar GA-sites. This is illustrated in Figure 4 showing the cleavage patterns of four different RNAs (7 - 10) with varying sequence composition in the single stranded region. In all experiments cleavage prevails at GA sites.

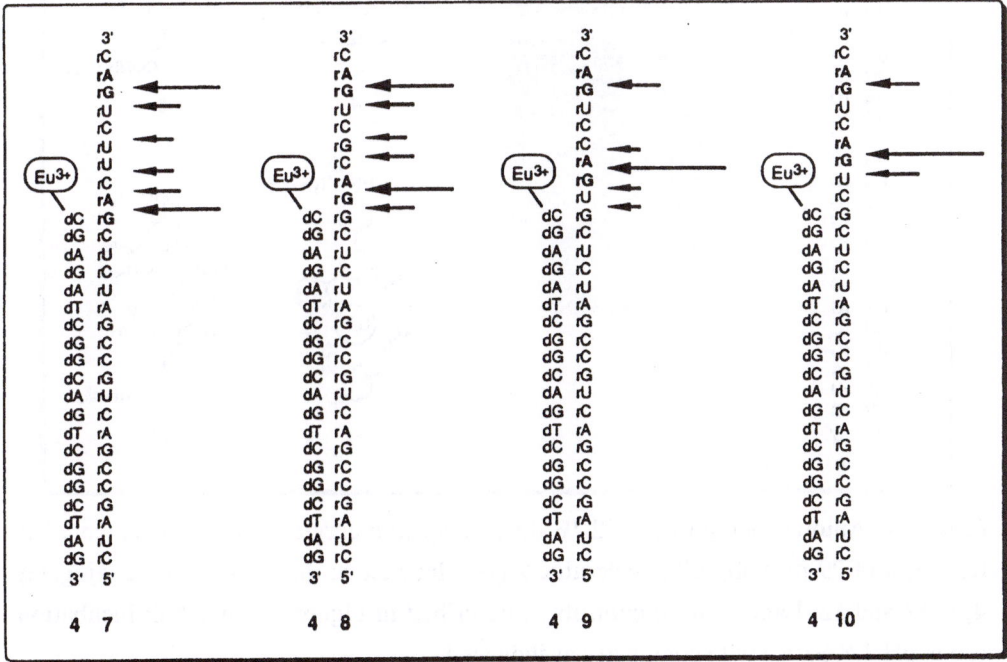

Figure 4. Cleavage pattern obtained with RNA's of varying base composition, illustrating a preference for cleavage by Eu(III)-complex DNA conjugates at GA-sites.

312

The cleavage efficiency of the conjugates depends on several factors, such as metal ion, type of linker, site of attachment and type of ligand. In order to investigate some of these factors in more detail, four different conjugates (4, 5, 11 and 12) were tested against target 9. Clearly, the two lanthanum derivatives 11 and 12 are much less active than the corresponding europium conjugates of which 5 in turn is significantly better than 4. Since the latter two compounds differ in more than one aspect (type of linker, site of attachment at the complex, substitution pattern of the ligand) further studies have to be carried out to examine this difference in activity more thoroughly.

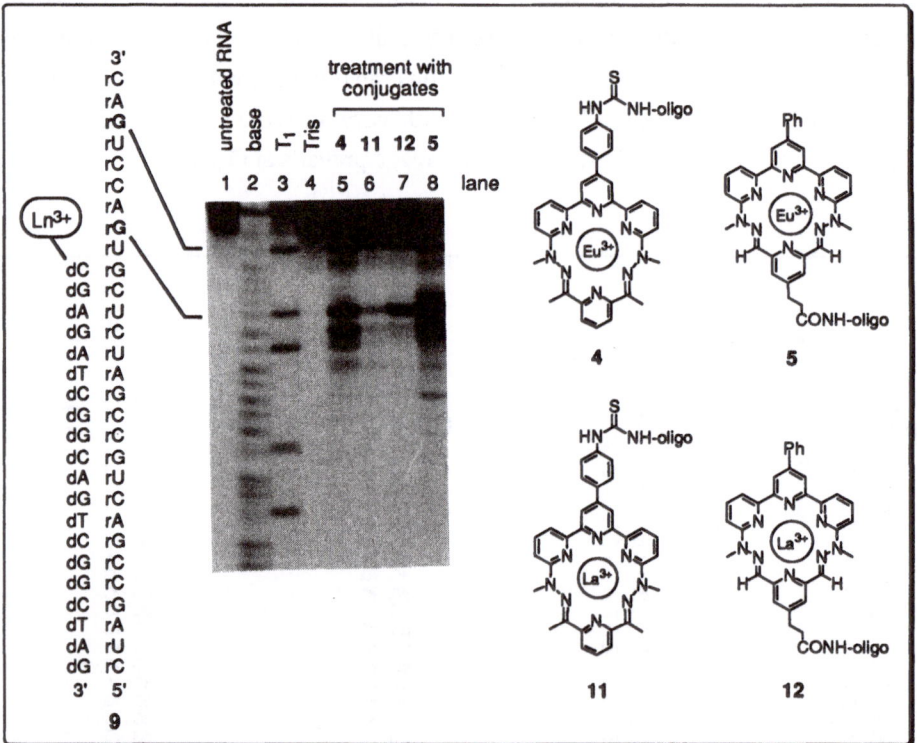

Figure 5. Autoradiograph of a 12 % denaturing polyacrylamide gel obtained after treatment of 29-mer oligoribonucleotide 9 (^{33}P-labelled at the 5'-end) with conjugates 4, 5, 11 and 12. Lanes 1 to 4: controls as described in Figure 3; lanes 5-8: incubation (16 h, pH 7.4, 37°C) with conjugates as indicated.

The results described above show that the terpyridine derived lanthanide complexes conjugates are efficient artificial ribonucleases, which can be used to cleave ribonucleic acid in a single strand. This is illustrated in Figure 6.

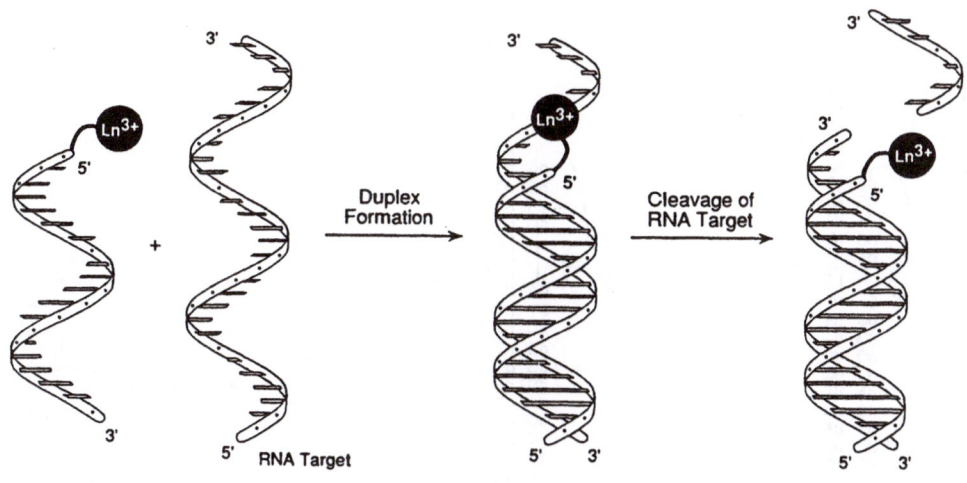

Figure 6. Schematic illustration of the sequence specific cleavage of a target oligoribonucleotide in a single strand region obtained with a lanthanide complex conjugated to a complementary oligonucleotide.

Cleavage of RNA at Bulged Nucleotides

In all the experiments described so far cleavage occurs only in the single strand region of the RNA. The fact that double stranded ribonucleic acids are largely protected against transesterification is well-known [2b, 19, 20]. Cleavage within a duplex region could, however, be effected by the introduction of a vulnerable site. Based upon earlier investigations we devised a strategy making use of bulged RNA residues [21]. Figure 7 shows the results obtained with targets **13** and **14**, which, after hybridisation with

314

conjugate **4**, give rise to bulged nucleotides. Target **13** leads to the formation of a single bulge and target **14** to a double bulge. In both cases cleavage takes place around the bulged residues. However, whereas in the case of a single bulged residue (**13**) cleavage also occurs in the single strand region, the GA-double bulge in **14** directs cleavage exclusively to the bulge site.

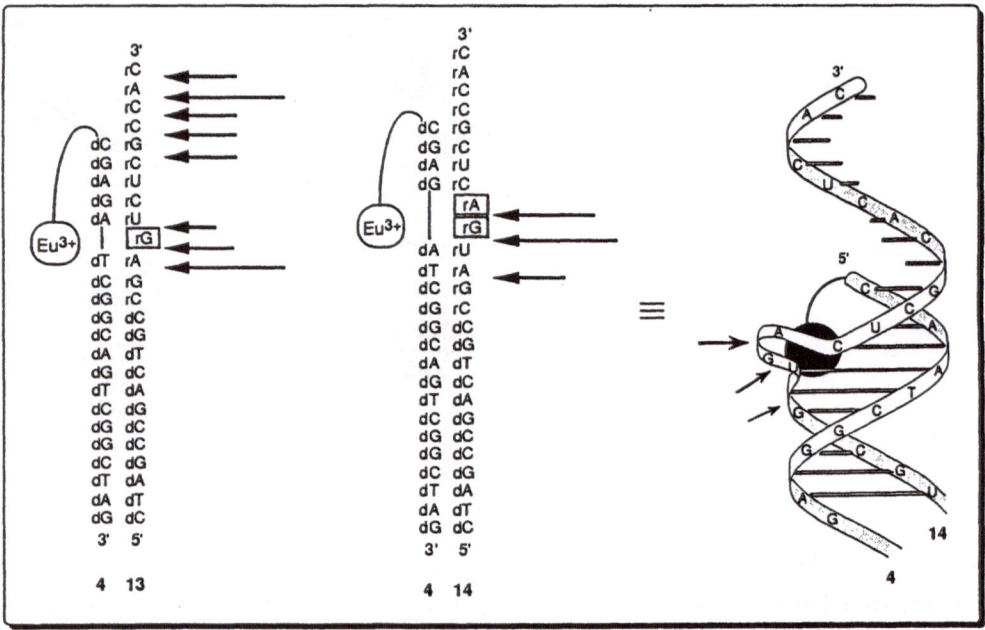

Figure 7. Cleavage pattern obtained by treatment of the target RNAs **13** and **14** with the conjugate **4**. The boxed residues indicate the sites of the bulges.

The cleavage experiments described above were all carried out in a standard way using 1μM conjugate at a total RNA concentration of 10-50nM. Investigation of the concentration dependence showed that cleavage efficiency remained unchanged down to concentrations as little as 100nM. Below this concentration loss of cleavage efficiency was observed, indicating that the cleavage takes place in a stochiometric way. Figure 8 shows the effect of the concentration of conjugate **4** on the cleavage of target **14**.

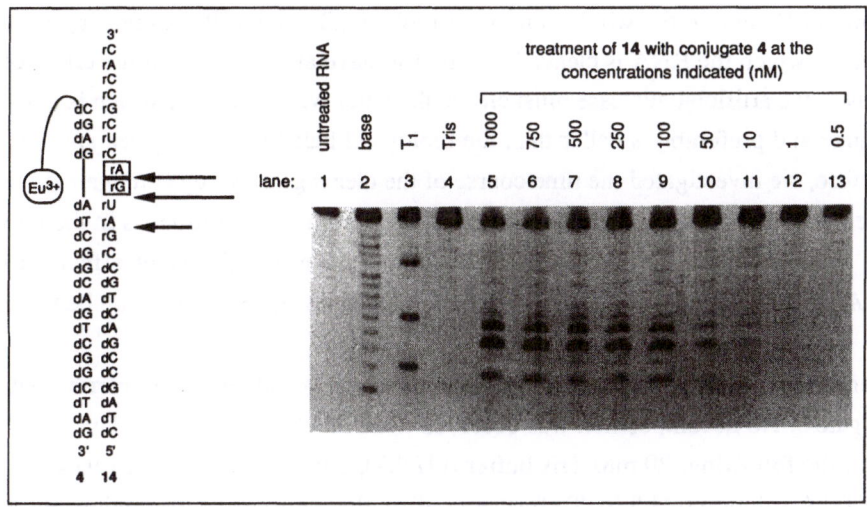

Figure 8. Dependence of the cleavage of target oligo **14** on the concentration of conjugate **4**. All reactions were carried out at 37°C (16 h, pH = 7.4).

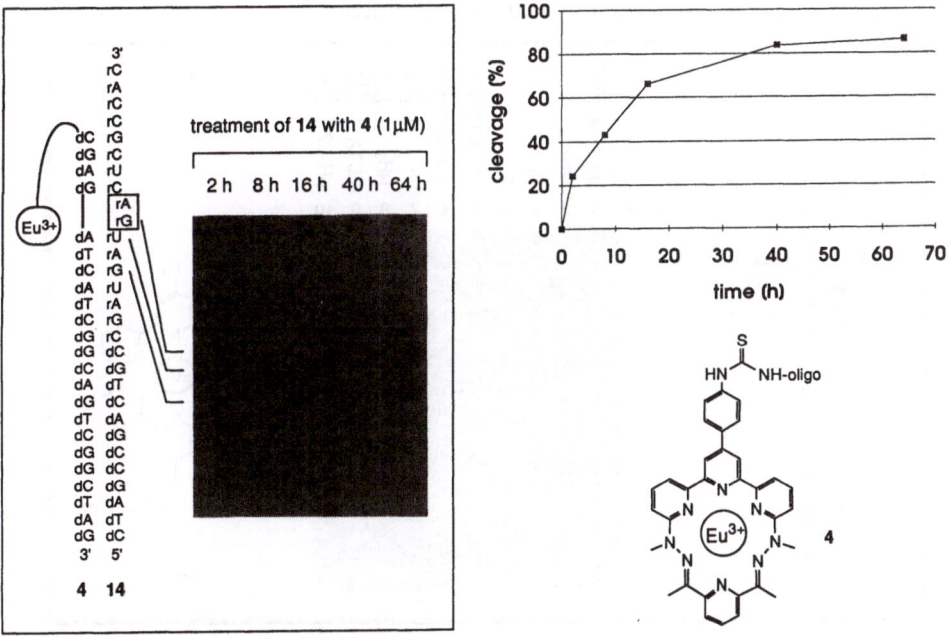

Figure 9. Time dependence of the cleavage of target oligo **14** by conjugate **4**. All reactions were carried out at 37°C (16 h, in 20 mM Tris buffer, pH = 7.4).

316

A major challenge in the whole context of biologically active RNA cleaving agents is the rate at which the RNA is cleaved. In order to have an influence on the effect of e.g. a mRNA, the artificial nuclease must cleave the latter with a half life which is at least in the range and preferably smaller than the biological half life of the given species [22]. Therefore, we investigated the time course of the cleavage of 14 by conjugate 4. As can be seen from Figure 9, the half life of the RNA under these conditions is in the range of 10 h. Whereas this is probably not sufficient to have a siginificant effect on most mRNAs, it is likely, that conjugates with increased cleavage efficiency can be found.

To demonstrate the stability of the metal complexes, several tests were carried out with conjugate 5. No decomposition was observed upon incubation during 9 days at 37°C in each of the following: 20 mM Tris buffer (pH 7.7), 2 mM acetate buffer (pH 4.7) and 5 mM EDTA solution (pH 8.0). To further confirm the inertness of the conjugates, 5 was re-isolated after treatment under the conditions mentioned above and subsequently used to cleave the target RNA 14 (see Figure 10, lanes 8-10). No loss of cleavage activity was observed compared to the non-treated conjugate (lane 7).

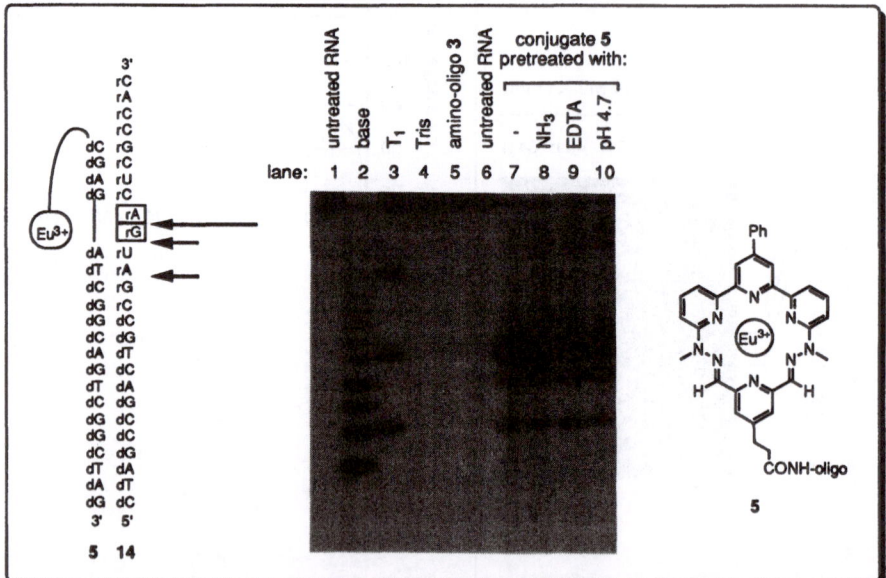

Figure 10. Cleavage of 14 with untreated conjugate 5 (lane 7) or with 5 which had been treated under the conditions indicated (lanes 8-10; see text for more information). All reactions were carried out at 37°C (16 h, pH = 7.4) using 1μM conjugate.

Summary

We have described the sequence-specific cleavage of different model RNAs through the use of chemically stable macrocyclic lanthanide(III) complexes covalently linked to oligodeoxynucleotides. The cleavage efficiency of the conjugates can vary greatly with the type of lanthanide metal used. Whereas lanthanum(III) complexes are rather weak cleavers, almost complete cleavage of the RNA target is effected within 16 h using europium(III) complexes. Preferred sites of cleavage are NA-sites. Transesterification of the RNA backbone, which readily takes place in single stranded regions, is much more difficult to achieve within a duplex. The presence of a bulge, however, renders the RNA susceptible to cleavage. Sequence specific cleavage within a duplex region at bulged sites was demonstrated using lanthanide complex conjugates. Strand scission takes place at or near the bulge.

References

1. (a) Sigman, D.S., Mazumder, A. and Perrin, D.M. (1993). Chemical Nucleases. *Chem. Rev.* **93**, 2295-2316. (b) Hélène, C., Thuong, N.T., Saison-Behmoaras, T. and François, J.-C. (1989). Sequence-specific artificial endonucleases.*Trends Biotechnol.* **7**, 310-315. (c) Manoharan, M. (1993). Designer Antisense Oligonucleotides: Conjugation Chemistry and Functionality Placement, in S.T. Crooke and B. Lebleu (eds.), *Antisense Research and Applications*, CRC Press, Florida, pp. 303-349. (d) Pyle, A.M. (1993). Ribozymes: A Distinct Class of Metalloenzymes. *Science* **261**, 709-714. (e) Christoffersen, R.E. and Marr, J.J. (1995). Ribozymes as Human Therapeutic Agents. *J. Med. Chem.* **38**, 2023-2037 and refs. cited therein.

2. (a) Morrow, J.R., Buttrey, L.A., Shelton, V.M. and Berback, K.A. (1992). Efficient Catalytic Cleavage of RNA by Lanthanide(III) Macrocyclic Complexes: Towards Synthetic Nucleases for in Vivo Applications. *J. Am. Chem. Soc.* **114**, 1903-1905. (b) Kolasa, K.A., Morrow, J.R. and Sharma, A.P. (1993). Trivalent Lanthanide Ions Do Not Cleave RNA in DNA-RNA Hybrids. *Inorg. Chem.* **32**, 3983-3984. (c)

Hayashi, N., Takeda, N., Shiiba, T., Yashiro, M., Watanabe, K. and Komiyama, M. (1993). Site-Selective Hydrolysis of tRNA by Lanthanide Metal Complexes. *Inorg. Chem.* **32**, 5899-5900. (d) Amin, S., Morrow, J.R., Lake, C.H. and Rowen Churchill, M. (1994). Lanthanide(III) Tetraamide Macrocyclic Complex as Synthetic Ribonucleases: Structure and Catalytic Properties of [La(tcmc)(CF3SO3)(EtOH)](CF3SO3)2. *Angew. Chem. Int. Ed. Engl.* **33**, 773-775. (e) Morrow, J.R.and Shelton, V.M. (1994). Toward the design of artificial ribonucleases: The effect of ligands and RNA bases on the cleavage of dinucleotides and dinucleosides by La^{3+}, Pb^{2+} and Zn^{2+}. *New J.Chem.* **18**, 371-375. (f) Chin, K.O.A., Morrow, J.R. (1994). RNA Cleavage and Phosphate Diester Transesterification by Encapsulated Lanthanide Ions: Traversing the Lanthanide Series with Lanthanum(III), Europium (III), and Lutetium(III) Complexes of 1,4,7,10-Tetrakis(2-hydroxyalkyl)-1,4,7,10-tetraazacyclododecane. *Inorg Chem.* **33**, 5036-5041.

3. Stern, M.K., Bashkin, J.K. and Sall, E.D. (1990). Hydrolysis of RNA by Transition-Metal Complexes. *J. Am. Chem. Soc.* **112**, 5357-5359. (b) Modak, A.S., Gard, J.K., Merriman, M.C., Winkeler, K.A., Bashkin, J.K. and Stern, M.K., (1991). Toward Chemical Ribonucleases. 2. Synthesis and Characterization of Nucleoside-Bipyridine Conjugates. Hydrolytic Cleavage of RNA by their Copper(II) Complexes. *J. Am. Chem. Soc.* **113**, 283-291.

4. Matsumoto, Y. and Komiyama, M. (1990). Efficient Cleavage of Adenylyl(3'-5')adenosine by Triethylenetetraminecobalt(III). *J. Chem. Soc., Chem. Commun.* 1050-1051.

5. Breslow., R. and Huang, D.-L. (1991). Effects of metal ions, including Mg^{2+} and lanthanides, on the cleavage of ribonucleotides and RNA model compounds. *Proc. Natl. Acad. Sci. USA* **88**, 4080-4083.

6. (a) Wall, M., Hynes, R.C. and Chin, J. (1993). Double Lewis Acid Activation in Phosphate Diester Cleavage. *Angew. Chem. Int. Ed. Engl.* **32**, 1633-1635. (b) Bashkin, J.K., Jenkins, L.A. (1993). Catalytic Hydrolysis of 2',3'-Cyclic Adenosine Monophosphate by Aqua(2,2':6',2"-terpyridine)copper(II): Breakdown of the Analogy Between Activated Phosphodiesters and RNA. *J. Chem. Soc. Dalton Trans.* 3631-3632. (c) Linkletter, B., Chin, J. (1995). Rapid Hydrolysis of RNA with a Cu[II] Complex. *Angew. Chem. Int. Ed. Engl.* **34**, 472-474.

7. Schneider, H.J., Rammo, J. and Hettich, R. (1993). Catalysis of the Hydrolysis of Phosphoric Acid Diesters by Lanthanide Ions and the Influence of Ligands. *Angew. Chem. Int. Ed. Engl.* **32**, 1716-1719.

8. Bashkin, J.K., Frolova, E.I. and Sampath, U. (1994). Sequence-Specific Cleavage of HIV mRNA by a Ribozyme Mimic. *J. Am. Chem. Soc.* **116**, 5981-5982.

9. Magda, D., Miller, R.A., Sessler, J.L. and Iverson, B.L. (1994). Site-Specific Hydrolysis of RNA by Europium(III) Texaphyrin Conjugated to a Synthetic Oligodeoxyribonucleotide. *J. Am. Chem. Soc.* **116**, 7439-7440.

10. Matsumura, K., Endo, M. and Komiyama, M. (1994). Lanthanide Complex-Oligo-DNA Hybrid for Sequence-selective Hydrolysis of RNA. *J. Chem. Soc., Chem. Commun.* 2019-2020.

11. Hall, J., Hüsken, D., Pieles, U., Moser, H.E., and Häner, R. (1994). Efficient sequence-specific cleavage of RNA using novel europium complexes conjugated to oligonucleotides. *Chemistry & Biology* **1**, 185-190.

12. (a) Uhlmann, E. and Peyman, A. (1990). Antisense Oligonucleotides: A New Therapeutic Principle. *Chem. Reviews* **90**, 543- 584. (b) Hélène, C. and Toulmé, J.-J. (1990). Specific regulation of gene expression by antisense, sense and antigene nucleic acids. *Biochim. Biophys. Acta* **1049**, 99-125. (c) Milligan, J.F., Matteucci, M.D. and Martin, J.C. (1993). Current Concepts in Antisense Drug Design. *J. Med. Chem.* **36**, 1923-1937. (d) S.T. Crooke and B. Lebleu (eds.), *Antisense Research and Applications*, CRC Press, Florida.

13. De Mesmaeker, A., Häner, R., Martin, P. and Moser, H.E. (1995). Antisense Oligonucleotides. *Acc. Chem. Res. in press.*

14. Huber, P.W. (1993). Chemical nucleases: their use in studying RNA structure and RNA-protein interactions. *FASEB* **7**, 1367-1375.

15. (a) Bamann, E. (1939). Über "phosphatatische" Wirkungen von Hydrogelen. *Angew. Chem.* **52**, 186-188. (b) Bamann, E., Trapmann, H. and Fischler, F. (1954). Verhalten und Spezifität von Cer und Lanthan als Phosphatase-Modelle gegenüber Nucleinsäuren und Mononucleotiden. *Biochem. Z.* **326**, 89-96.

16. Butcher, W.W. and Westheimer, F.H. (1955). The Lanthanum Hydroxide Gel Promoted Hydrolysis of Phosphate Esters. *J. Am. Chem. Soc.* **77**, 2420-2424.

17. Eichhorn, G.L. and Butzow, J.J. (1965). Interactions of Metal Ions with Polynucleotides and Related Compounds. III. Degradation of Polyribonucleotides by Lanthanum Ions. *Biopolymers* **3**, 79-94.

18. Constable, E.C. and Holmes, J.M. (1988). The Preparation and Coordination Chemistry of 2,2':6',2"-Terpyridine Macrocycles - VI. Planar Hexadentate Macrocycles Incorporating 2,2':6',2"-Terpyridine. *Polyhedron* **7**, 2531-2536.

19. Usher, D.A. and McHale, A.H. (1976). Hydrolytic stability of helical RNA: A selective advantage for the natural 3',5'-bond. *Proc. Nat. Acad. Sci. USA* **73**, 1149-1153.

20. Gornicki, P., Baudin, F., Romby, P., Wiewiorowski, M., Kryzosiak, W., Ebel, J.P., Ehresmann, C. and Ehresmann, B. (1989). Use of Lead(II) to probe the Structure of Large RNA's. Conformation of the 3' Terminal Domain of *E.coli* 16S rRNA and its Involvement in Building the tRNA Binding Sites. *J. Biomol. Struct. Dyn.* **6**, 971-984.

21. Hüsken, D., Goodall, G., Hall, J., Häner, R. and Moser, H.E. *in preparation.*

22. Sachs, A.B. (1993). Messenger RNA Degradation in Eukaryotes. *Cell* **74**, 413-421.

SYNERGISM OF TWO METAL IONS FOR THE HYDROLYSIS OF DNA AND RNA

The Most Active Catalyst Ever for DNA Hydrolysis

MAKOTO KOMIYAMA, * **NAOYA TAKEDA, MAKOTO IRISAWA AND MORIO YASHIRO**
Department of Chemistry and Biotechnology, Graduate School of Engineering, University of Tokyo, Hongo, Tokyo 113, Japan

Abstract

Enormously active catalysts for the hydrolysis of DNA and RNA are prepared by use of the synergetic cooperation of two metal ions (the combinations of lanthanide/lanthanide, lanthanide/non-lanthanide, and non-lanthanide/non-lanthanide ions). The most active catalysts for DNA hydrolysis ever have been obtained by combining the Ce(IV) ion with either the Pr(III) or Nd(III) ion (the Pr(III) and Nd(III) ions themselves are inactive). The activities of these combinations are 10 times as great as that of the Ce(IV) ion, hitherto the most active catalyst, so that linear DNA is hydrolyzed for the first time under the physiological conditions. DNA as well as RNA is also efficiently hydrolyzed by the combination of La(III) with either Fe(III) or Sn(IV), although neither of the components is eminent when used separately. Furthermore, RNA is promptly hydrolyzed by the combinations of two non-lanthanide ions (Zn(II)/Sn(IV) and Zn(II)/In(III)). All these catalyses involve mixed hydroxide clusters of the two metal ions, in which the water molecules bound to the metal ions show effective acid-base cooperation. These bimetallic combinations provide valuable tools for biotechnology, molecular biology, and therapy in the future.

B. Meunier (ed.), DNA and RNA Cleavers and Chemotherapy of Cancer and Viral Diseases, 321–335.
© *1996 Kluwer Academic Publishers. Printed in the Netherlands.*

1. Introduction

Non-enzymatic hydrolysis of the phosphodiester linkages in DNA and RNA under the physiological conditions has been one of the most challenging topics. In 1992, we succeeded in the first non-enzymatic hydrolysis of DNA by use of a series of lanthanide ions: the cerium ion is especially competent [1-6].

Later, the active species for the DNA hydrolysis was shown to be the Ce(IV) ion, produced *in situ* by the oxidation of Ce(III) by ambient oxygen [7-10]. Quite importantly, no molecular oxygen is required for the catalytic process (the hydrolysis is effective even without oxygen, when Ce(IV) salts are used in place of Ce(III) salts [7,8,10]). Thus, the DNA hydrolysis proceeds via the nucleophilic attack by the hydroxide ion bound to the Ce(IV) ion [10], not via the attack by the peroxide-like species as proposed by Chin *et al* [9]. In addition, we efficiently hydrolyzed RNA by use of either lanthanide ions [6,11] or oligoamines [12].

Furthermore, artificial nucleases and ribonucleases, which selectively hydrolyze the target phosphodiester linkages in DNA [13-15] and in RNA [15-18], were prepared by the attachment of these catalytic residues to synthetic DNA oligomers. Since both the selectivities and the reaction rates are great and in addition the scission manner is identical with those in enzymatic scission, these artificial enzymes are highly potent for applications to molecular biology, biotechnology, and therapy. Typical examples of them as well as their scission patterns are presented in *Figure 1*.

One of the next goals is to design still more active catalysts. In nature, some of phosphoesterases take advantage of the cooperation of two metal ions at the active sites (e.g., Mg(II), Zn(II), and Mn(II)). In fact, activated phosphodiesters and their analogs were hydrolyzed by sophisticated enzyme mimics [19]. However, none of them could hydrolyze the phosphodiester linkages in DNA and RNA themselves.

Here we present several bimetallic systems which hydrolyze the phosphodiester linkages in nucleic acids under the physiological conditions. Especially, novel catalysts for DNA hydrolysis, which greatly exceed the Ce(IV) ion in the activity, have been prepared by use of the bimetallic synergism.

1). Artificial Hydrolytic Nuclease

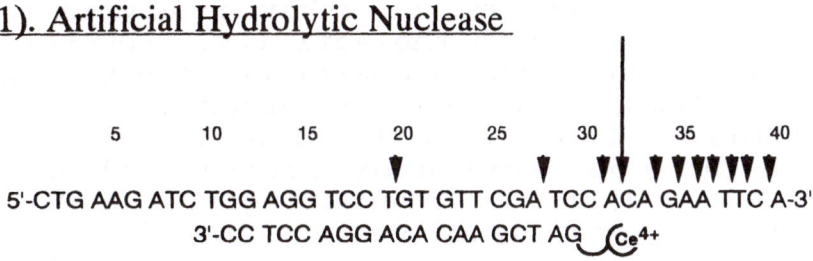

2). Artificial Hydrolytic Ribonuclease

(a)

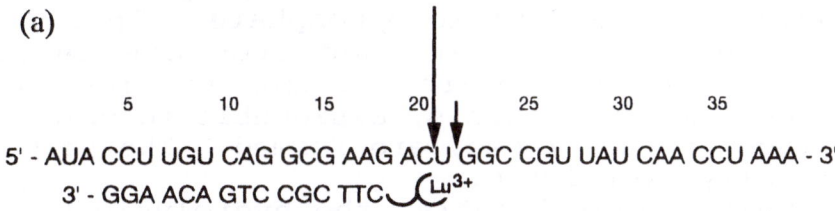

(b)

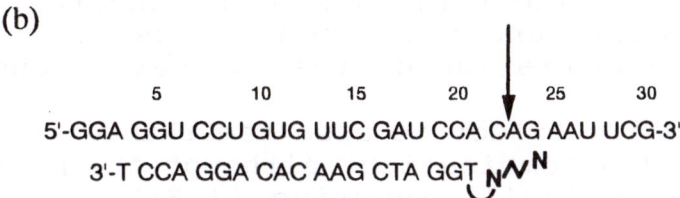

Figure 1. Artificial enzymes for the site–selective scission of DNA and RNA as well as their scission patterns. The catalytic residues are the Ce(IV) ion for 1 (refs. 13 and 14), the Lu(III) ion for 2a (ref. 16), and ethylenediamine for 2b (ref. 17).

2. Bimetallic Synergism for the Hydrolysis of DNA and RNA

2.1. COOPERATION OF TWO LANTHANIDE IONS (THE MOST ACTIVE CATALYST FOR DNA HYDROLYSIS EVER WHICH GREATLY EXCEEDS CERIUM(IV) IN THE ACTIVITY!) [20]

Before the present study, the most active catalyst for DNA hydrolysis was the Ce(IV) ion. In spite of many efforts, no other catalysts could beat the metal ion in the activity. However, we have succeeded in the preparation of novel catalytic systems, which are one order of magnitude more active than the Ce(IV) ion, by use of the synergetic cooperation of Ce(IV) with either Pr(III) or Nd(III). Interestingly, the Pr(III) and Nd(III) ions themselves are inactive for DNA hydrolysis.

By use of 2:1 combination of $Ce(NH_4)_2(NO_3)_6$ and $PrCl_3$ at pH 7 and 50°C, more than half of thymidylyl(3'→5')thymidine (TpT) is hydrolyzed to thymidine (Thd) within 30 min (*Figure 2* (a)). Thymidine 3'- and 5'-monophosphates (Tp and pT) as the hydrolysis intermediates are rapidly converted to Thd and are not significantly accumulated. No by-products, assignable to oxidative cleavage of the ribose, are formed. The scission is totally hydrolytic.

Quite significantly, the hydrolysis by the Ce(IV)/Pr(III) combination is 10 fold faster than that by the Ce(IV) ion alone (see *Figure 2* (b) and TABLE 1). The Pr(III) ion is inactive when used separately under the conditions (c). The synergetic cooperation of the two metal ions is conclusive.

Synergetic effect is also remarkable for the Ce(IV)/Nd(III) combination, although the Nd(III) ion is intrinsically inactive (TABLE 1). The magnitudes of the synergetic acceleration (over the activity of the Ce(IV) ion) are 7 fold at 50°C and 10 fold at 30°C. In contrast, other lanthanide ions and non-lanthanide ions suppressed the activity of the Ce(IV) ion. The synergism with the Ce(IV) ion is specific to the Pr(III) and Nd(III) ions.

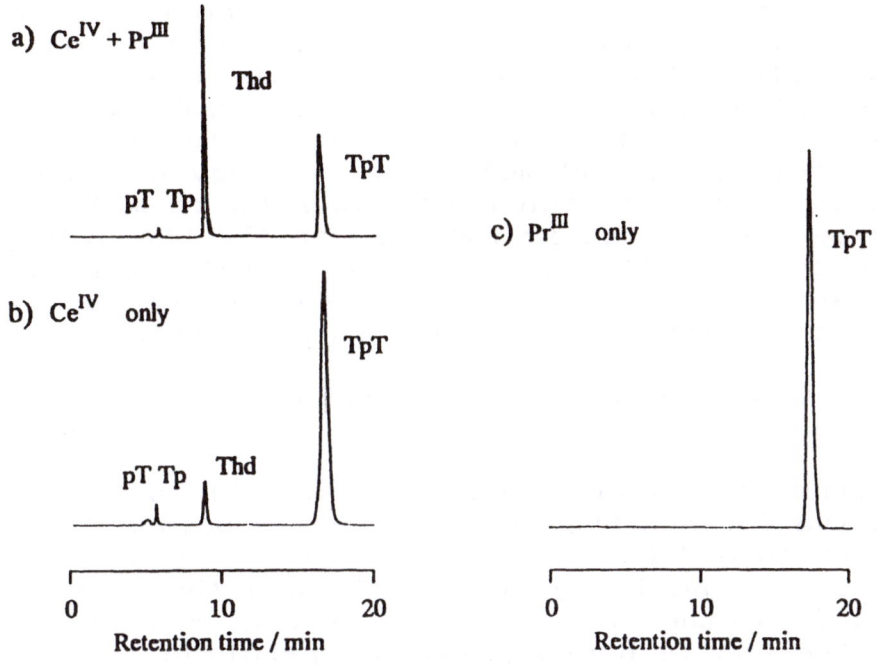

Figure 2. Reversed–phase HPLC profiles for the hydrolysis of TpT at pH 7 and 50°C for 30 min: (a) the combination of Ce(IV) and Pr(III); (b) Ce(IV) only; (c) Pr(III) only. [Ce(IV)] = 20 *mM* and [Pr(III)] = 10 *mM*.

TABLE 1. The pseudo first–order rate constants for the hydrolysis of TpT at pH 7 by the 2:1 Ce(IV)/Pr(III) and Ce(IV)/Nd(III) combinations [a]

Metal ions	Rate constant (h^{-1})	
	at 50°C	at 30°C
Ce(IV) + Pr(III)	1.7	0.11
Ce(IV) only	0.17	0.011
Pr(III) only	0.0	0.0
Ce(IV) + Nd(III)	1.1	0.11
Nd(III) only	0.0	0.0

a. [Ce(IV)] = 20 *mM* and [Pr(III)] = 10 *mM*.

With the present bimetallic combinations, DNA can be hydrolyzed under the physiological conditions (pH 7 and 30°C) within 1 day. When the Ce(IV) ion is used alone, however, it takes nearly two weeks. Thus, truly useful tools for biotechnology and molecular biology are now in hand. The reaction does not require molecular oxygen at all, as is the case in the DNA hydrolysis by the Ce(IV) ion [7,8,10]. This is another advantage for applications, especially *in vivo*, since the interior of cells is anaerobic.

2.2. COOPERATION OF A LANTHANIDE ION WITH A NON-LANTHANIDE ION FOR HYDROLYSIS OF DNA AND RNA [21]

Neither the La(III) nor the Fe(III) ion is very active for ApA hydrolysis, when is used separately (*Figure 3* (b) and (c)). However, a remarkable catalytic activity comes out when they are combined together (a). In only 5 min, more than 70% of the ApA is hydrolyzed. Cooperative catalysis is also evident for the combination of the La(III) and the Sn(IV) ions (TABLE 2).

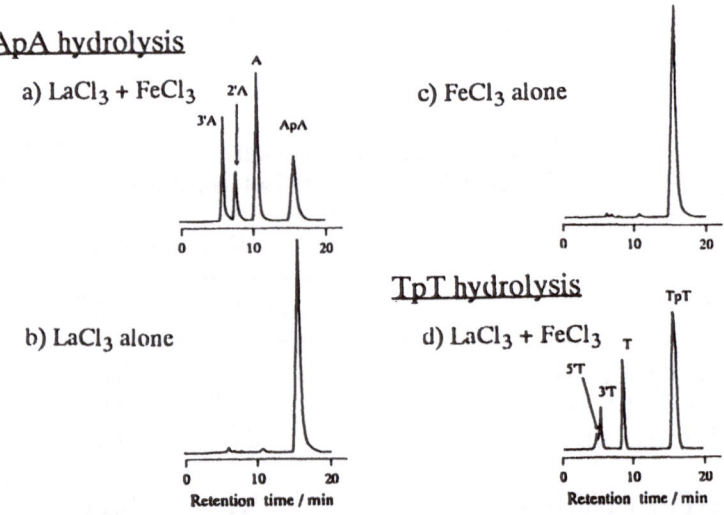

Figure 3. Reversed–phase HPLC profiles of the hydrolysis of ApA at pH 7.0 and 50°C for 5 min (a–c), and of TpT at pH 7.5 and 70°C for 24 h (d): (a) and (d), $LaCl_3$ (10 *mM*) + $FeCl_3$ (10 *mM*); (b), $LaCl_3$ alone (10 *mM*); (c), $FeCl_3$ alone (10 *mM*).

TABLE 2. Hydrolysis of ApA and TpT by the combination of the La(III) ion (10 mM) and non–lanthanide ion (10 mM)

Substrate	Lanthanide ion	Non–lanthanide ion	Conversion (mole%)
ApA [a]	La(III)	Fe(III)	72
	La(III)	-----	1.4
	------	Fe(III)	1.3
	La(III)	Sn(IV)	81
	------	Sn(IV)	0.7
TpT [b]	La(III)	Fe(III)	36
	La(III)	-----	2.1
	------	Fe(III)	0.0
	La(III)	Sn(IV)	26
	------	Sn(IV)	0.1

a. pH 7.0 and 50°C for 5 min.
b. pH 7.5 and 70°C for 24 h.

It is noteworthy that these bimetallic combinations are also applicable to DNA hydrolysis: they are the second non-enzymatic catalysts for the hydrolysis of linear DNA (the first one is the Ce(IV) ion). TpT is efficiently hydrolyzed by both the La(III)/Fe(III) and La(III)/Sn(IV) combinations at pH 7.5 and 70°C (*Figure 3* (d)). The conversions at 24 h are 36 and 26 mole%, respectively. However, the activity of each of the La(III), Fe(III), and Sn(IV) ions is marginal to nil (see the lower part of TABLE 2). All the products are hydrolytic ones, and thymine and other byproducts are not formed at all. The combination of the La(III) and In(IV) ions also shows a synergetic acceleration (the conversion at 24 h is 9 mole%, although In(IV) itself is inactive).

The hydrolysis of ApA and TpT, induced by the Eu(III) ion, was also promoted by the addition of the Fe(III) ion (by 2-3 fold when [Fe(III)]/[Eu(III)] = 1).

2.3. COOPERATION OF TWO NON-LANTHANIDE IONS [22,23]

It is now clear that quite active catalysts for the hydrolysis of DNA and RNA can be prepared by combining a lanthanide ion with either another lanthanide ion or non-lanthanide ion, as long as the combination is appropriate. We have further extended the findings and have shown that efficient catalysts for the purpose can be also obtained even by use of only two non-lanthanide metal ions.

As depicted in *Figure 4* (a), ApA is promptly hydrolyzed by the combination of $ZnCl_2$ (10 *mM*) and $SnCl_4$ (10 *mM*) at pH 7 and 50°C. In 3 h, about half of ApA is hydrolyzed to adenosine (Ado) and its 2'- and 3'-monophosphates (2'A and 3'A). A small amount of 2',3'-cyclic monophosphate of adenosine (A>p) as the hydrolysis intermediate is accumulated. However, either $ZnCl_2$ or $SnCl_4$ is virtually inactive when used separately ((b) and (c)). The synergetic catalysis is again evident. The pseudo first-order rate constants and the magnitudes of cooperative acceleration are presented in TABLE 3.

Synergetic catalysis is also notable for the Zn(II)/In(III), Zn(II)/Fe(III), and Zn(II)/Al(III) combinations (TABLE 3). The Zn(II)/Ni(II), Zn(II)/Co(II), and Co(II)/Fe(III) combinations exhibit 2-3 fold cooperative acceleration. In contrast, no measurable synergetic effects were observed when Zn(II) was combined with Mg(II), Ca(II), Pb(II), Na(I), K(I), and Rb(I). The combinations of Fe(III) with Sn(IV), Al(III), Cr(III), In(III), and Ni(II) (as well as with alkali metal and alkaline earth metal ions) were not cooperative either. Only the specific combinations of two metal ions give rise to remarkable catalysis.

Thus, RNA is for the first time promptly hydrolyzed by the bimetallic combinations of only non-lanthanide metal ions. The use of lanthanide ion as a component is not a definite requisite any more for the efficient hydrolysis of phosphoesters.

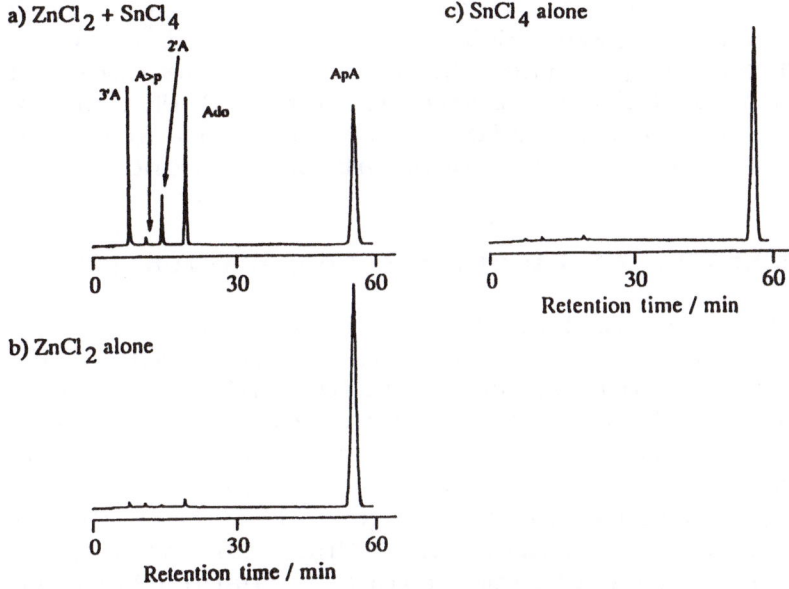

Figure 4. Reversed–phase HPLC profiles for the hydrolysis of ApA at pH 7 (Hepes buffer) and 50°C for 3 h: (a) ZnCl$_2$ (10 *mM*) + SnCl$_4$ (10 *mM*), (b) ZnCl$_2$ alone (10 *mM*), and (c) SnCl$_4$ alone (10 *mM*).

TABLE 3. Cooperation of Zn(II) (10 *mM*) with Sn(IV), In(III), Fe(III), and Al(III) (10 *mM*) for the hydrolysis of ApA at pH 7 and 50°C

Catalyst	First–order rate constant (10^{-2} h^{-1})	Magnitude of cooperative acceleration [a]
Zn(II) + Sn(IV)	21	26
Zn(II)	0.6	
Sn(IV)	0.2	
Zn(II) + In(III)	14	4.0
In(III)	2.9	
Zn(II) + Fe(III)	31	3.2
Fe(III)	9.1	
Zn(II) + Al(III)	4.6	2.6
Al(III)	1.2	

a. Ratio of the rate constant for the catalysis by the bimetallic combination to the sum of the values for each of the metal ions.

Furthermore, a dinuclear zinc(II) complex which hydrolyzes RNA at pH 7 and 30°C has been prepared [24]. The structure of the complex has been concretely characterized by ^{1}H-NMR spectroscopy. The monomeric analog has no activity so that the bimetallic cooperation is conclusive.

3. Mechanism of Bimetallic Cooperation [20-23]

In the present bimetallic synergetic catalyses, the two metal ions form mixed hydroxide clusters in the reaction mixtures (*vide infra*). According to the kinetic and spectroscopic analyses, the molar ratios of the metal ions in the mixed clusters are 2:1 for the Ce(IV)/Pr(III) and the Ce(IV)/Nd(III) systems, and 1:1 for the other bimetallic combinations. Thus the two kinds of metal ions are located close to each other therein and can show cooperative catalysis, as is the case in natural phosphoesterases.

For example, $FeCl_3$ is readily precipitated as polymeric aggregates of hydroxide clusters in neutral to alkaline solutions. However, the precipitation is greatly suppressed by the addition of $ZnCl_2$. When $[ZnCl_2]/[FeCl_3] > 1$, homogeneous mixtures can be prepared. Apparently, the specific interaction between the two metal ions prevents the formation of polymeric aggregates. Similar results were obtained for the other bimetallic synergetic systems, confirming the formation of the mixed clusters.

3.1. MECHANISM OF RNA HYDROLYSIS

The most plausible mechanism for the RNA hydrolysis by the Zn(II)/Sn(IV) combination is schematically depicted in *Figure 5*. Here, the hydroxide ion bound to the Zn(II) in the mixed hydroxide cluster promotes, as a general base catalyst, the intramolecular attack of the 2'-OH towards the phosphorus atom. The reaction is further assisted by the acid catalysis of the water, bound to the Sn(IV) in the mixed cluster. The assignment of the two types of catalyses to the metal ions is based on the smaller pK_a of the Sn(IV)-bound water than that of the Zn(II)-bound water (-0.6 vs. 8.2). Under the reaction conditions, the coordination water molecules of Sn(IV), except for one, exist in the form of Sn-OH$_2$. These water molecules function as active acid catalysts, since they release protons when the metal-bound hydroxide is protonated in the rapid proton-exchange process. Furthermore, the positive charges of the metal ions electrostatically stabilize the negatively charged transition states for the hydrolysis. The cooperation renders the stable phosphodiester linkages susceptible to hydrolysis.

The mechanism is supported by the following results. (1) An Sn(IV) ion, which is the most eminent for the synergism with Zn(II), has the most acidic coordination water. (2) The Fe(III)/Sn(IV) combination shows no synergetic catalysis, in spite of the remarkable catalysis by the Zn(II)/Sn(IV) system. The Fe(III)-bound hydroxide is much poorer general base catalyst than the Zn(II)-bound one, as estimated from the pK_a value (2.5-3.1) of the coordination water.

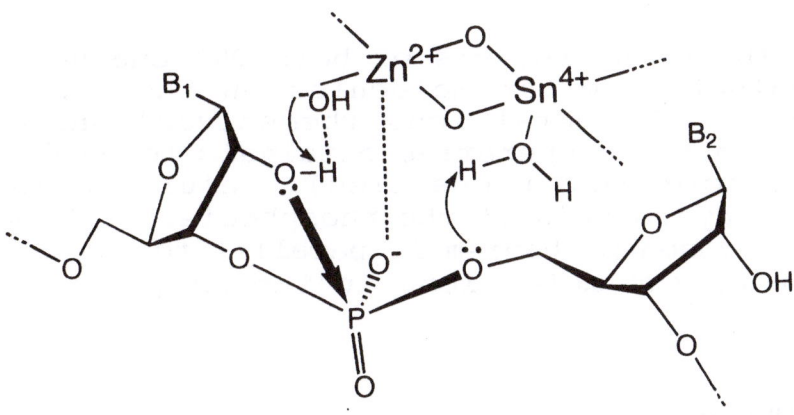

Figure 5. Proposed mechanism for the hydrolysis of RNA by the Zn(II)/Sn(IV) combination.

3.2. MECHANISM OF DNA HYDROLYSIS

The hydrolysis of DNA by the Ce(IV)/Pr(III), Ce(IV)/Nd(III), and La(III)/Fe(III) combinations proceeds in the same way as the RNA hydrolysis, except for the fact that the metal-bound hydroxide directly attacks the phosphorus atom as a nucleophile, rather than functions as a general base catalyst (*Figure 6*).

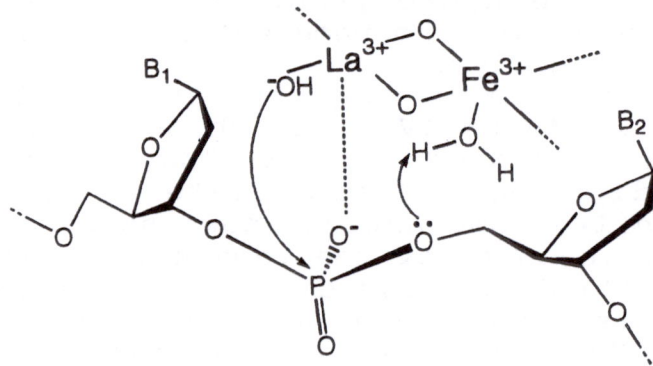

Figure 6. Proposed mechanism for the hydrolysis of DNA by the La(III)/Fe(III) combination.

In the hydrolysis of both DNA and RNA, the coordination water molecules on the two metal ions (or the metal ions themselves) share the roles in the cooperative catalysis (depending on their acid-base properties), resulting in the prompt hydrolysis of the phosphodiester linkage. The proposed mechanism is parallel to the cooperation of two metal ions in the enzymatic reactions.

4. Conclusion

By use of the cooperation of two metal ions, enormously active catalysts for the hydrolysis of

DNA and RNA are now in hand. The typical combinations are as follows: (1) Ce(IV)/Pr(III) and Ce(IV)/Nd(III) for DNA hydrolysis (the most active catalysts ever reported), (2) La(III)/Fe(III) and La(III)/Sn(IV) for the hydrolysis of DNA and RNA, and (3) Zn(II)/Sn(IV) and Zn(II)/In(III) for RNA hydrolysis. Their catalytic activities can be regulated by the modulation of the mutual orientation of the two metal ions. These findings pave the way to the further development of the field. A study on dinuclear complexes of these two metal ions is currently under way in our laboratory.

Acknowledgements

This work was partially supported by a Grant-in-Aid for Scientific Research on Priority Areas "New Development of Rare Earth Complexes" No. 06241103 and 06241107 from The Ministry of Education, Science, and Culture.

REFERENCES

1. Matsumoto, Y. and Komiyama, M. (1992) DNA hydrolysis by rare–earth metal ions, *Nucleic Acids, Symp. Ser.*, **27**, 33–34.
2. Matsumoto, Y. and Komiyama, M. (1992) DNA hydrolysis by rare earth metal ions, *Chem. Express* **7**, 785–788.
3. Komiyama, M., Matsumura, K., Yonezawa, K., and Matsumoto, Y. (1993) Consecutive catalysis by cerium(III) ion for complete hydrolysis of phosphodiester linkage in DNA, *Chem. Express*, **8**, 85–88.
4. Komiyama, M., Matsumoto, Y., Hayashi, N., Matsumura, K., Takeda, N., and Watanabe, K. (1993) Hydrolysis of oligoDNAs by lanthanide metal(III) chloride, *Polym. J.* **25**, 1211–1214.
5. Shiiba, T., Yonezawa, K., Takeda, N., Matsumoto, Y., Yashiro, M., and Komiyama, M. (1993) Lanthanide metal complexes for the hydrolysis of linear DNAs, *J. Mol. Catal.* **84**, L21–L25.
6. Komiyama, M., Yashiro, M., Matsumoto, Y., Sumaoka, J., and Matsumura, K. (1993) Lanthanide metal ions for remarkably efficient hydrolysis of DNA, RNA, and adenosine 3',5'–cyclic phosphate, *J. Chem. Soc. Jpn.*, 411–417.

7. Komiyama, M., Kodama, T., Takeda, N., Sumaoka, J., Shiiba, T., Matsumoto, Y., and Yashiro, M. (1994) Catalytically active species for $CeCl_3$–induced DNA hydrolysis, *J. Biochem.* **115**, 809–810.

8. Komiyama, M., Shiiba, T., Kodama, T., Takeda, N., Sumaoka, J., and Yashiro, M. (1994) DNA hydrolysis by cerium(IV) does not involve either molecular oxygen or hydrogen peroxide, *Chem. Lett.*, 1025–1028.

9. Takasaki, B.K. and Chin, J. (1994) Cleavage of the phosphate diester backbone of DNA with cerium(III) and molecular oxygen, *J. Am. Chem. Soc.*, **116**, 1121–1122.

10. Komiyama, M., Takeda, N., Uchida, H., Takahashi, Y., Shiiba, T., Kodama, T., and Yashiro, M. (1994) Efficient and oxygen–independent hydrolysis of single–stranded DNA by Ce(IV) ion, *J. Chem. Soc., Perkin Trans. 2*, 269–274.

11. Komiyama, M., Matsumura, K., and Matsumoto, Y. (1992) Unprecedentedly fast hydrolysis of the RNA dinucleoside monophosphates ApA and UpU by rare earth metal ions, *J. Chem. Soc., Chem. Commun.*, 640–641.

12. Yoshinari, K., Yamazaki, K., and Komiyama, M. (1991) Oligoamines as simple and efficient catalysts for RNA hydrolysis, *J. Am. Chem. Soc.*, **113**, 5899–5901.

13. Komiyama, M., Takeda, N., Shiiba, T., Takahashi, Y., Matsumoto, Y., and Yashiro, M. (1994) Rare earth metal ions for DNA hydrolyses and their use to artificial nucleases, *Nucleosides and Nucleotides*, **13**, 1297–1309.

14. Komiyama, M., Shiiba, T., Takahashi, Y., Takeda, N., Matsumura, K., and Kodama, T. (1994) Cerium(IV) complex–oligoDNA hybrid as highly selective artificial nuclease, *Supramol. Chem.*, **4**, 31–34.

15. Komiyama, M. (1995) Sequence–specific and hydrolytic scission of DNA and RNA by lanthanide complex–oligoDNA hybrids, *J. Biochem.*, **118**, 665–670 (JB Review).

16. Matsumura, K., Endo, M., and Komiyama, M. (1994) Lanthanide complex–oligo DNA hybrid for sequence–selective hydrolysis of RNA, *J. Chem. Soc., Chem. Commun.*, 2019–2020

17. Komiyama, M., Inokawa, T., and Yoshinari, K. (1995) Ethylenediamine–oligoDNA hybrid as sequence–selective artificial ribonuclease, *J. Chem. Soc., Chem. Commun.*, 77–78.

18. Inokawa, T. and Komiyama, M. (1994) Selective hydrolysis of tRNA by ethylenediamine bound to a DNA oligomer, *J. Biochem.*, **116**, 719–720.

19. Tsubouchi, A. and Bruice, T.C. (1995) Remarkable (10^{13}) rate enhancement in phosphonate ester hydrolysis catalyzed by two metal ions, *J. Am. Chem. Soc.*, **116**, 11614–11615 and references therein.

20. Takeda, T., Imai, T., Irisawa, M., Sumaoka, J., Yashiro, M., and Komiyama, M. (1995) Synergism of Ce(IV) and Pr(III) for the fastest non–enzymatic DNA hydrolysis ever, *Nucleic Acids, Symp. Ser.*, in press.
21. Irisawa, M., Takeda, N., and Komiyama, M. (1995) Synergetic catalysis by two non–lanthanide metal ions for hydrolysis of diribonucleotides, *J. Chem. Soc., Chem. Commun.*, 1221–1222.
22. Irisawa, M. and Komiyama, M. (1995) Hydrolysis of DNA and RNA through cooperation of two metal ions: A novel mimic of phosphoesterases, *J. Biochem.*, **117**, 465–466.
23. Takeda, N., Irisawa M., and Komiyama, M. (1994) Cooperation of lanthanum ion and non–lanthanide metal ion for the hydrolysis of bis(4–nitrophenyl)phosphate, *J. Chem. Soc., Chem. Commun.*, 2773–2774.
24. Yashiro, M., Ishikubo, A. and Komiyama, M. (1995) Dinuclear zinc(II) complex for the efficient hydrolysis of the phosphodiester linkage in a diribonucleotide, *J. Chem. Soc., Chem. Commun.*, in press.

TEXAPHYRIN-BASED NUCLEASE ANALOGUES. RATIONALLY DESIGNED APPROACHES TO THE CATALYTIC CLEAVAGE OF RNA AND DNA TARGETS.

D. MAGDA, R.A. MILLER, M. WRIGHT, J. RAO
Pharmacyclics, Inc.
Sunnyvale, CA 94086

J.L. SESSLER, B.L. IVERSON, P.I. SANSOM
University of Texas at Austin
Austin, Texas 78712

1. Introduction

The concept of using DNA or DNA analogues as chemotherapeutics has origins dating back to the 1960's.[1] This approach did not gain serious recognition, however, until the pioneering efforts of Ts'o and Miller[2] and Zamecnik[3,4] and their coworkers in the mid-1970's. Since the emergence of automated synthesizers,[5] which made DNA more available, research in this area has increased at an exponential rate. Several comprehensive reviews of this literature are available.[6-12]

As originally conceived, the DNA or its analogues would bind to the complementary RNA (or, later, DNA) target and act as a competitive inhibitor of gene expression. In theory, it seemed apparent that if the exquisite specificity of base pairing could be harnessed, these agents might truly perform as "magic bullets". The ability to target at the genotypic, as opposed to the phenotypic level, suggested that this approach would have special significance in both antiviral and anticancer areas, where the discrimination between healthy and diseased cells is more difficult. In addition, one would anticipate broad application, as such agents could presumably be made to bind to nearly any complementary nucleic acid target merely through altering their sequence. Although much experimentation has demonstrated that internal[13] and higher order[14,15] structural factors can impose serious limitations on this simplistic view of hybridization, the general idea remains sound.

It rapidly became apparent in practice that a number of interrelated properties, such as nuclease stability, lipophilicity, and length, were tied to efficacy in cell culture. The particular site targeted also seemed important. A further complication was non-sequence specific activity observed with certain DNA analogues. Even taking these considerations into account, it was obvious that the understanding of why certain oligonucleotides showed efficacy and others did not was incomplete. Gradually, there emerged a generalized understanding that the initial assumptive mechanism of competitive inhibition was relatively unimportant in a chemotherapy context,[16] and that a different mechanism was responsible for the observed activities of antisense agents.

B. Meunier (ed.), DNA and RNA Cleavers and Chemotherapy of Cancer and Viral Diseases, 337–353.
© *1996 Kluwer Academic Publishers. Printed in the Netherlands.*

Depending on the cell culture examined, an activity termed RNase H is present to a greater or lesser extent.[17] This enzyme catalyzes hydrolysis of the RNA portion of a RNA/DNA heteroduplex, functionally removing RNA primers during replication. RNase H is also produced during retroviral infection, where it abets the process of reverse transcription. The significance of this activity to antisense chemotherapy is that it can act as an adjuvant with extraneous oligodeoxynucleotides, degrading the RNA to which they hybridize.[18,19] The presence of RNase H thus explains why DNA and certain analogues, such as the phosphorothioates,[20] can display potent activity in cell culture. Other analogues, which may be of identical sequence and length, may be ineffectual as they are unable to act in concert with this enzyme. In retrospect, it seems unsurprising that short oligomers at low concentration should be incapable of competitively inhibiting a large macromolecule such as a ribosome. *Catalytic cleavage* of the RNA, however, which in effect serves to "tear up the track", cannot fail to block expression of the RNA.

The recent application of ribozyme chemistry to gene therapy also has as its basis the catalytic cleavage of RNA. Ribozymes were originally discovered as intervening sequences of RNA capable of self-splicing by means of an intramolecular transesterification reaction.[21] The study and alteration of these sequences has led to shortened forms which can recognize and hydrolyze complementary RNA sequences in an intermolecular reaction.[22-25] These sequences can be expressed in cells transfected with a ribozyme-coding vector, and have been demonstrated to result in the specific elimination of the complementary RNA's targeted.[26-29] Further development of ribozyme derivatives may lead to agents which can be administered as a chemotherapeutic. Such an approach could lead to drugs which would effect the catalytic cleavage of RNA in a manner independent of RNase H activity.

The efficacy of RNase H and ribozyme based methods validates the principle of using catalytic cleavage of RNA to control gene expression. Both approaches, however, have drawbacks which may limit their application in practice. DNA in itself is insufficiently stable in vivo to be of use as a chemotherapeutic.[30] The phosphorothioates are stable,[31] but have been demonstrated to bind polymerases[32-34] and in general exhibit a substantial background inhibition of gene expression at fairly low doses.[20,31,35] In addition, there is an inherent limit to the sequence specificity of the RNase H-dependent agents, since regions of partial complementarity can serve as substrate for the enzyme.[36] These problems have been only partly solved through the use of "chimeric" analogues, in which outlying regions of the agent have a different backbone modification which is not recognized by the enzyme.[37-39] On the other hand, at present the ribozyme approach is limited to gene therapy,[26-29] and it is not clear whether chemotherapeutic agents will be derived which are stable to environmental nucleases, are able to traverse the barriers to cell penetration, and still retain activity. In view of the substantial higher order structure required for catalysis, the transport barrier may prove to be the most formidable obstacle to their development.

In this paper a third approach to the catalytic cleavage of RNA is described, in which the ability to hydrolyze RNA is incorporated into the antisense agent by virtue of covalent attachment of a lanthanide(III) texaphyrin complex. We have termed such agents *ribozyme analogues*, as they would possess specificity and cleavage properties similar to those of the ribozymes. Inclusion of the catalytic metal core as an integral component of the agent, however, would functionally replace such structural architecture of ribozymes required in order to sequester Mg(II) ion(s) within the active site.[40-42] DNA analogue-lanthanide(III) texaphyrin conjugates of this type would as a consequence need only be of sufficient length to recognize specifically the RNA sequence targeted, ie., 12-15 bases.[36,43] As they would be expected to act independently of cellular RNase H, this would permit any of the available DNA backbone analogues to be incorporated into their design. These properties of ribozyme analogues suggest that they may avoid the major deficiencies of the other catalytic cleavage approaches in having improved cellular uptake, potent activity, and reduced toxicity. The development of such agents could therefore represent an important advance in the chemotherapeutic treatment of disease using the antisense strategy.

2. Sequence-Specific Hydrolysis of RNA

Lanthanide[44-59] and transition metal[60-79] cations and their complexes are known to catalyze RNA hydrolysis. In particular, the 1:1 lanthanide(III) complexes of the hexa-aza macrocycle (HAM, Figure 1A) have formed the basis for a number of studies,[52,53,55] exhibiting rates which may be considered state-of-the-art for phosphate ester hydrolysis. We speculated that the lanthanide(III) complexes of the texaphyrin ligand (Figure 1B), which have the advantage of being stable under physiologic conditions, might also serve as catalysts. As the total charge differs between the two types of complex (3+ for the HAM, 2+ for the texaphyrins), it was anticipated that different catalytic properties might also be observed.

Figure 1

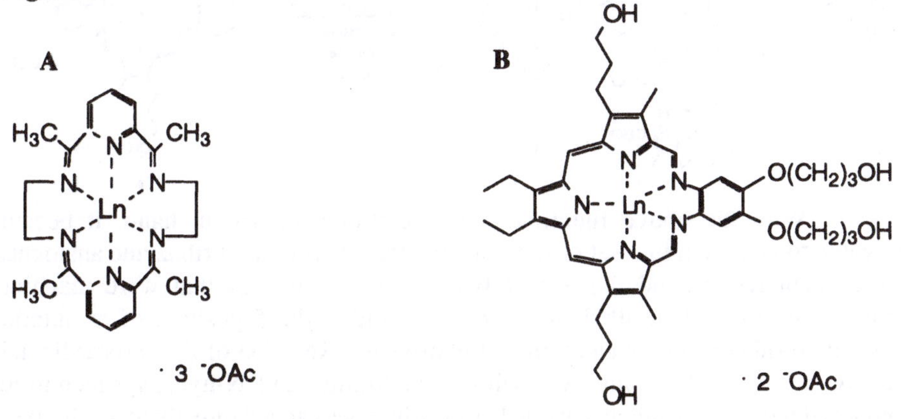

Ln = Gd(III), Eu(III), or Lu(III)

Initial studies focused on comparing the bimolecular RNA hydrolysis activity of lanthanide(III) texaphyrin and HAM complexes. Towards this end, an assay was selected that involved incubating a ca. 2 Kb run-off RNA transcript in the presence of metal complex at 37 °C and observing the resulting degradation of this fragment by electrophoresis. Such experiments indicated that lanthanide(III) texaphyrins do indeed show cleavage activity, with europium(III) texaphyrin (EuTx) catalyzing roughly similar amounts of RNA degradation as free Eu(III) or EuHAM. Of the three cations studied initially as their texaphyrin adducts, (Gd(III), Eu(III), Lu(III)), Eu(III) was found to be most active, and lutetium(III) the least. Equimolar EDTA inhibited the activity of the free lanthanide(III) cations but not that of the texaphyrin or HAM complexes.

Having established that EuTx catalyzes the hydrolysis of RNA in an intermolecular reaction at least as well as EuHAM, the next step consisted of synthesizing a EuTx derivative that could be covalently attached to an oligodeoxynucleotide. This was done via the preparation of the monocarboxylic acid substituted texaphyrin complex as outlined in Scheme 1. Here, condensation of the appropriate aromatic diamine, formed in situ, with a diformyltripyrrane, generated the hydrochloride salt of the macrocyclic texaphyrin precursor. Metallation and concomitant oxidation produced the fully aromatized EuTx complex bearing an acetoxy substituent. Zeolite was used during workup to remove excess Eu(III), while axial acetate ligands, which could interfere with the subsequent DNA conjugation, were replaced via ion exchange chromatography.

Scheme 1

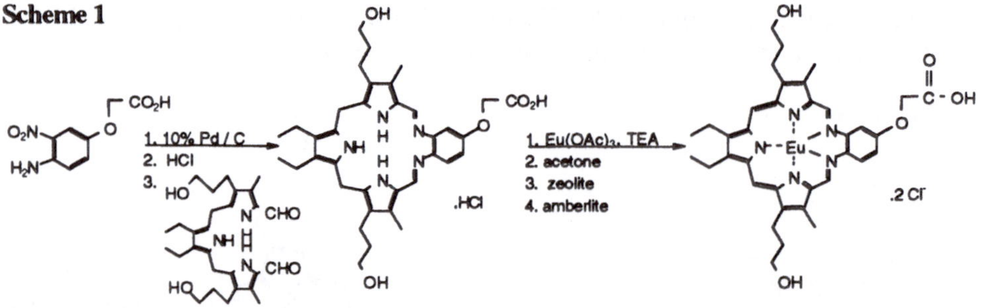

With the above functionalized metallotexaphyrin in hand, it became possible to prepare a first set of four putative texaphyrin-based ribozyme analogues. These centered around the use of two 20-mer sequences that were machine-synthesized to contain alkylamine groups at either the 5-position of an internal thymine residue or the 5'-end terminal phosphate. Reaction of the carboxylic acid functionalized EuTx complex 1 with carbodiimide and N-hydroxysuccinimide produced the corresponding activated ester, which was added directly to a solution of the chosen oligodeoxynucleotide amine. The resulting DNA-EuTx conjugates (2-5, Figure 2) were purified by electrophoresis. The presence of the EuTx complex resulted in an observable reduction in gel mobility, and could be affirmed by virtue of its characteristic UV-vis absorbances at ca. 460 and 760 nm.

Figure 2

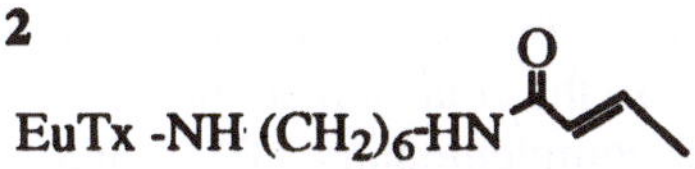

EuTx -NH·(CH₂)₆-HN—(structure with C=O)

$$\text{EuTx -NH·(CH}_2)_6\text{-HN}-\overset{\displaystyle O}{\overset{\|}{C}}\diagup\diagdown$$

5'- CAT CTG TGA GCC GGG TGT TG-3'   **2**

5'-EuTx -NH(CH₂)₆·PO₄–CAT CTG TGA GCC GGG TGT TG-3'   **3**

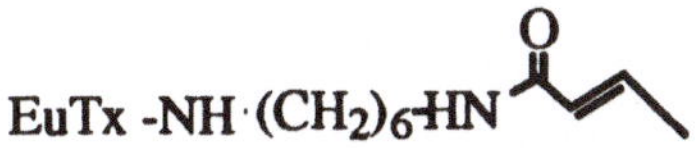

EuTx -NH·(CH₂)₆-HN—(structure with C=O)

5'- CTC GGC CAT AGC GAA TGT TC-3'   **4**

5'-EuTx -NH (CH₂)₆·PO₄–CTC GGC CAT AGC GAA TGT TC-3'   **5**

5'- H₂N·(CH₂)₆–PO₄–CTC GGC CAT AGC GAA TGT TC-3'   **6**

We next attempted the sequence specific hydrolysis of an RNA target. Towards this end, a synthetic RNA 30-mer (7, Figure 3) was prepared as substrate, with a sequence selected from a unique site within the gene transcript for multiple drug resistance (MDR). The 3'-^{32}P-labeled substrate was incubated with an excess of oligodeoxynucleotide conjugate at 37 °C for 18-24 h in a buffered salt solution, ethanol precipitated, and assayed on a 20% denaturing polyacrylamide gel. As illustrated schematically in Figure 3 and in more explicit detail in an earlier-published report,[44] a ca. 30% cleavage occurred near the expected location of the europium (III) texaphyrin complex upon hybridization with conjugate **5**. The corresponding cleavage bands were not observed when this same substrate was incubated with oligonucleotides **2-4**, or **6** that were non-complementary in sequence, unmodified, or were modified internally with the complex. This last result was expected, as duplex regions of RNA are known to resist hydrolysis due to conformational restrictions imposed by the secondary structure.[53] Other control reactions indicated that ambient light, calf thymus DNA, type of buffer (Tris acetate or HEPES, EDTA) or oxygen had no apparent effect on cleavage efficiency.

The cleavage fragments produced by the EuTx co-migrate with bands in sequencing lanes produced by incubation of the 3'-labeled substrate under alkaline conditions or subjected to partial digestion with a series of base-specific ribonucleases.[25,80] This observation is consistent with a hydrolytic mechanism, as 3'-labeled fragments generated via oxidative damage would remain phosphorylated on their 5'-end and show corresponding altered mobility on the 20% polyacrylamide gel employed.

Maximal cleavage activity of the DNA-EuTx was observed down to 25 nM conjugate. Slightly decreased cleavage was observed when the conjugate concentration was reduced to 2.5 nM. At this lower conjugate level, the reduction in conjugate-mediated cleavage is considered to arise from incomplete hybridization (as evidenced by increased background cleavage of the target RNA, which is present at ca. 1 nM). By means of comparison, the free EuTx complex 1 non-specifically hydrolyzed the RNA substrate at 25 μM. In the control reaction containing both free EuTx complex and the non-derivatized complementary DNA oligomer, cleavage occurred predominantly in the single stranded region, although still at lower efficiency than the Eu-Tx-DNA conjugate at 2.5 nM. Thus, attachment of the EuTx to the DNA probe increased its effective concentration more than 10,000-fold.

Figure 3

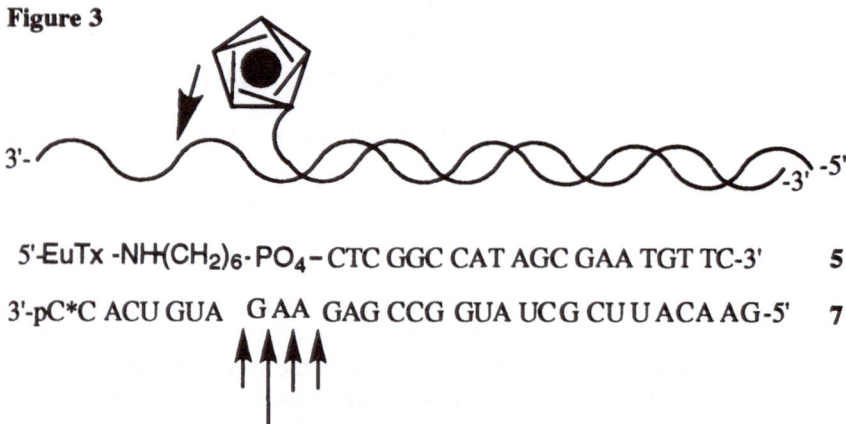

5'-EuTx -NH(CH$_2$)$_6$·PO$_4$ – CTC GGC CAT AGC GAA TGT TC-3'   **5**

3'-pC*C ACU GUA G AA GAG CCG GUA UCG CU U ACA AG-5'   **7**

The sequence specific hydrolysis of RNA "in trans" validates the concept of using lanthanide(III) texaphyrins as ribozyme analogues. A major concern was that the attachment of the complex to the DNA would inhibit its ability to bind to and cleave the RNA, but it is apparent that the ligand exchange rate of the EuTx is sufficiently rapid to allow recomplexation of the RNA backbone. It is important to note that the assay was carried out under salt and pH conditions which approximate those found in vivo, and that calf thymus DNA did not interfere. These results augur well for the use of such ribozyme analogues in an antisense context and, as importantly, inspired us to try our hand at preparing yet-improved hydrolysis systems.

As a first step towards improving the cleavage efficiency of the ribozyme analogues, dysprosium(III) texaphyrin (DyTx) analogues of the 20-mers used in the original study were prepared (**8** and **9**, Figure 4). The RNA 36-mer **10**, complementary to the control oligomers used in the original study (also a unique site in human MDR), was also prepared, in order to demonstrate further the sequence specificity of the agents. Cleavage studies, resulting from incubation of the two RNA substrates with the two new DyTx conjugates as compared to their EuTx analogues **3** and **5**, revealed that replacement of EuTx by DyTx approximately doubles the cleavage yield, to ca. 60% after 18 h at 37 °C. As anticipated, the DyTx ribozyme analogues, like their EuTx congeners, only hydrolyze their appropriate complementary RNA substrate.

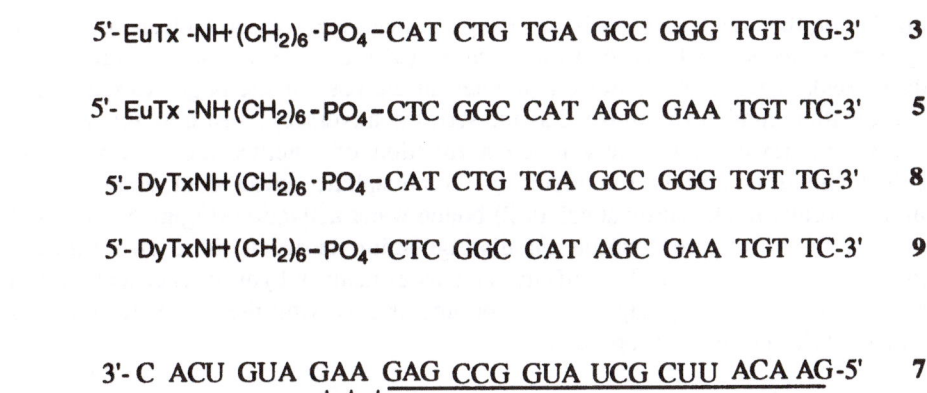

Figure 4. Synthetic RNA substrates **7** and **10** used as targets for sequence-specific hydrolysis by complementary EuTx-DNA conjugates **3** and **5** and DyTx-DNA conjugates **8** and **9**. Hybridized regions of RNA are underlined. The arrows indicate positions of hydrolysis.

The above "compare and contrast" results are significant on at least two levels. First, they provide a direct indication that the choice of the central chelated metal center is critical and that, at least in the case of the slower EuTx systems, cleavage efficiency is not necessarily limited by inherent features associated with either the texaphyrin macrocycle (size, charge, etc.) or conjugate-target binding affinity. Second, they provide support for the contention that this same texaphyrin conjugate approach could be made more efficacious simply by optimizing the choice of metal. In other words, the fact that EuTx and analogous DyTx conjugates were found to cleave identical targets at different rates meant that a full survey of the Lewis acidic lanthanide(III) cations was in order.

To date, the following trivalent lanthanide cations have been attached to RNA-targeting oligodeoxynucleotide sequences: Nd, Eu, Gd, Tb, Dy, Ho, Er, Tm, Lu. With these in hand, it became possible to ascertain relative cleavage efficiencies for the resulting Ln(III) texaphyrin DNA conjugates. This was done by comparing the absolute cleavage of two complementary RNA targets analogous to those described above after a defined incubation time. In this way, the following efficacy order was obtained for trivalent lanthanide cations when encapsulated within the texaphyrin core: Nd<<Eu<Gd<Tb ≈ Dy > Ho > Er > Tm > Lu. This order stands in marked contrast to what is observed for the free Ln(III) aquo ions when tested as dinucleotide hydrolysis catalysts in an intermolecular sense.[49]

The exact reasons for the observed texaphyrin conjugate-based metal dependence remain recondite. However, the clear periodic trend observed as the lanthanide series is traversed, wherein efficiency increases initially, reaches a maximum, then decreases, leads us to suggest that metal-centered efficacy in these systems reflects a balance of Lewis acidity (increased as the size of the cation decreases), steric accessibility (augmented in the case of the larger cations), and coordination number (diminished as the series is traversed). Just how this balance plays out, however, is likely to be a function of whether the observed rate enhancement is largely the result of 1) phosphate anion complexation and intramolecular nucleophilic attack or 2) bound water activation (Figure 5). Current efforts, therefore, are focused on distinguishing between these two limiting mechanistic possibilities. Towards this end, assesments of hydrolysis efficiency as a function of pH are ongoing, as are experiments involving the use of texaphyrins with modified electronic properties.

Figure 5

General Base Mechanism

Lewis Acid Mechanism

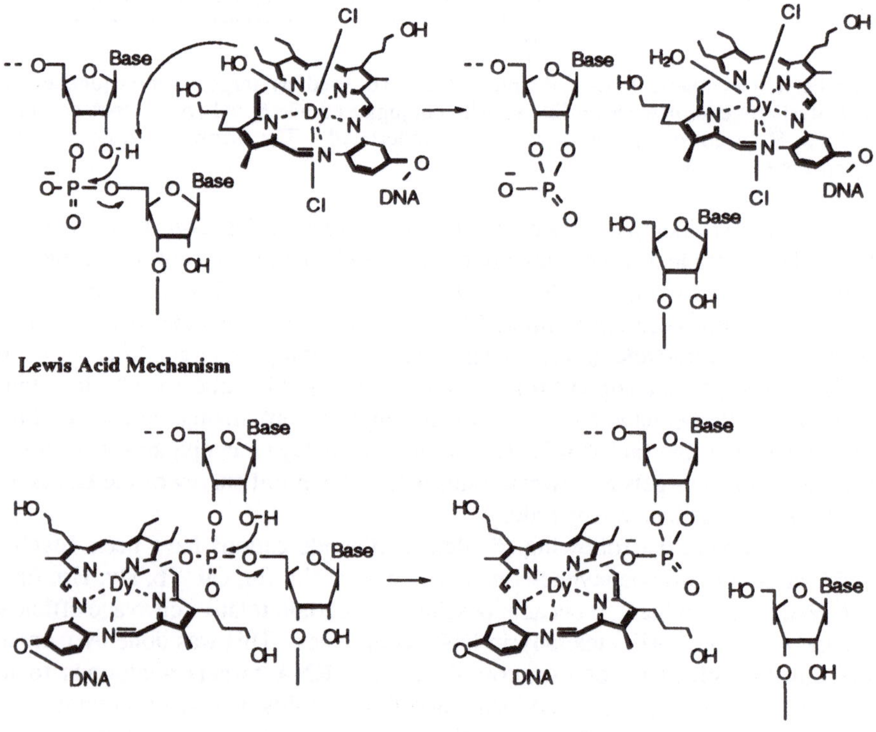

3. Photolytic Approaches

As mentioned above, the lutetium(III) texaphyrin (LuTx) complex hydrolyzed RNA in the intermolecular assay at a slower rate than all of the other metals studied. This result also held for LuTx-DNA conjugates bound to RNA. This, in turn, leads to the rather obvious conclusion that this particular metal cation is less than ideal as far as *hydrolysis*-based cleavage applications are concerned. However, it is to be appreciated that lutetium(III) differs from other trivalent lanthanide cations in that it is diamagnetic, which gives rise to unique photochemical properties. In particular, this particular diamagnetic texaphyrin displays a long-lived triplet state and can efficiently produce singlet oxygen upon irradiation.[81] In conjunction with the strong absorbance at ca. 740 nm characteristic of the texaphyrins, this property has led to development of a free LuTx complex as an agent for photodynamic therapy.[82] Studies of RNA cleavage by LuTx in the absence of light have shown that background level hydrolysis is the only significant reaction occurring. Intermolecular treatment of both RNA and DNA with 25 μM LuTx, however, was found to result in substantial light dependent photomodification. The paramagnetic EuTx and DyTx complexes, by contrast, produced no discernable modification of DNA either in the presence or absence of light.

These studies led us to consider the use of LuTx-antisense conjugates as site-specific photocleavers of nucleic acids. In order to test this idea, LuTx derivatives of DNA 15-mers and 2'-O-methyl RNA 15-mers were prepared (Figure 6).[83] Synthetic DNA 36-mers **11** and **12** were selected as substrates, each complementary in sequence to only one of the paired sets of the LuTx DNA and 2'-O-methyl RNA conjugates, respectively (see Figure 6). The 5'-^{32}P-labeled 36-mers were allowed to hybridize with an excess (50 nM) of conjugate in the dark, whereupon samples were irradiated for 15 minutes using a dye laser tuned to 732 nm. No cleavage was detected in the absence of light. The presence of the complementary LuTx conjugates, but not the control non-complementary conjugates, led to photomodification of each of the DNA targets, as assayed by treatment with piperidine. Cleavage products were found to co-migrate exclusively with bands generated by the Maxim-Gilbert sequencing (G) reaction. This observation is consistent with a mechanism involving singlet oxygen generation, which is known to lead to the depurination of guanine.[84-86] The overall cleavage of DNA totaled 70-90% depending on the specific choice of substrate and conjugate.

Interestingly, the patterns of photocleavage differ for the two types of conjugate backbone employed. The sites of reaction primarily lie across the minor groove from the position of LuTx attachment within the duplex, with additional cleavage occurring at sites across the adjacent major groove in reactions involving the use of the 2'-O-methyl backbones. These differences likely reflect the different conformational types of duplex formed between ssDNA target and its ssDNA and RNA complements (B-form vs. A-form, respectively).

Similar experiments were conducted using RNA substrates of identical sequence. In general, these gave results analogous to the above. However, the observed levels of cleavage were found to be lower. This we consider to be a consequence of having substituted aniline for piperidine in the reaction workup (required by the sensitivity of RNA to the latter reagent), rather than intrinsic differences in per photon reactivity of RNA and DNA.

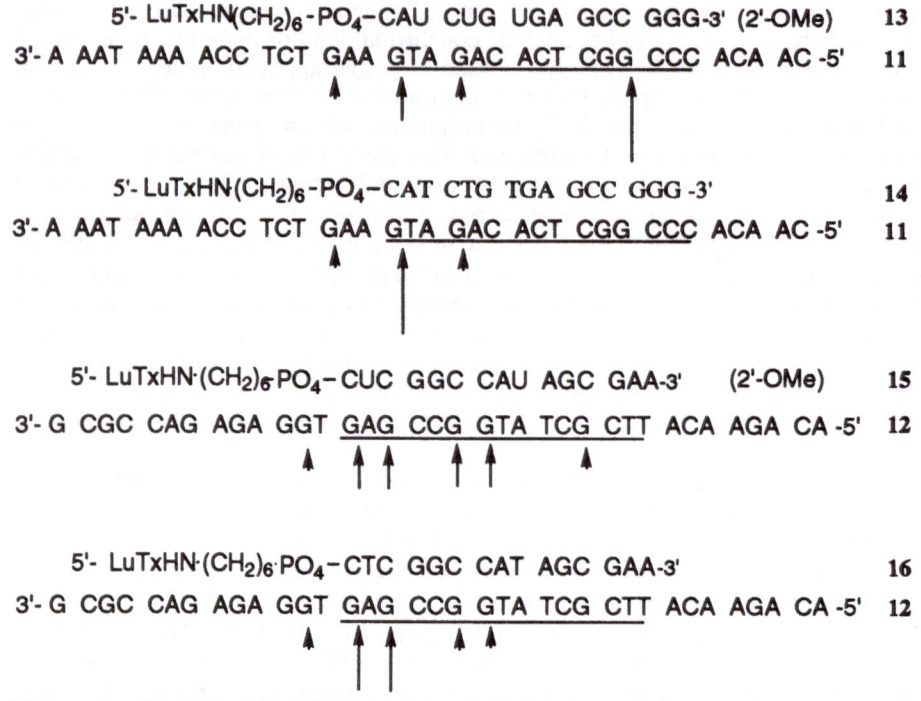

Figure 6. Synthetic DNA 36-mers **11** and **12** used as targets for sequence-specific photocleavage and complementary LuTx-2'-O-methyl RNA conjugates **13** and **15**. and LuTx-DNA conjugates **14** and **16**. The arrows indicate positions of strong, intermediate, or weak DNA modification as exposed by treatment with piperidine.[87]

4. Combined Approaches

The above success in achieving independently both RNA hydrolysis and DNA cleavage, led us to consider whether a texaphyrin-based conjugate could be prepared that would allow for DNA photo-modification *and* RNA hydrolysis using the *same* antisense reagent. Conceptually, this could be achieved using a metallotexaphyrin conjugate that incorporated a diamagnetic (and hence photo-active) metal center with greater hydrolysis-inducing capabilities than Lu(III). Since, however, this latter cation forms the only diamagnetic lanthanide(III) texaphyrin complex, we were led to investigate the use of other, non-lanthanide metal centers. We were fortunate in obtaining some success using the first cation surveyed for this purpose, namely the rare earth cation Y(III).

As above, hydrolysis was demonstrated by incubating target RNA sequences in the presence of complementary and non-complementary YTx conjugates (Cf., Figure 7). In this way, it was determined that the hydrolysis-enhancing activity of YTx is roughly equivalent to that of EuTx. However, unlike this latter europium(III) conjugate, the YTx conjugates were found to be photo-active with regard to both RNA and DNA photo-modification at efficiency levels that are essentially the same as those of the LuTx conjugates. These compounds, which represent a new class of potential antisense agents, could play an important role in situations where a combined hydrolytic/ photolytic approach to antisense is in order.

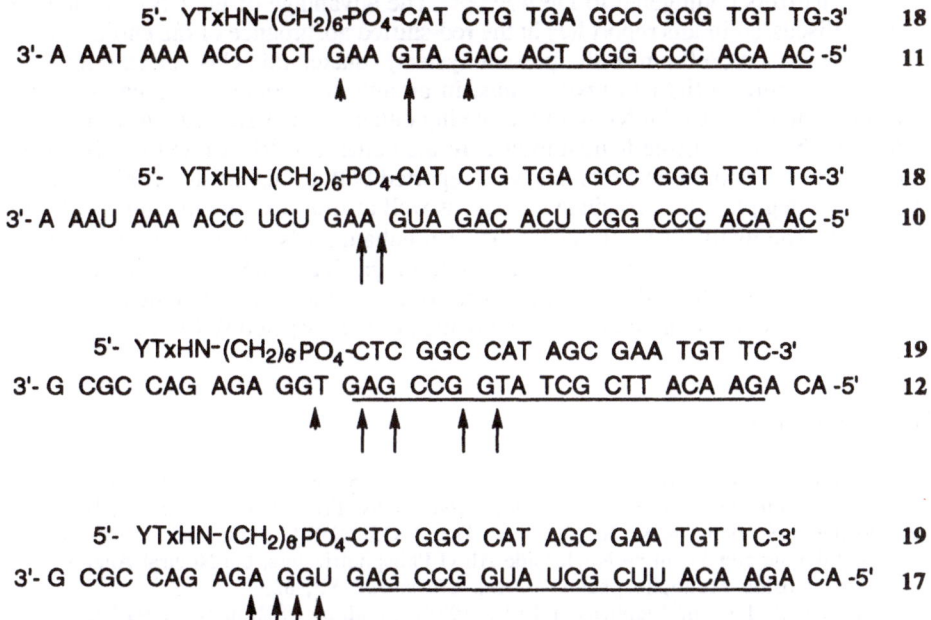

Figure 7. Synthetic DNA 36-mers **11** and **12** and analogous RNA 36-mers **10** and **17** used as targets for sequence-specific photocleavage and and hydrolysis, respectively, by complementary YTx-DNA conjugates **18** and **19**. The arrows indicate positions of DNA modification as exposed by treatment with piperidine or RNA hydrolysis.

5. Future Directions

While the present results are considered very exciting, it is nonetheless clear that greater rates of hydrolysis are required before the full potential of ribozyme analogues can be realized. In this context, it is important to appreciate that in the current system the rates of hydrolysis are not at present limited by the binding of the conjugate to the RNA, but by the intrinsic reactivity of the RNA/DNA complex (K_{cat}). This idea is supported by a study of the RNA hydrolysis rates of EuTx and DyTx conjugates made from DNA oligomers of different length: Conjugates of 12-mers were found to effect hydrolysis at the same rate as the 20-mers studied above, while conjugates of 9-mers were considerably less efficient. Thus, assuming a minimum conjugate chain length is maintained, it is to be expected that augmentations in the texaphyrin catalyst efficiency will translate directly into improved RNA hydrolyzing efficacy. Accordingly, a range of new, modified texaphyrins are being prepared in our laboratories and are being tested vis-a-vis their hydrolysis-enhancing abilities.

Turning to a discussion of the photo-active systems, it is to be noted that DNA and RNA photocleavage had been effected previously using a range of photosensitizers conjugated to DNA.[88-99] The advantage of the DNA-texaphyrin system discussed in this report lies in the red-shifted absorbance of the chromophore. This absorbance (760 nm in the present system) makes this class of chromophore more accessible to light in vivo. Thus, in an antiviral context, one can envision treatment of a localized infection by targeting either single stranded DNA or double standed DNA via a triple helical motif. In the context of RNA photomodification, treatment with photosensitizers has been reported both to inactivate mRNA towards acting as a template for protein synthesis, as well as diminish the infectivity of viral RNA.[84] The availability of agents that can enhance RNA degradation via either hydrolysis or photolysis is considered to be of particular advantage, as treatment protocols which alternatively exploit first one and then the other mechanism of action may insure a more complete inactivation of the targeted RNA species.

6. References

1. Zon, G. (1993) History of Antisense Drug Discovery, in S. T. Crooke and B. Lebleu (Eds.), *Antisense Research and Applications*, CRC Press, Boca Raton, FL, 1-5.
2. Miller, P.S., Fang, K.N., Kondo, N.S., Ts'o, P.O.P. (1971) Synthesis and Properties of Adenine and Thymine Nucleoside Alkyl Phosphotriesters, the Neutral Analogs of Dinucleoside Monophosphates *J. Am. Chem. Soc.* **93**, 6657.
3. Zamecnick, P.C.and Stephenson, M.L. (1978) Inhibition of Rous Sarcoma Virus Replication and Cell Transformation by a Specific Oligodeoxynucleotide *Proc. Natl. Acad. Sci. U.S.A.* **75**, 280.
4. Stephenson, M.L., Zamecnick, P.C.(1978) Inhibition of Rous Sarcoma Viral RNA Translation by a Specific Oligodeoxynucleotide *Proc. Natl. Acad. Sci. U.S.A.* **75**, 285.
5. Alvarado-Urbina, G., Sathe, G.M., Liu, W-C, Gillen, M.F., Duck, P.D., Bender, R., Ogilvie, K.K. (1981) Automated Synthesis of Gene Fragments *Science* **214**, 270.
6. Uhlmann, E.and Peyman, A. (1990) Antisense Oligonucleotides: A New Therapeutic Principle *Chemical Reviews* **90**, 543.
7. Zamecnik, P. C. (1991) Introduction: Oligonucleotide Base Hybridization as a Modulator of Genetic Message Readout, in E. Wickstrom (Ed.), *Prospects for Antisense Nucleic Acid Therapy of Cancer and AIDS*, Wiley-Liss, New York, 1-6.
8. Ts'o, P.O.P., Miller, P.S., Aurelian, L., Murakami, A., Agris, C., Blake, K.R., Lin, S-B, Lee, B.L., Smith, C.C. (1993) "An Approach to Chemotherapy Based on Base Sequence Information and Nucleic Acid Chemistry" *Ann. N. Y. Acad. Sci.* **14**, 220.
9. Cazenave, C., Helene, C. (1991) Antisense Oligonucleotides, in J.N.M. Mol and A.R. van der Krol (Eds.), *Antisense Nucleic Acids and Proteins*, Marcel Dekker, Inc., New York, 47.
10. English, U.and Gauss, D.H. (1991) Chemically Modified Oligonucleotides as Probes and Inhibitors *Angew. Chem. Int. Ed. Engl.* **30**, 613.

11. Tidd, D.M. (1990) A Potential Role for Antisense Oligonucleotide Analogues in the Development of Oncogene Targeted Cancer Chemotherapy *Anticancer Research* **10**, 1169.

12. Zon, G. (1988) "Oligonucleotide Analogues as Potential Chemotherapeutic Agents" *Pharm. Res.* **5**, 539.

13. Lima, W.F., Monia, B.P., Ecker, D.J., Freier, S.M. (1992) "Implication of RNA Structure on Antisense Oligonucleotide Hybridization Kinetics" *Biochemistry* **31**, 12055.

14. Rittner, K., Burmester, C., Sczakiel, G. (1993) In Vitro Selection of Fast-Hybridizing and Effective Antisense RNAs Directed against the Human Immunodeficiency Virus Type 1 *Nucleic Acids Res.* **21**, 1381.

15. Chastain, M. and Tinoco, Jr., I. (1993) RNA Structure as Related to Antisense Drugs in S.T. Crooke and B. Lebleu (Eds.), *Antisense Research and Applications*, CRC Press, Boca Raton, FL, 55.

16. Walder, R.Y., Walder, J.A. (1988) Role of RNase H in Hybrid-Arrested Translation by Antisense Oligonucleotides *Proc. Natl. Acad. Sci. U.S.A.* **85**, 5011.

17. Crouch, R.J.and Dirksen, M.-L. (1982) Ribonuclease H in S. M. Linn and R. J. Roberts (Eds.), *Nucleases*, Cold Spring Harbor Laboratory Press, Cold Spring Harbor, 211.

18. Shuttleworth, J. and Colman, A. (1988) Antisense Oligodeoxyribonucleotide-directed Cleavage of Maternal mRNA in *Xenopus* Oocytes and Eggs *EMBO J.* **7**, 427.

19. Shuttleworth, J., Matthews, G., Dale, L., Baker, C., Colman, A. (1988) Antisense Oligodeoxyribonucleotide-directed Cleavage of Maternal mRNA in *Xenopus* Oocytes and Embryos" *Gene* **72**, 267.

20. Wagner, R.W., Matteucci, M.D., Lewix, J.G., Gutierrez, A.J., Moulds, C., Froehler, B.C. (1993) Antisense Gene Inhibition by Oligonucleotides Containing C-5 Propyne Pyrimidines *Science* **260**, 1510.

21. Cech, T.R.and Bass, B.L. (1986) Biological Catalysis by RNA *Ann. Rev. Biochem.* **55**, 599.

22. Zaug, A., Been, M., Cech, T. (1986) The *Tetrahymena* Ribozyme Acts Like an RNA Restriction Endonuclease *Nature* **324**, 429.

23. Perrotta, A.T.and Been, M.D. (1992) Cleavage of Oligoribonucleotides by a Ribozyme Derived from the Hepatitis δ Virus RNA Sequence" *Biochemistry* **1992**, *31*, 16.

24. Fu, D.-J., Benseler, F., McLaughlin, L.W. (1994) Hammerhead Ribozymes Containing Non-Nucleoside Linkers are Active RNA Catalysts *J. Am. Chem. Soc.* **116**, 4591.

25. Dange, V., Van Atta, R.B., Hecht, S.M. (1990) A Mn^{2+}-Dependent Ribozyme *Science* **248**, 458.

26. Uhlenbeck, O.C. (1993) Using Ribozymes to Cleave RNAs, in S.T. Crooke and B. Lebleu (Eds.), *Antisense Research and Applications*, CRC Press, Boca Raton, FL, 83.

27. Sullenger, B.A.and Cech, T.R. (1993) Tethering Ribozymes to a Retroviral Packaging Signal for Destruction of Viral RNA *Science* **262**, 1566.

28. Sarver, N., Cantin, E.M., Chang, P.S., Zaia, J.A., Ladne, P.A., Stephens, D.A., Rossi, J.J. (1990) Ribozymes as Potential Anti-HIV-1 Therapeutic Agents *Science* **247**, 1222.

29. Yu, M., Ojwang, J., Yamada, O., Hampel, A., Rapapport, J., Looney, D., Wong-Staal, F. (1993) A Hairpin Ribozyme Inhibits Expression of Diverse Strains of Human Immunodeficiency Virus Type 1 *Proc. Natl. Acad. Sci. U.S.A.* **90**, 6340.

30. Cazenave, C., Chevrier, M., Thuong, N.T., Helene, C. (1987) Rate of Degradation of [α]- and [β]-Oligodeoxynucleotides in *Xenopus* Oocytes. Implications for Anti-messenger Strategies" *Nucleic Acids Res.* **15**, 10507.

31. Morvan, F., Porumb, H., Degols, G., Lefebvre, I., Pompon, A., Sproat, B. S., Rayner, B., Malvy, C. (1993) Comparative Evaluation of Seven Oligonucleotide Analogues as Potential Antisense Agents *J. Med. Chem.* **36**, 280.

32. Majumdar, C., Stein, C.A., Cohhen, J.S., Broder, S., Wilson, S.H. (1989) Stepwise Mechanism of HIV Reverse Transcriptase: Primer Function of Phosphorothioate Oligodeoxynucleotide *Biochemistry* **28**, 1340.

33. Cohen, J.S. (1993) Phosphorothioate Oligodeoxynucleotides, in S.T. Crooke and B. Lebleu (Eds.), *Antisense Research and Applications*, CRC Press, Boca Raton, FL, 205.

34. Gao, W., Stein, C.A., Cohen, J.S., Dutschman, G.E., Cheng, Y.-C. (1989) Effect of Phosphorothioate Homo-oligodeoxynucleotides on Herpes Simplex Virus Type 2-Induced DNA Polymerase" *J. Biol. Chem.* **264**, 11521.

35. Stein, C.A., Cheng, Y.-C. (1993) Antisense Oligonucleotides as Therapeutic Agents-Is the Bullet Really Magical? *Science* **261**, 1004.

36. Woolf, T.M., Melton, D.A., Jennings, C.G.B. (1992) Specificity of Antisense Oligonucleotides In Vivo *Proc. Natl. Acad. Sci. U.S.A.* **89**, 7305.

37. Giles, R.V., Spiller, D.G., Tidd, D.M. (1993) Chimeric Oligodeoxynucleotide Analogues: Enhanced Cell Uptake of Structures which Direct Ribonuclease H with High Specificity *Anti-Cancer Drug Design* **8**, 33.

38. Giles, R.V.and Tidd, D.M. (1992) Increased Specificity for Antisense Oligodeoxynucleotide Targeting of RNA Cleavage by RNase H using Chimeric Methylphosphonodiester/Phosphodiester Structures *Nucleic Acids Res.* **20**, 763.

39. Shibahara, S., Mukai, S., Nishihara, T., Inoue, H., Ohtsuka, E., Morisawa, H. (1987) Site-Directed Cleavage of RNA" *Nuc. Acids Research* **15**, 4403.

40. Dahm, S.C., Uhlenbeck, O.C. (1991) Role of Divalent Metal Ions in the Hammerhead RNA Cleavage Reaction *Biochemistry* **30**, 9464.

41. Pley, H.W., Flaherty, K.M., McKay, D.B. (1994) Three-Dimensional Structure of a Hammerhead Ribozyme, *Nature* **372**, 68.

42. Scott, W.G., Finch, J.T., Klug, A. (1995) The Crystal Structure of an All-RNA Hammerhead Ribozyme: A Proposed Mechanism for RNA Catalytic Cleavage, *Cell* **81**, 991.

43. Helene, C., Toulme, J.-J. (1989) Control of Gene Expression by Oligodeoxynucleotides Covalently Linked to Intercalating Agents and Nucleic Acid-Cleaving Reagents, in J.S. Cohen (Ed.),*Oligodeoxynucleotides. Antisense Inhibitors of Gene Expression*, Macmillan Press, London, 137.

44. Magda, D.; Miller, R.A.; Sessler, J.L.; Iverson, B.L. (1994) Site-Specific Hydrolysis of RNA by Europium(III) Texaphyrin Conjugated to a Synthetic Oligodeoxyribonucleotide *J. Am. Chem. Soc.* **116**, 7439-40.

45. Hall, J., Husken, D., Pieles, U., Moser, H.E., Haner, R. (1994) Efficient Sequence-Specific Cleavage of RNA Using Novel Europium Complexes Conjugated to Oligonucleotides *Chemistry & Biology* **1**, 185-190.

46. Takasaki, B. K.and Chin, J. (1994) Cleavage of the Phosphate Diester Backbone of DNA with Cerium(III) and Molecular Oxygen *J. Am. Chem. Soc.* **116**, 1121.

47. Takasaki, B.K., Chin, J. (1995) La(III)-Hydrogen Peroxide Cooperativity in Phosphate Diester Cleavage: A Mechanistic Study *J. Am. Chem. Soc.* **117**, 8582-8585.

48. Breslow, R.and Huang, D.-L. (1991) Effects of Metal Ions, Including Mg^{2+} and Lanthanides, on the Cleavage of Ribonucleotides and RNA Model Compounds" *Proc. Natl. Acad. Sci. U.S.A.* **88**, 4080.

49. Komiyama, M., Matsumura, K., Matsumoto, Y. (1992) Unprecendentedly Fast Hydrolysis of the RNA Dinucleoside Monophosphates ApA and UpU by Rare Earth Metal Ions *J. Chem. Soc., Chem. Commun.* 640.

50. Matsumura, K., Endo, M., Komiyama, M. (1994) Lanthanide Complex-Oligo-DNA Hybrid for Sequence-selective Hydrolysis of RNA *J. Chem. Soc., Chem. Commun.* 2019-2020.

51. Sumaoka, J., Yashiro, M., Komiyama, M. (1992) Remarkably Fast Hydrolysis of 3',5'-Cyclic Adenosine Monophosphate by Cerium(III) Hydroxide Cluster *J. Chem. Soc., Chem. Commun.* 1707.

52. Morrow, J.R.; Buttrey, L.A., Shelton, V.M., Berback, K.A. (1992) Efficient Catalytic Cleavage of RNA by Lanthanide(III) Macrocyclic Complexes: Toward Synthetic Nucleases for in Vivo Applications" *J. Am. Chem. Soc.* **114**, 1903.

53. Kolasa, K.A., Morrow, J.R., Sharma, A.P. (1993) Trivalent Lanthanide Ions Do Not Cleave RNA in DNA-RNA Hybrids *Inorg. Chem.* **32**, 3983.

54. Morrow, J.R., Amin, S., Lake, C.H., Churchill, M.R. (1993) Synthesis, Structure, and Dynamic Properties of the Lanthanum(III) Complex of 1,4,7,10-tetrakis(2-carbamoylethyl)-1,4,7,10-tetraazacyclododecane*Inorg. Chem.* **32**, 4566.

55. Morrow, J.R., Buttrey, L.A., Shelton, V.M., Berback, K.A. (1992) Efficient Catalytic Cleavage of RNA by Lanthanide(III) Macrocyclic Complexes: Toward Synthetic Nucleases for In Vivo Applications *Chemtracts* **4**, 262.

56. Amin, S., Morrow, J.R., Lake, C.H., Churchill, M.R. (1994) Lanthanide(III) Tetraamide Macrocyclic Complexes as Synthetic Ribonucleases: Structure and Catalytic Properties of [La(tcmc)(CF$_3$SO$_3$)(EtOH)](CF$_3$SO$_3$)$_2$" *Angew. Chem. Int. Ed. Engl.* **33**, 773.

57. Hayashi, N., Takeda, N., Shiiba, T., Yashiro, M., Watanabe, K., Komiyama, M. (1993) Site-Selective Hydrolysis of tRNA by Lanthanide Metal Complexes*Inorg. Chem.* **32**, 5899.

58. Schneider, H-J, Rammo, J., Hettich, R. (1993) Catalysis of the Hydolysis of Phosphoric Acid Diesters by Lanthanide Ions and the Influence of Ligands *Angew. Chem. Int. Ed. Engl.* **32**, 1716.

59. Hashimoto, S. and Nakamura, Y. (1995) Nuclease Activity of a Hydroxamic Acid Derivative in the Presence of Various Metal Ions *J. Chem. Soc., Chem. Commun.* 1413-1414.

60. Bashkin, J.K., Frolova, E.I., Sampath, U. (1994) Sequence-Specific Cleavage of HIV mRNA by a Ribozyme Mimic *J. Am. Chem. Soc.* **116**, 5981.

61. Modak, A.S., Gard, J.K., Merriman, M.C., Winkeler, K.A., Bashkin, J. K., Stern, M.K. (1991) Toward Chemical Ribonucleases. 2. Synthesis and Characterization of Nucleoside-Bipyridine Conjugates. Hydrolytic Cleavage of RNA by Their Copper(II) Complexes *J. Am. Chem. Soc.* **113**, 283.

62. Stern, M.K., Bashkin, J.K., Sall, E.D. (1990) Hydrolysis of RNA by Transition-Metal Complexes *J. Am. Chem. Soc.* **112**, 5357.

63. Baker, B.F. (1993) 'Decapitation' of a 5'-Capped Oligoribonucleotide by o-Phenanthroline:Cu(II) *J. Am. Chem. Soc.* **115**, 3378.

64. Kim, J.H.and Chin, J. (1992) Dimethyl Phosphate Hydrolysis at Neutral pH" *J. Am. Chem. Soc.* **114**, 9792.

65. Chin, J.and Banaszczyk, M. (1989) Highly Efficient Hydrolytic Cleavage of Adenosine Monophosphate Resulting in a Binuclear Co(III) Complex with a Novel Doubly Bidentate μ^4-Phosphato Bridge *J. Am. Chem. Soc.* **111**, 4103.

66. Chin, J., Banaszczyk, M., Jubian, V., Zou, X. (1989) Co(III) Complex Promoted Hydrolysis of Phosphate Diesters: Comparison in Reactivity of Rigid *cis*-Diaquotetraazacobalt(III) Complexes *J. Am. Chem. Soc.* **111**, 186.

67. Chin, J.and Zou, X. (1987) Catalytic Hydrolysis of cAMP *Can. J. Chem.* **65**, 1882.

68. Hendry, P. and Sargeson, A.M. (1989) Metal Ion Promoted Phosphate Ester Hydrolysis. Intramolecular Attack of Coordinated Hydroxide Ion *J. Am. Chem. Soc.* **111**, 2521.

69. Hendry, P. and Sargeson, A.M. (1990) Metal Ion Promoted Reactions of Phosphate Derivatives, in S.J. Lippard (Ed.),*Progress in Inorganic Chemistry: Bioinorganic Chemistry*, John Wiley & Sons, Inc. 201.

70. Vance, D.H.and Czarnik, A.W. (1993) Functional Group Convergency in a Binuclear Dephosphorylation Reagent *J. Am. Chem. Soc.* **115**, 12165.

71. Chung, Y., Akkaya, E.U., Venkatachalam, T.K., Czarnik, A.W. (1990) Synthesis and Characterization of a Reactive Binuclear Co(III) Complex. Cooperative Promotion of Phosphodiester Hydrolysis *Tetr. Lett.* **31**, 5413.

72. Menger, F.M., Gan, L.H., Johnson, E., Durst, D.H. (1987) Phosphate Ester Hydrolysis Catalyzed by Metallomicelles *J. Am. Chem. Soc.* **109**, 2800.

73. Visscher, J.and Schwartz, A.W. (1992) Selective Cleavage of Pyrophosphate Linkages *Nucleic Acids Res.* **20**, 5749.

74. Koike, T.and Kimura, E. (1991) Roles of Zinc(II) Ion in Phosphatases. A Model Study with Zinc(II)-Macrocyclic Polyamine Complexes *J. Am. Chem. Soc.* **113**, 8933.

75. Browne, K.A.and Bruice, T.C. (1992) Chemistry of Phosphodiesters, DNA and Models. 2. The Hydrolysis of Bis(8-hydroxyquinoline) Phosphate in the Absence and Presence of Metal Ions *J. Am. Chem. Soc.* **114**, 4951.

76. Ikenaga, K.and Inoue, Y. (1974) Metal(II) Ion Catalyzed Transphosphorylation of Four Homodinucleotides and Five Pairs of Dinucleotide Sequence Isomers *Biochemistry* **13**, 577.

77. Eichhorn, G., Tarien, E., Butzow, J.J. (1971) Interaction of Metal Ions with Nucleic Acids and Related Compounds. XVI. Specific Cleavage Effects in the Depolymerization of Ribonucleic Acids by Zinc(II) Ions *Biochemistry* **10**, 2014.

78. Butzow, J.J.and Eichhorn, G.L. (1971) Interaction of Metal Ions with Nucleic Acids and Related Compounds. XVII. On the Mechanism of Degradation of Polyribonucleotides and Oligoribonucleotides by Zinc(II) Ions *Biochemistry* **10**, 2019.

79. Breslow, R., Huang, D.-L., Anslyn, E. (1989) On the Mechanism of Action of Ribonucleases: Dinucleotide Cleavage Catalyzed by Imidazole and Zn^{2+} *Proc. Natl. Acad. Sci. U.S.A.* **86**, 1746.

80. England, T.E. and Uhlenbeck, O.C. (1978) Enzymatic Oligoribonucleotide Synthesis with T4 DNA Ligase*Biochemistry* **17**, 2069-2076.

81. Maiya, B.G., Harriman, A., Sessler, J.L., Hemmi, G, Murai, T., Mallouk, T.E. (1989) Ground- and Excited-State Spectral and Redox Properties of Cadmium(II) Texaphyrin *J. Phys. Chem.* **93**, 8111.

82. Sessler, J.L., Hemmi, G., Mody, T.D., Murai, T., Burrell, A., Young, S.W. (1994) Texaphyrins: Synthesis and Applications *Acc. Chem. Res.* **27**, 43.

83. Magda, D., Wright, M., Miller, R.A., Sessler, J.L., Sansom, P.I. (1995) Sequence-Specific Photocleavage of DNA by an Expanded Porphyrin with Irradiation above 700 nm *J. Am. Chem. Soc.* **117**, 3629-3630.

84. Spikes, J.D. and Straight, R. (1967) Sensitized Photochemical Processes in Biological Systems *Ann. Rev. Phys. Chem.* **18**, 409-436.

85. Foote, C. S. (1968) Mechanisms of Photosensitized Oxidation *Science* **162**, 963-970.

86. Rosenthal, I. and Pitts, Jr., J. N. (1971) Reactivity of Purine and Pyrimidine Bases toward Singlet Oxygen *Biophys. J.* **11**, 963-996.

87. Sambrook, J., Fritsch, E.F.; Maniatis, T. (1989) *Molecular Cloning, A Laboratory Manual, 2nd Ed.*; Cold Spring Harbor Laboratory Press: New York.

88. Le Doan, T., Praseuth, D., Perrouault, L., Chassignol, M., Thuong, N.T., Helene, C. (1990) Sequence-Targeted Photochemical Modification of Nucleic Acids by Complementary Oligonucleotides Covalently Linked to Porphyrins *Bioconjugate Chem.* **2**, 108-113.

89. Fedorova, O.S., Savitskii, A.P., Shoikhet, K.G., Ponomarev, G.V. (1990) Palladium(II)-coproporphyrin I as a Photoactivatable Group in Sequence-Specific Modification of Nucleic Acids by Oligonucleotide Derivatives *FEBS Lett.* **259**, 335-337.

90. Vlasssov, V.V., Deeva, E.A., Ivanova, E.M., Knorre, D.G., Maltseva, T.V., Frolova, E.I. (1991) Photoactivatable Porphyrin Oligonucleotide Derivatives for Sequence Specific Chemical Modification and Cleavage of DNA *Nucleosides & Nucleotides* **10**, 641-643.

91. Mastruzzo, L., Woisard, A., Ma, D.D., Rizzarelli, E., Favre, A., Le Doan, T. (1994) Targeted Photochemical Modification of HIV-Derived Oligoribonucleotides by Antisense Oligodeoxynucleotides Linked to Porphyrins *Photochem. Photobiol.* **60**, 316-322.

92. Perrouault, L., Asseline, U., Rivalle, C., Thuong, N.T., Bisagni, E., Giovannangeli, C., Le Doan, T., Helene, C. (1990) Sequence-Specific Artificial Photoinduced Endonucleases Based on Triple Helix-Forming Oligonucleotides *Nature* **344**, 358-360.

93. Le Doan, T., Perrouault, L., Praseuth, D., Habhoub, N., Decout, J. L., Thuong, N.T., Lhomme, J., Helene, C. (1987) Sequence-Specific Recognition, Photocrosslinking and Cleavage of the DNA Double Helix by an Oligo-α-Thymidylate Covalently Linked to an Azidoproflavine Derivative *Nucleic Acids Res.* **15**, 7749-7760.

94. Le Doan, T., Perrouault, L., Asseline, U., Thuong, N.T., Rivalle, C., Bisagni, E., Helene, C. (1991) Recognition and Photoinduced Cleavage and Crosslinking of Nucleic Acids by Oligonucleotides Covalently Linked to Ellipticine *Antisense Res. Dev.* **1**, 43-54.

95. Bhan, P., Miller, P.S. (1990) Photocrosslinking of Psoralen Derivatized Oligonucleoside Methylphosphonate to Single-Stranded DNA *Bioconjugate Chem.* **1**, 82-88.

96. Levina, A.S., Berezovskii, M.V., Venjaminova, A.G., Dobrikov, M.I., Repkova, M.N., Zarytova, V.F. (1993) Photomodification of RNA and DNA fragments by oligonucleotide reagents bearing arylazide groups *Biochimie* **75**, 25-27.

97. Pieles, U.and Englisch, U. (1989) Psoralen covalently linked to oligodeoxyribonucleotides: synthesis, sequence specific recognition of DNA and photo-cross-linking to pyrimidine residues of DNA *Nucleic Acids Res.* **17**, 285-299.

98. Takasugi, M., Guendouz, A., Chassignol, M., Decout, J.L., Lhomme, J., Thuong, N.T., Helene, C. (1991) Sequence-specific photo-induced cross-linking of the two strands of double-helical DNA by a psoralen covalently linked to a triple-gelix-forming oligonucleotide *Proc. Natl. Acad. Sci. U.S.A.* **88**, 5602-5606.

99. Praseuth, D., Le Doan, T., Chassignol, M., Decout, J.-L., Habhoub, N., Lhomme, J., Thuong, N.T., Helene, C. (1988) Sequence-targeted photosensitized reactions in nucleic acids by oligo-α-deoxynucleotides and oligo-β-deoxynucleotides covalently linked to proflavin *Biochemistry* **27**, 3031-3038.

RIBOZYME MIMICS FOR CATALYTIC ANTISENSE STRATEGIES

JAMES K. BASHKIN,* JIN XIE, ANDREW T. DANIHER,
LISA A. JENKINS AND GEORGE C. YEH
Department of Chemistry, Washington University
One Brookings Drive, Campus Box 1134,
St. Louis, Missouri 63130-4899

Abstract.   We have designed ribozyme mimics that consist of a deoxyoligonucleotide (for molecular recognition) and a covalently incorporated RNA hydrolysis catalyst. These mimics can cleave RNA in a sequence-specific fashion, *via* a biocompatible hydrolytic reaction, and may serve as catalytic antisense drugs. Our initial design used terpyridylCu(II) as the hydrolytic agent and delivered this metal complex across the major groove of the RNA/DNA duplex that forms with the substrate RNA. As part of our goal to optimize the activity of ribozyme mimics, we report DNA building blocks that can deliver hydrolytic agents across the minor groove. We also describe mechanistic studies of RNA transesterification and hydrolysis, and a new type of RNA substrate, termed embedded RNA. Embedded RNA (embRNA) contains one RNA residue in a deoxyoligonucleotide, and combines advantages of both dimeric and polymeric RNA substrates.

1.   Introduction

Since 1986 we have been working on a class of molecules that we describe as ribozyme mimics [1-9]. These ribozyme mimics are designed to serve as catalytic antisense [10] reagents for the destruction of harmful messenger (or genomic) RNA. There are two important structural domains in the ribozyme mimics: a recognition domain and a catalytic domain. As with some natural ribozymes [11-13], recognition is provided by Watson-Crick base pairing. In common with all known ribozymes, catalysis is provided by one or more metal ions that are bound at specific sites. In natural ribozymes, the catalytic and recognition domains are contiguous, and not necessarily separable. Metal binding sites are formed by a complex folding of RNA that requires the participation of certain 2'-hydroxyl groups and many other RNA functional groups.

B. Meunier (ed.), DNA and RNA Cleavers and Chemotherapy of Cancer and Viral Diseases, 355–366.
© *1996 Kluwer Academic Publishers. Printed in the Netherlands.*

We have chosen to separate the catalytic and recognition domains by incorporating metal binding groups as side chains in a DNA sequence. This design is represented by the cartoon in Figure 1.

Differences between our ribozyme mimics and their natural counterparts include: (1) Ribozymes require RNA residues at some positions in order to fold into the active, metal-bound form. Our mimics can be constructed of DNA or DNA analogs, including molecules designed for cellular uptake and nuclease resistance, since the 2'-hydroxyl group is not a necessary

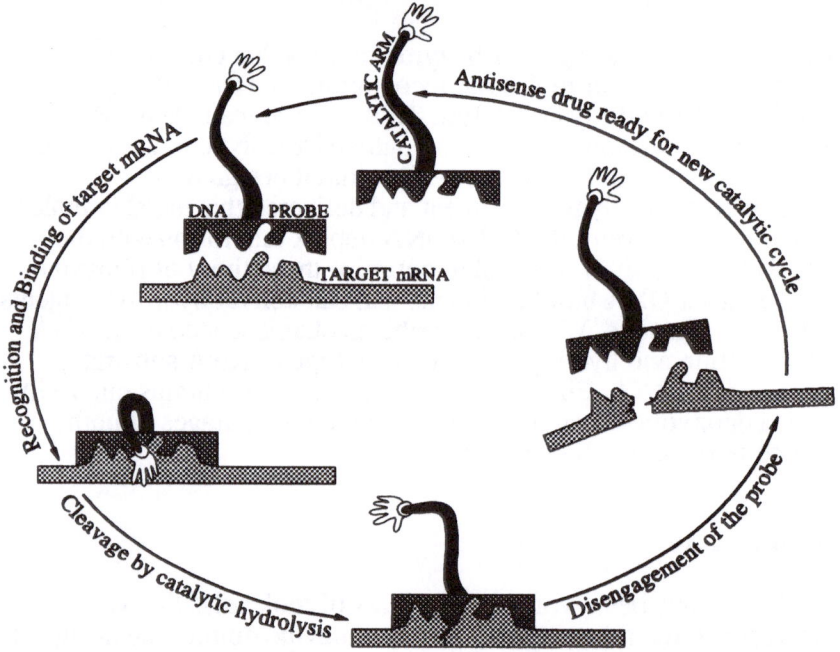

Figure 1. Schematic View of Catalytic Antisense Drugs

component of an active ribozyme mimic. (2) The molecular weight of natural ribozymes is relatively high, since the active, tertiary structure requires a long primary sequence. Ribozyme mimics replace this tertiary structure with a small molecule catalyst, thereby decreasing the MW and perhaps improving cellular uptake.

Here I emphasize a few additional design aspects that distinguish our ribozyme mimics from their naturally-occurring counterparts and from other chemical approaches to reactive antisense probes. First, we rely on hydrolytic cleavage to destroy the target RNA. This reaction is highly chemoselective for RNA over DNA, so our mimics do not self-destruct. Hydrolytic cleavage, or transesterification, does not require ancillary redox equivalents, and is potentially free of harmful side effects of *in vivo* redox chemistry. Second, Figure 1 depicts a ribozyme mimic whose catalytic domain is attached at an internal site on the DNA probe, or recognition domain. This is a key element of our strategy to achieve catalytic turnover. As Figure 1 shows, cleavage at an internal position of the RNA-DNA duplex disrupts the duplex. The RNA-DNA binding constant is thereby decreased and the product fragments can be displaced by a new substrate molecule. In contrast, cleaving outside the duplex fails to change the DNA-RNA binding constant, the product fragments are not released, and product inhibition occurs. For this reason, our program is focused on reactive DNA building blocks that allow hydrolytic agents to be incorporated at internal sites in DNA sequences. Other approaches to the sequence-specific hydrolysis of RNA have been reported recently [14-17].

2. Development Of The First Ribozyme Mimic.

Implicit in Figure 1 is the ability to identify RNA cleavage agents and to incorporate them into DNA without overly compromising their activity. We chose two different routes to solve this problem. One was the use of imidazole, a known RNA hydrolysis catalyst, as a side chain in oligonucleotides. We have reported an extensive series of nucleosides and nucleotides with pendent imidazole groups [1, 2, 5]. Cohen proposed a similar approach [18, 19], and both he [20] and Lönnberg [21] have recently reported studies in this area.

The main focus of our work has been on inorganic catalysts. Many metal ions were known to hydrolyze RNA, e.g. from the work of Dimroth [22] and Eichhorn [23, 24]. However, we needed well defined metal complexes that could be covalently incorporated into DNA. We reported the first examples of such complexes, along with an HPLC assay that used true RNA substrates to evaluate the hydrolytic activity of potential catalysts [9]. Among the active complexes we identified were bipyridine (bipy) and terpyridine (terpy) complexes of Cu(II), and a Zn macrocycle complex. These results are summarized in Figure 2 and Table 1. Trogler and Morrow had reported the hydrolysis of the activated substrate bis(*p*-nitrophenyl)phosphate by Cu(bipy) [25, 26]. Interestingly, TRogler reported (and we confirmed) that Cu(terpy) is inactive for the hydrolysis of bis(*p*-nitrophenyl)phosphate. Chin has recently reported the hydrolytic cleavage of RNA dimers by terpy and bipy complexes of Cu(II) [27].

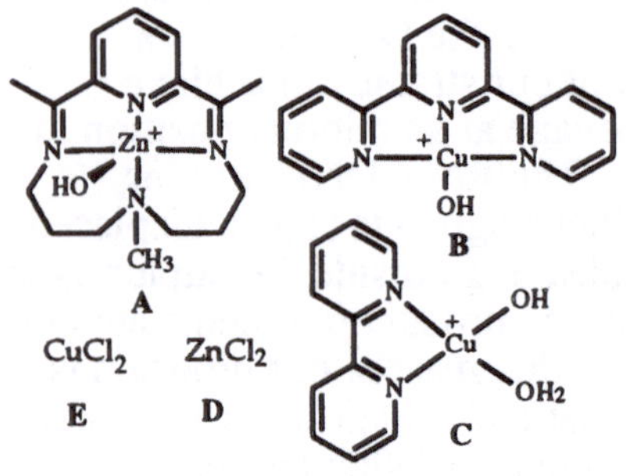

Table 1. Extent of $(A)_{12-18}$ hydrolysis by transition metal complexes. Conditions: 37°C, 20mM HEPES buffer, pH = 7.1, reaction time = 20h, [complex] = 0.160 mM, [RNA] = 0.06 mM.

Complex	% of Substrate Hydrolyzed
A	70
B	97
C	61
D	72
E	31
none*	5

* control

Figure 2. Identification of metal complexes that cleave RNA hydrolytically-, using the substrate $(A)_{12-18}$. (a) 2',3'-Cyclic AMP was formed from the cleavage of $(A)_{12-18}$ by complexes A and B. (b) No nucleic acid cleavage was observed with deoxy $(A)_{12-18}$ in the presence of B. (c) Complexes A, B and C all produced the same product signature as determined by anion exchange HPLC.

We then began the design and synthesis of DNA building blocks containing bipy and terpy ligands. At the same time, we initiated mechanistic studies on the transesterification and hydrolysis of RNA by Cu(terpy).

Many early examples [8] of our reactive DNA building blocks used 2'-deoxyuridine as a starting point, and relied on functionalization of the uracil C-5 position . This point of attachment delivers the metal complex across the major groove of the RNA/DNA duplex that forms between the ribozyme mimic and the substrate. Synthesis of the first ribozyme mimic is shown in Figures 3 and 4 [7]. A terpyridine derivative was linked to C-5, and a conventional DNA phosphoramidite was prepared. Automated DNA synthesis gave the desired compound, a 17-mer deoxyoligonucleotide with one pendent terpy. This particular reagent was complementary to a section of mRNA of the HIV *gag* gene. Using this approach, we achieved sequence-specific cleavage of the target RNA, within the duplex region. The resulting cleavage pattern is shown in Fig. 5. Under the reaction conditions used for hydrolytic cleavage, the complementary DNA target sequence was not cleaved. However, adding dithiothreitol (DTT) changed the mechanism from hydrolytic to oxidative. Under oxidation conditions, cleavage of both DNA and RNA substrates occurred, over a wide range of positions.

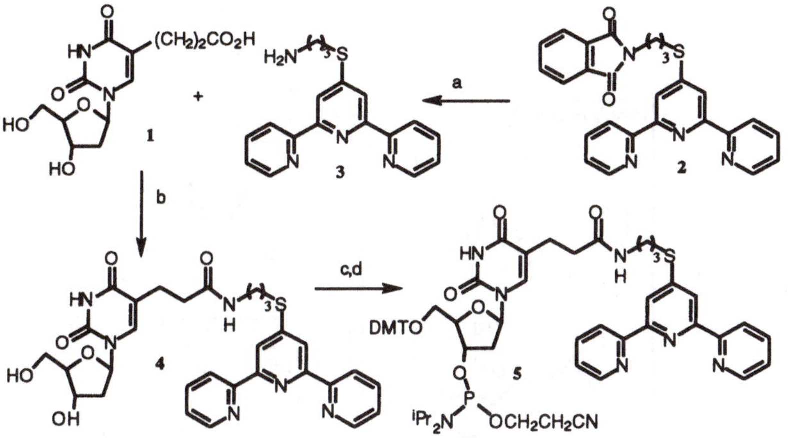

Figure 3. Synthesis of terpyridine-containing phosphoramidite. a. Hydrazine, EtOH, 60
°C, yield 96%; b. Ethyldimethylaminopropyl carbodiimide HCl, DMSO, room
temperature, 24 h, yield 77%; c. Pyridine, DMTCl, 24 h, yield 65%; d. iPr2NEt, THF,
4° C, chloro-β-cyanoethyl-N,N-diisopropyl phosphoramidite, yield 86%.

Thus, we are able to cleave a target RNA molecule at a specific sequence.
This proved the concept that a hydrolytically-active metal complex can be
made sequence-specific by incorporating it into DNA. It further
demonstrated that hydrolytic cleavage can be directed to an internal site in
an RNA/DNA duplex, in keeping with our design for catalytic turnover.
However, we have not yet demonstrated catalytic turnover in this system,
although we know that Cu(terpy) is a capable catalyst under other
conditions [4]. Furthermore, improvements are necessary, since the
cleavage reaction was rather slow, and Cu(terpy) is too labile to survive in
a living cell.

3. Optimization Of Ribozyme Mimics

Our strategies for optimization include determining the mechanism of
RNA cleavage by Cu(terpy) and bipy reagents. We are also exploring the
synthesis of alternative catalytic moieties for incorporation into new
ribozyme mimics. One promising approach is to attack the target RNA *via*
the minor groove, where the 2'-hydroxyl groups of RNA in an A-form
RNA/DNA duplex are located. In addition, we hope to achieve
bifunctional catalysis by simultaneous major- and minor-groove attack.
Figures 6 and 7 show some of our recent work [28] on the synthesis of
reagents that target the minor groove. For these syntheses, the metal
complex is attached either at C1', following methods reported by Hovinen
and Ono [29, 30] , or at C2' using a method reported by Manoharan [31].

360

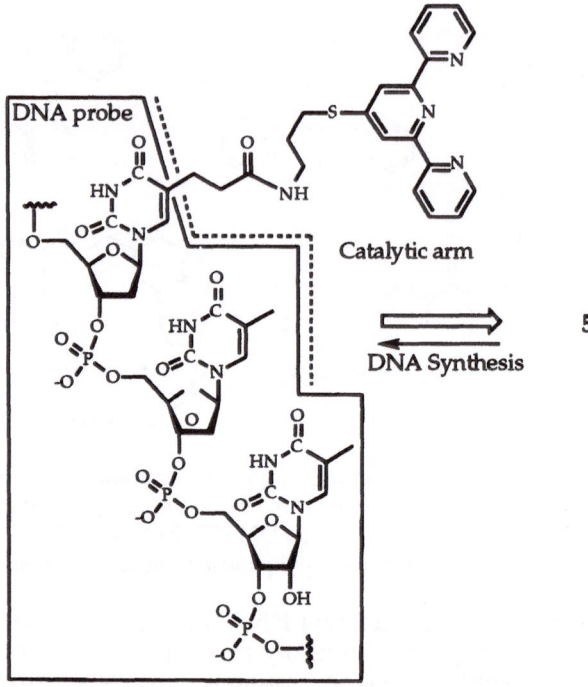

Figure 4. Preparation of ribozyme mimics: Incorporation of the terpyridine group into DNA by automated synthesis using phosphoramidite 5.

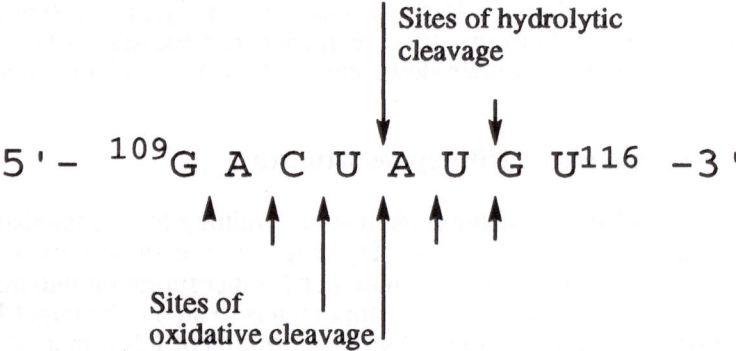

Figure 5. Sequence-specific cleavage of the 159-mer RNA target by a terpyridine-based ribozyme mimic: sites of cleavage.

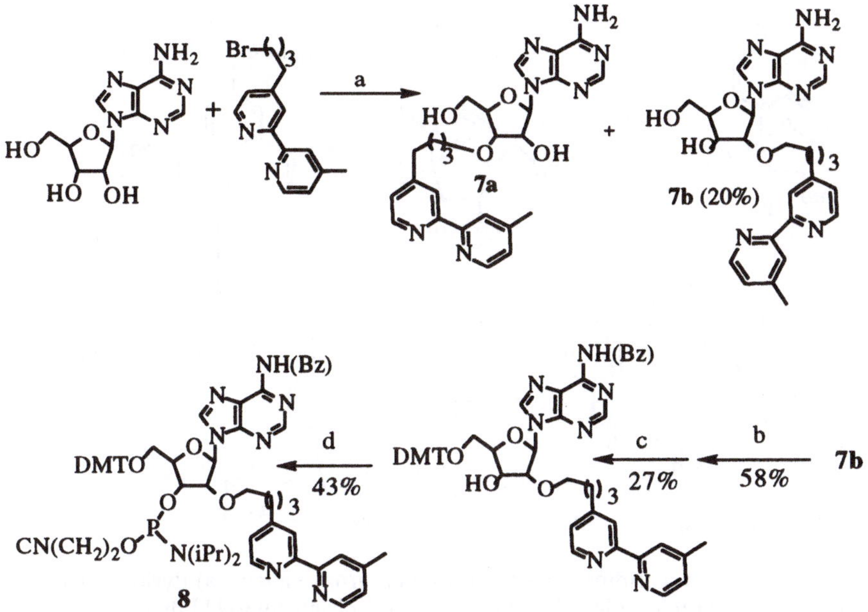

Figure 6. Synthesis of an adenosine-bipyridine reagent for minor-groove attack. a) NaH, DMF, 24 h, -5 °C; b) DMT-Cl, Pyridine, overnight; c) i. TMS-Cl; ii. Bz-Cl; iii. 2M NH₄OH; d) iPr₂NEt, THF, chloro-ß-cyanoethyl-N,N-diisopropyl phosphoramidite. Isolated yields are given

We have also developed a new assay for studying the transesterification of RNA [32]. In principle, dimers like ApA and ApUp provide the simplest substrates for studying RNA cleavage. However, when studying the hydrolysis of RNA by metal ions, Eichhorn and Inoue noted the extraordinary inertness of dimers relative to long RNA molecules [23, 33]. Short sequences (< 4 nucleotides) may not provide sufficient binding sites for multiple metal ions to bind and catalyze cleavage. However, long RNA molecules also suffer some limitations as substrates for kinetic studies. For example, they contain heterogeneous sites for reaction and each set of reaction products can serve as substrates for additional cleavage events. Thus, the kinetics can be complicated, especially beyond the initial rate limit.

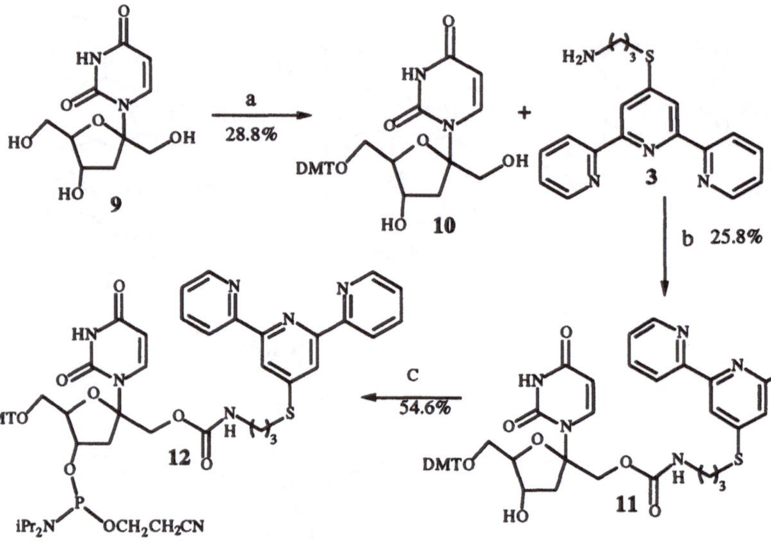

Figure 7. Synthesis of a uridine analogue for minor groove attack. a) Pyridine, DMTCl, 24 h. b) Carbonyl diimidazole, CH_2Cl_2. c)chloro-β-cyanoethyl-N,N-diisopropyl phosphoramidite, 30 min. Isolated yields are given.

We have therefore developed a new RNA substrate which we call embedded RNA (embRNA) [32]. This new class of substrate has only one cleavage site, like an RNA dimer, but it has multiple metal binding sites, like a polymer. EmbRNA has one RNA residue incorporated into a deoxyoligonucleotide. With embRNA, we can study the sequence context of RNA hydrolysis by changing the flanking sequences. One can also hope to identify specific phosphates and other functional groups that bind catalytically active metal ions by the systematic use of methyl phosphonates, phosphorothioates and other chemical modifications. Our first embRNA, $T_{11}UT_7A$, is shown in Figure 8. Kinetic results on the transesterification of $T_{11}UT_7A$ by Cu(terpy) are shown in Figure 9. Figure 9 illustrates the dependence of the initial rate of transesterification on the concentration of substrate. A preliminary fit of the data to a Michaelis-Menten rate equation is also shown.

4. Conclusions

We have developed a systematic approach to ribozyme mimics, molecules that cleave target RNA in a sequence-specific manner, using a biocompatible hydrolytic pathway. We have explored the chemical and kinetic mechanisms of metal-catalyzed RNA transesterification and hydrolysis, and we have prepared new building blocks for ribozyme mimics that are designed to attack RNA *via* the minor groove. These results are directed toward exploring the potential of catalytic drugs to fight cancer, viral infections and any diseases that may be described as the overproduction of dangerous proteins.

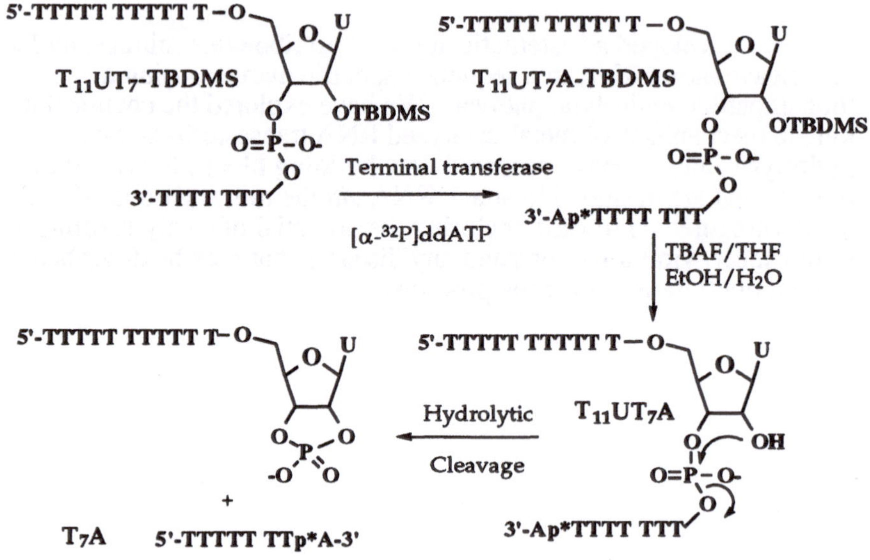

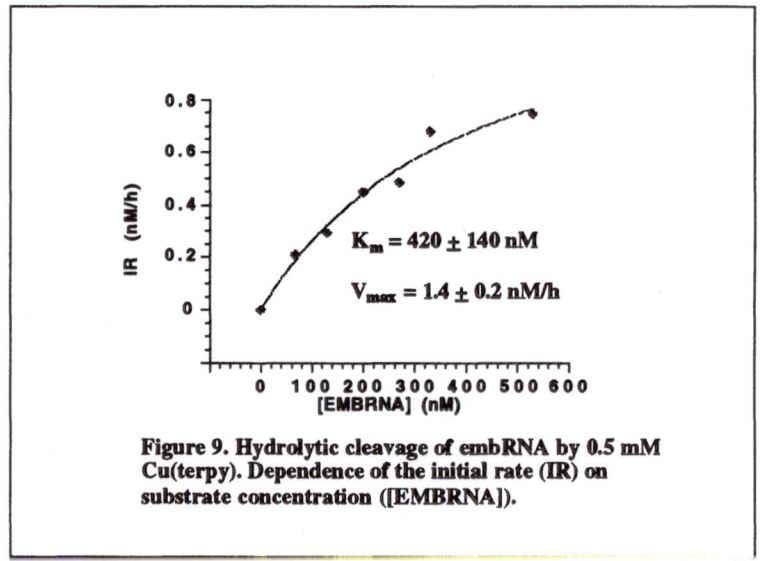

5'-TTTTT TTTTT T-O
T11UT7-TBDMS
OTBDMS
O=P-O-
3'-TTTT TTT
Terminal transferase
[α-32P]ddATP
5'-TTTTT TTTTT T-O
T11UT7A-TBDMS
OTBDMS
O=P-O-
3'-Ap*TTTT TTT
TBAF/THF
EtOH/H2O
5'-TTTTT TTTTT T-O
T11UT7A
OH
O=P-O-
3'-Ap*TTTT TTT
Hydrolytic
Cleavage
5'-TTTTT TTTTT T-O
+
T7A
5'-TTTTT TTp*A-3'

Figure 8. Preparation, deprotection and transesterification of an embedded RNA (embRNA) substrate.

IR (nM/h)
0.8
0.6
0.4
0.2
0
Km = 420 ± 140 nM
Vmax = 1.4 ± 0.2 nM/h
0 100 200 300 400 500 600
[EMBRNA] (nM)

Figure 9. Hydrolytic cleavage of embRNA by 0.5 mM Cu(terpy). Dependence of the initial rate (IR) on substrate concentration ([EMBRNA]).

Acknowledgments

This work was supported in part by NSF Grant CHE-9318581 and a grant from the U.S. Department of Education. Acknowledgment is also made to the donors of The Petroleum Research Fund, administered by the ACS, for partial support of this research.

References

1. Bashkin, J. K., Gard, J. K. and Modak, A. S. (1990) *J. Org. Chem* **55**, 5125.
2. Bashkin, J. K., McBeath, R. J., Modak, A. S., Sample, K. R. and Wise, W. B. (1991) *J. Org. Chem.* **56**, 3168.
3. Bashkin, J. K. (1993) in *Bioinorganic Chemistry of Copper*, eds. Karlin, K. D. and Tyeklar, Z. (Chapman and Hall, New York), pp. 132-139.
4. Bashkin, J. K. and Jenkins, L. A. (1993) *J. Chem. Soc. Dalton Trans.* 3631.
5. Bashkin, J. K., Sondhi, S. M., Sampath, U., d'Avignon, D. A. and Modak, A. S. (1994) *New J. Chem.* **18**, 305-316.
6. Bashkin, J. K. and Jenkins, L. A. (1994) *Comments on Inorganic Chemistry* **16**, 77-93.
7. Bashkin, J. K., Frolova, E. I. and Sampath, U. (1994) *J. Am. Chem. Soc.* **116**, 5981-5982.
8. Modak, A. S., Gard, J. K., Merriman, M. C., Winkeler, K. A., Bashkin, J. K. and Stern, M. K. (1991) *J. Am. Chem. Soc.* **113**, 283.
9. Stern, M. K., Bashkin, J. K. and Sall, E. D. (1990) *J. Am. Chem. Soc.* **112**, 5357.
10. Uhlmann, E. and Peyman, A. (1990) *Chemical Reviews* **90**, 544.
11. Cech, T. R. (1987) *Science* **236**, 1532.
12. Cech, T. R. (1988) *JAMA* **260**, 3030-3034.
13. Altman, S. (1990) *Angew. Chem. Int. Ed. Engl.* **29**, 749-758.
14. Magda, D., Miller, R. A., Sessler, J. L. and Iverson, B. L. (1994) *J. Am. Chem. Soc.* **116**, 7439-7440.
15. Hall, J., Huesken, D., Pieles, U., Moser, H. E. and Haener, R. (1994) *Chemistry and Biology* 185-190.
16. Komiyama, M., Inokawa, T. and Yoshinari, K. (1995) *J. Chem. Soc., Chem. Commun.* 77-78.
17. Matsumura, K., Endo, M. and Komiyama, M. (1994) *J. Chem. Soc., Chem. Commun.* 2019-2020.
18. Cohen, J. S. (1989) in *Oligodeoxynucleotides: Antisense Inhibitors of Gene Expression* (CRC Press, Boca Raton).
19. Stein, C. A. and Cohen, J. S. (1988) *Cancer Res.* **48**, 2659-2668.
20. Polushin, N. P., Chen, B.-C., Anderson, L. W. and Cohen, J. S. (1993) *J. Org. Chem.* **58**, 4606.
21. Hovinen, J., Guzaev, A., Azhayeva, E., Azhayev, A. and Lonnberg, H. (1995) *J. Org. Chem.* **60**, 2205-2209.
22. Dimroth, K., Witzel, H. and Mirbach, W. (1975) *Annalen* **620**, 94.
23. Butzow, J. J. and Eichhorn, G. L. (1971) *Biochemistry* **10**, 2019-2027.
24. Butzow, J. J. and Eichhorn, G. L. (1975) *Nature* **254**, 358.
25. Morrow, J. R. and Trogler, W. C. (1988) *Inorg. Chem.* **27**, 3387-94.
26. Morrow, J. R. and Trogler, W. C. (1992) *Inorg. Chem.* **31**, 1544.

27. Linkletter, B. and Chin, J. (1995) *Angew. Chem. Int. Ed. Engl.* **34,** 472-474.

28. Bashkin, J. K., Xie, J., Sampath, U. S., Daniher, A. T. and Kao, J. (1995) *Manuscript in preparation.*

29. Hovinen, J., Azhayev, A., Guzaev, A. and Lonnberg, H. (1995) *Nucleosides and Nucleotides* **14,** 329-332.

30. Ono, A., Dan, A. and Matsuda, A. (1993) *Bioconj. Chem.* **4,** 499-508.

31. Manoharan, M., Johnson, L. K., Tivel, K. L., Springer, R. H. and Cook, P. D. (1993) *BioMed. Chem. Lett.* **3,** 2765.

32. Jenkins, L. A., Bashkin, J. K. and Autry, M. E. (1995) *Submitted for publication.*

33. Ikenaga, H. and Inoue, Y. (1974) *Biochemistry* **13,** 577-582.

SUBJECT INDEX

—A—

—B—

—C—

368

—D—

—E—

—F—

—G—

—H—

—I—

—K—

—L—

—M—

—N—

—O—

—P—

—R—

—S—

—T—

—X—

—Y—